Applications of Radio Frequency Heating in Food Processing

Applications of Radio Frequency Heating in Food Processing

Editors

Shaojin Wang
Rui Li

MDPI • Basel • Beijing • Wuhan • Barcelona • Belgrade • Manchester • Tokyo • Cluj • Tianjin

Editors
Shaojin Wang
Northwest A&F University
China

Rui Li
Northwest A&F University
China

Editorial Office
MDPI
St. Alban-Anlage 66
4052 Basel, Switzerland

This is a reprint of articles from the Special Issue published online in the open access journal *Foods* (ISSN 2304-8158) (available at: https://www.mdpi.com/journal/foods/special_issues/radio_frequency_heating).

For citation purposes, cite each article independently as indicated on the article page online and as indicated below:

LastName, A.A.; LastName, B.B.; LastName, C.C. Article Title. *Journal Name* **Year**, *Volume Number*, Page Range.

ISBN 978-3-0365-7234-5 (Hbk)
ISBN 978-3-0365-7235-2 (PDF)

© 2023 by the authors. Articles in this book are Open Access and distributed under the Creative Commons Attribution (CC BY) license, which allows users to download, copy and build upon published articles, as long as the author and publisher are properly credited, which ensures maximum dissemination and a wider impact of our publications.

The book as a whole is distributed by MDPI under the terms and conditions of the Creative Commons license CC BY-NC-ND.

Contents

About the Editors vii

Shaojin Wang and Rui Li
Applications of Radio Frequency Heating in Food Processing
Reprinted from: *Foods* 2023, 12, 1133, doi:10.3390/foods12061133 1

Yuxiao Mao, Yujun Hao, Xiangyu Guan, Penghao Wang and Shaojin Wang
Temperature and Moisture Dependent Dielectric and Thermal Properties of Walnut Components Associated with Radio Frequency and Microwave Pasteurization
Reprinted from: *Foods* 2022, 11, 919, doi:10.3390/foods11070919 3

Sandro M. Goñi, Matteo d'Amore, Marta Della Valle, Daniela F. Olivera, Viviana O. Salvadori and Francesco Marra
Effect of Load Spatial Configuration on the Heating of Chicken Meat Assisted by Radio Frequency at 40.68 MHz
Reprinted from: *Foods* 2022, 11, 1096, doi:10.3390/foods11081096 21

Ke Wang, Lisong Huang, Yangting Xu, Baozhong Cui, Yanan Sun, Chuanyang Ran, Hongfei Fu, et al.
Evaluation of Pilot-Scale Radio Frequency Heating Uniformity for Beef Sausage Pasteurization Process
Reprinted from: *Foods* 2022, 11, 1317, doi:10.3390/foods11091317 39

Karn Chitsuthipakorn and Sa-nguansak Thanapornpoonpong
Effect of Large-Scale Paddy Rice Drying Process Using Hot Air Combined with Radio Frequency Heating on Milling and Cooking Qualities of Milled Rice
Reprinted from: *Foods* 2022, 11, 519, doi:10.3390/ foods11040519 53

Chang-Yi Huang, Yu-Huang Cheng and Su-Der Chen
Hot Air-Assisted Radio Frequency (HARF) Drying on Wild Bitter Gourd Extract
Reprinted from: *Foods* 2022, 11, 1173, doi:10.3390/foods11081173 71

Xiangqing Chen, Yu Liu, Ruyi Zhang, Huacheng Zhu, Feng Li, Deyong Yang and Yang Jiao
Radio Frequency Drying Behavior in Porous Media: A Case Study of Potato Cube with Computer Modeling
Reprinted from: *Foods* 2022, 11, 3279, doi:10.3390/foods11203279 83

Yu Gao, Xiangyu Guan, Ailin Wan, Yuan Cui, Xiaoxi Kou, Rui Li and Shaojin Wang
Thermal Inactivation Kinetics and Radio Frequency Control of *Aspergillus* in Almond Kernels
Reprinted from: *Foods* 2022, 11, 1603, doi:10.3390/foods11111603 103

Yu-Fen Yen and Su-Der Chen
Influence of Radio Frequency Heating on the Pasteurization and Drying of Solid-State Fermented *Wolfiporia cocos* Products
Reprinted from: *Foods* 2022, 11, 1766, doi:10.3390/foods11121766 117

Ke Wang, Chuanyang Ran, Baozhong Cui, Yanan Sun, Hongfei Fu, Xiangwei Chen, Yequn Wang, et al.
Sterilizing Ready-to-Eat Poached Spicy Pork Slices Using a New Device: Combined Radio Frequency Energy and Superheated Water
Reprinted from: *Foods* 2022, 11, 2841, doi:10.3390/foods11182841 129

Zhenna Zhang, Bin Zhang, Lin Zhu and Wei Zhao
Microstructure, Digestibility and Physicochemical Properties of Rice Grains after Radio Frequency Treatment
Reprinted from: *Foods* **2022**, *11*, 1723, doi:10.3390/foods11121723 . . . . **145**

Ting-Yu Lian and Su-Der Chen
Developing Radio-Frequency Roasting Protocols for Almonds Based on Quality Evaluations
Reprinted from: *Foods* **2022**, *11*, 1885, doi:10.3390/foods11131885 . . . . **159**

Jiwei Jiang, Fen Zhou, Caining Xian, Yuyao Shi and Xichang Wang
Effects of Radio Frequency Tempering on the Texture of Frozen Tilapia Fillets
Reprinted from: *Foods* **2021**, *10*, 2663, doi:10.3390/foods10112663 . . . . **171**

Rong Han, Jialing He, Yixuan Chen, Feng Li, Hu Shi and Yang Jiao
Effects of Radio Frequency Tempering on the Temperature Distribution and Physiochemical Properties of Salmon (*Salmo salar*)
Reprinted from: *Foods* **2022**, *11*, 893, doi:10.3390/foods11060893 . . . . **185**

About the Editors

Shaojin Wang

Shaojin Wang received his bachelors and master degrees from Zhejiang University (P.R. China), while he received his PhD from Gembloux Agricultural University (Belgium). He worked at the Department of Biological Systems Engineering at Washington State University (WSU) for 11 years, and has been at the College of Mechanical and Electronic Engineering at Northwest A&F University since 2011. He is currently a Full Professor at this college and an Adjunct Professor at WSU. His major research areas include the thermal death kinetics of microorganisms, the thermal/dielectric properties of food materials, and radio frequency heating for pasteurization as well as disinfestation. He is an Editorial Board Member of seven SCI journals, such as *Transactions of the ASABE*, the *International Journal of Agricultural and Biological Engineering*, and so on. He has published more than 274 SCI papers, has more than 11296 citations, has an h-index of 61, and has given more than 20 invited keynote presentations. He is the PI of more than 12 projects, primarily supported by USDA-NIFA, the NSF of China, and Key R&D projects of the Chinese Ministry of Science and Technology. He has been listed as a high-citation scientist in the agriculture and biology fields of China for 8 years by Elsevier.

Rui Li

Rui Li received her bachelors degree from China Agricultural University, her masters degree from Hokkaido University (Japan), and her PhD from Northwest A&F University (P.R. China). She has worked at the College of Mechanical and Electronic Engineering at Northwest A&F University since 2012; she is a Senior Scientific Researcher (equal to an Associate Professor) at this university. Her major research areas include radio frequency heating for pasteurization, disinfestation, drying, the thermal death kinetics of microorganisms, and the non-destructive testing of agricultural product quality. She has published more than 34 SCI papers. She is a reviewer of SCI journals, such as the *Journal of Food Engineering and the Journal of Food Process Engineering*. She is the PI of more than six projects, primarily supported by Key R&D projects of the Shaanxi Provincial Department of Science and Technology, the China Postdoctoral Foundation, and the Yulin Science and Technology Bureau.

Editorial

Applications of Radio Frequency Heating in Food Processing

Shaojin Wang [1,2,*] and Rui Li [1]

1. College of Mechanical and Electronic Engineering, Northwest A&F University, Xianyang 712100, China
2. Department of Biological Systems Engineering, Washington State University, Pullman, WA 99164-6120, USA
* Correspondence: shaojinwang@nwafu.edu.cn

Citation: Wang, S.; Li, R. Applications of Radio Frequency Heating in Food Processing. *Foods* **2023**, *12*, 1133. https://doi.org/10.3390/foods12061133

Received: 27 February 2023
Accepted: 7 March 2023
Published: 8 March 2023

Copyright: © 2023 by the authors. Licensee MDPI, Basel, Switzerland. This article is an open access article distributed under the terms and conditions of the Creative Commons Attribution (CC BY) license (https://creativecommons.org/licenses/by/4.0/).

Considering safety concerns regarding postharvest agricultural products or foods, environmental pollution caused by chemical fumigations, and increased international regulations to limit the use of fumigants, it is an extremely urgent task to develop novel and environmentally friendly physical alternatives to the postharvest control of insect pests and pathogens. Radio-frequency (RF) treatment has been identified as a novel physical heating method, providing fast and volumetric heating. The purpose of this Special Issue is to focus on the recent developments and applications of RF heating in food and agricultural product processing, such as disinfestations, drying, pasteurization, sterilization, roasting, temping and thawing. This Special Issue aims to present the most significant methods, research strategies and protocols used in the development of environmentally friendly food processing based on RF energy.

This Special Issue collates 13 papers related to RF processing technology. The thermal and dielectric properties of agricultural products [1] are important for understanding RF heating principles and necessary for establishing a computer simulation model to improve RF heating uniformity in foods [2,3]. RF drying has been widely used to replace conventional dehydration methods for improving product quality and storage stability [4–6]. Based on thermal inactivation kinetics, RF energy has been applied to control pests and pathogens [7–9] in foods and agricultural products, shortening treatment time but maintaining product quality [10]. RF heating could also be used in the food industry to roast nuts since RF-roasted almonds were found to have a better flavor, texture, and overall preferability compared to commercial almonds [11]. Finally, RF heating has been successfully used for the tempering and thawing the frozen foods by reducing the drip and micronutrient losses [12,13].

In this Special Issue, we hope to establish a sound basis for the further development of thermal methods for the RF control of pests and pathogens, as well as the drying, roasting and thawing of foods and harvested agricultural commodities. This Special Issue will form an important resource for readers who are interested in the knowledge, methods and strategies used in the development of environmentally friendly RF processes. This Special Issue may also be suitable for researchers who work on heating technologies relevant to RF all over the world.

The editorial team would like to express its gratitude to the contributors for sharing their novel ideas, new knowledge, and innovative findings. Each paper has been handled by a qualified editorial team and reviewed by two international experts in RF fields. Therefore, we thank those reviewers for helping us maintain a high standard in the Special Issue. We hope to maintain a strong understanding, collaboration, and friendship with our colleagues across disciplinary, institutional, and country border lines.

Author Contributions: S.W. wrote the first version of the manuscript; R.L. revised the manuscript. All authors have read and agreed to the published version of the manuscript.

Funding: This research was funded by National Foreign Expert Project, Ministry of Science and Technology, China (DL2022172013L, G2022172003L).

Institutional Review Board Statement: Not applicable.

Informed Consent Statement: Not applicable.

Data Availability Statement: Not applicable.

Conflicts of Interest: The authors declare no conflict of interest.

References

1. Mao, Y.; Hao, Y.; Guan, X.; Wang, P.; Wang, S. Temperature and Moisture Dependent Dielectric and Thermal Properties of Walnut Components Associated with Radio Frequency and Microwave Pasteurization. *Foods* **2022**, *11*, 919. [CrossRef]
2. Goñi, S.; d'Amore, M.; Della Valle, M.; Olivera, D.; Salvadori, V.; Marra, F. Effect of Load Spatial Configuration on the Heating of Chicken Meat Assisted by Radio Frequency at 40.68 MHz. *Foods* **2022**, *11*, 1096. [CrossRef] [PubMed]
3. Wang, K.; Huang, L.; Xu, Y.; Cui, B.; Sun, Y.; Ran, C.; Fu, H.; Chen, X.; Wang, Y.; Wang, Y. Evaluation of Pilot-Scale Radio Frequency Heating Uniformity for Beef Sausage Pasteurization Process. *Foods* **2022**, *11*, 1317. [CrossRef] [PubMed]
4. Chitsuthipakorn, K.; Thanapornpoonpong, S. Effect of Large-Scale Paddy Rice Drying Process Using Hot Air Combined with Radio Frequency Heating on Milling and Cooking Qualities of Milled Rice. *Foods* **2022**, *11*, 519. [CrossRef] [PubMed]
5. Huang, C.; Cheng, Y.; Chen, S. Hot Air-Assisted Radio Frequency (HARF) Drying on Wild Bitter Gourd Extract. *Foods* **2022**, *11*, 1173. [CrossRef] [PubMed]
6. Chen, X.; Liu, Y.; Zhang, R.; Zhu, H.; Li, F.; Yang, D.; Jiao, Y. Radio Frequency Drying Behavior in Porous Media: A Case Study of Potato Cube with Computer Modeling. *Foods* **2022**, *11*, 3279. [CrossRef]
7. Gao, Y.; Guan, X.; Wan, A.; Cui, Y.; Kou, X.; Li, R.; Wang, S. Thermal Inactivation Kinetics and Radio Frequency Control of Aspergillus in Almond Kernels. *Foods* **2022**, *11*, 1603. [CrossRef] [PubMed]
8. Yen, Y.; Chen, S. Influence of Radio Frequency Heating on the Pasteurization and Drying of Solid-State Fermented Wolfiporia cocos Products. *Foods* **2022**, *11*, 1766. [CrossRef] [PubMed]
9. Wang, K.; Ran, C.; Cui, B.; Sun, Y.; Fu, H.; Chen, X.; Wang, Y.; Wang, Y. Sterilizing Ready-to-Eat Poached Spicy Pork Slices Using a New Device: Combined Radio Frequency Energy and Superheated Water. *Foods* **2022**, *11*, 2841. [CrossRef] [PubMed]
10. Zhang, Z.; Zhang, B.; Zhu, L.; Zhao, W. Microstructure, Digestibility and Physicochemical Properties of Rice Grains after Radio Frequency Treatment. *Foods* **2022**, *11*, 1723. [CrossRef] [PubMed]
11. Lian, T.; Chen, S. Developing Radio-Frequency Roasting Protocols for Almonds Based on Quality Evaluations. *Foods* **2022**, *11*, 1885. [CrossRef] [PubMed]
12. Jiang, J.; Zhou, F.; Xian, C.; Shi, Y.; Wang, X. Effects of Radio Frequency Tempering on the Texture of Frozen Tilapia Fillets. *Foods* **2021**, *10*, 2663. [CrossRef] [PubMed]
13. Han, R.; He, J.; Chen, Y.; Li, F.; Shi, H.; Jiao, Y. Effects of Radio Frequency Tempering on the Temperature Distribution and Physiochemical Properties of Salmon (*Salmo salar*). *Foods* **2022**, *11*, 893. [CrossRef] [PubMed]

Disclaimer/Publisher's Note: The statements, opinions and data contained in all publications are solely those of the individual author(s) and contributor(s) and not of MDPI and/or the editor(s). MDPI and/or the editor(s) disclaim responsibility for any injury to people or property resulting from any ideas, methods, instructions or products referred to in the content.

Article

Temperature and Moisture Dependent Dielectric and Thermal Properties of Walnut Components Associated with Radio Frequency and Microwave Pasteurization

Yuxiao Mao [1], Yujun Hao [1], Xiangyu Guan [1], Penghao Wang [1] and Shaojin Wang [1,2,*]

1. College of Mechanical and Electronic Engineering, Northwest A & F University, Yangling 712100, China; maoyuxiao@nwafu.edu.cn (Y.M.); haoyujun@nwafu.edu.cn (Y.H.); xiangyuguan@nwafu.edu.cn (X.G.); wangpenghao@nwafu.edu.cn (P.W.)
2. Department of Biological Systems Engineering, Washington State University, Pullman, WA 99164-6120, USA
* Correspondence: shaojinwang@nwsuaf.edu.cn; Tel.: +86-29-8709-2391; Fax: +86-29-8709-1737

Abstract: To provide necessary information for further pasteurization experiments and computer simulations based on radio frequency (RF) and microwave (MW) energy, dielectric and thermal properties of walnut components were measured at frequencies between 10 and 3000 MHz, temperatures between 20 and 80 °C, and moisture contents of whole walnuts between 8.04% and 20.01% on a dry basis (d.b.). Results demonstrated that dielectric constants and loss factors of walnut kernels and shells decreased dramatically with raised frequency within the RF range from 10 to 300 MHz, but then reduced slightly within the MW range from 300 to 3000 MHz. Dielectric constant, loss factor, specific heat capacity, and thermal conductivity increased with raised temperature and moisture content. Dielectric loss factors of kernels were greater than those of shells, leading to a higher RF or MW heating rate. Penetration depth of electromagnetic waves in walnut components was found to be greater at lower frequencies, temperatures, and moisture contents. The established regression models with experimental results could predict both dielectric and thermal properties with large coefficients of determination ($R^2 > 0.966$). Therefore, this study offered essential data and effective guidance in developing and optimizing RF and MW pasteurization techniques for walnuts using both experiments and mathematical simulations.

Keywords: dielectric properties; thermal properties; walnut components; pasteurization; radio frequency; microwave

1. Introduction

Walnuts (*Juglans regia* L.) have experienced fast-growing harvest areas and annual yields due to their unique and favorite flavor and taste, and a rapid increase in the value of production. The largest producer of in-shell walnuts in 2022 was China, where over 30% (1.10 Mt) of the global production (3.32 Mt) was produced [1]. Fresh walnuts are susceptible to decay during storage due to microbial spoilage. Many bacterial pathogens, especially *Salmonella*, have been detected in walnuts [2,3]. In addition, aflatoxin contamination in walnuts during storage is the primary cause of food safety problems and mainly results from the infection of *Aspergillus flavus*, *A. niger*, and *Penicillium citrinum* [4]. For in-shell walnuts, the contamination of pathogenic bacteria is firstly on shells, and usually transfers to kernels during processing and storage [5]. To minimize the influence of potential spoilage on quality deterioration and reduce the population of bacterial pathogens and mold to a level qualified for consumption, pasteurization treatments for in-shell walnuts are crucial.

Some traditional methods have been investigated as pasteurization techniques for agricultural products. However, pasteurization by hot water at 76 °C for 3 min only achieves a reduction of about 1 log CFU/g of total plate count (TPC) on the surface of cantaloupe and is not suitable for dried products [6]. Steam provides a short pasteurizing duration of 25 s and a 5 log CFU/g reduction of *Salmonella* serotype Enteritidis [7], but the high processing temperature might lead to quality deterioration. Therefore, exploring alternative advanced technologies is desirable to develop an effective, practical, and economically viable method for pasteurizing in-shell walnuts.

Dielectric heating, including radio frequency (RF) and microwave (MW) treatments, raises temperatures of materials quickly and volumetrically since the heat generates inside products by dipole rotation and ionic migration as subjected to an alternating electromagnetic (EM) field [8]. Owing to the rapid heating, shortened processing duration, maintained product quality, and no chemical residues, the RF and MW treatments have been studied as novel pasteurization methods for nuts and other agricultural products, such as hazelnuts [9], green beans [10], walnuts [11], and paprika [12]. RF wave is more suitable for the treatment of bulk and thick materials because of the more uniform heating caused by the larger penetration depth as compared with MW processing [13]. By contrast, MW wave is commonly used for pasteurizing thin layer materials owing to the higher heating rates at larger frequencies [14]. Dielectric properties (DPs) of walnuts are essential data used to analyze the efficiency of MW and RF pasteurization processes. Mathematically, DPs are expressed in the form of relative complex permittivity (ε) as follows:

$$\varepsilon = \varepsilon' - j\varepsilon'' \qquad (1)$$

where j is the imaginary unit, $\sqrt{-1}$. The real part, ε', is dielectric constant, representing the capacity of materials to store the electric energy. The imaginary part, ε'', is dielectric loss factor and refers to the ability of transforming electric energy into thermal energy. During RF or MW heating for agricultural products, thermal properties (TPs) containing specific heat capacity (C_p, J/(kg·K)) and thermal conductivity (k, W/(m·K)) are also vital factors influencing the absorption and conduction of the thermal energy [15]. Specific heat capacity indicates the amount of heat required to raise the temperature of 1 kg of materials by 1 K without any phase or chemical changes, while thermal conductivity refers to the rate of heat transfer within materials. Nevertheless, both dielectric and thermal properties are mainly affected by some factors, such as temperature, moisture content, and frequency. Consequently, a comprehensive understanding of dielectric and thermal properties of walnuts is important for the effective design and improvement of pasteurization processes using RF or MW energy [16].

Several studies have investigated the dielectric and thermal properties of agricultural products as subjected to different ranges of frequency, moisture content (MC), and temperature for various purposes [17–20]. Specifically, Zhu et al. (2014) studied the dielectric properties of hazelnut kernels within a frequency band from 10 to 4500 MHz, a temperature range of 20–60 °C, and a MC range of 4.6–20.3% on a wet basis (w.b.) using an open-ended coaxial method [21]. The results showed that dielectric constant and loss factor of hazelnut kernels raised with increasing temperature, MC and falling frequency. Boldor et al. (2004) found that dielectric properties of both in-shell and shelled peanuts increased with increasing density and applied dielectric theory mixture equations to correlate these two parameters [22]. In terms of thermal properties, Huang et al. (2016) reported that both specific heat capacity and thermal conductivity of soybeans increased with raised temperature from 20 to 80 °C and MC from 4.64% to 7.86% (w.b.) [23]. Similarly, Perussello et al. (2014) chose Okara as an object, which is the residue of the soy beverage and tofu production and observed that thermal properties of Okara decreased with reduced moisture [24]. In the nut processing industry, shelling is a procedure after the pasteurization, so in-shell walnuts are common materials used for pasteurization treatments. Walnut kernels contain mainly fat and protein whereas lignocellulose in shells accounts for more than 80%, leading to a large difference between their dielectric properties [25,26]. In addition, to reduce the thermal resistance of bacteria, the MC of walnuts is commonly raised by adding distilled water before the pasteurization and removed by drying after the pasteurization. Therefore, dielectric properties of walnuts vary greatly as the moisture is evaporated continuously during the pasteurization and subsequent drying. However, to the authors' knowledge,

DPs of walnuts were only determined by Wang et al. (2003) at a single MC of 3% (w.b.) for kernels [27]. Furthermore, the information about dielectric and thermal properties of walnut components including kernels and shells is indispensable for further investigations in dielectric pasteurization treatments but unavailable at various MCs, temperatures, and frequencies.

Therefore, the objectives of this study were (1) to determine the dielectric properties of walnut kernels and shells within a frequency range of 10–3000 MHz at temperatures varying from 20 to 80 °C and four MCs of whole walnuts (8%, 12%, 16%, and 20% d.b.), (2) to measure the thermal properties of kernels and shells at a temperature range of 20–80 °C and the four MC levels of whole walnuts, (3) to provide empirical equations describing dielectric and thermal properties of walnut kernels and shells as affected by MC and temperature, and (4) to calculate the penetration depth of electromagnetic energy into kernels and shells of walnuts at the four representative frequencies (27, 40, 915, and 2450 MHz).

2. Materials and Methods

2.1. Materials and Sample Preparation

Dried and raw in-shell walnuts (*Juglans regia* L.) were purchased from a local farm market in Yangling, Shaanxi, China. The integrity of shells was adopted as the criterion of selecting walnuts for experiments. The selected walnuts with intact shells were then stored in polyethylene bags inside a refrigerator at 4 °C until further handling to avoid moisture loss and inhibit quality deterioration in ambient conditions [28]. For obtaining walnut kernels and shells with different levels of MC for further determining dielectric properties, four MC levels of 8%, 12%, 16%, and 20% (d.b.) were chosen as target MCs of whole walnuts based on the pasteurization treatment and a MC of 8%, which is qualified for long-term storage of walnuts in the ambient environment [11,29]. Distilled water with different pre-calculated quantities was added to the polyethylene bags with in-shell walnuts to adjust the MCs to the target ones. These in-shell walnuts were subsequently sealed in polyethylene bags at 4 °C for the complete absorption of water for 7 d when the bags were shaken twice a day to ensure the uniform distribution and complete absorption of the added water. Finally, four MCs of 8.04 ± 0.28%, 12.11 ± 0.08%, 15.91 ± 0.26%, and 20.01 ± 0.82% (d.b.) were obtained for in-shell walnuts, while the MCs of walnut components were adjusted to 4.21 ± 0.32%, 8.08 ± 0.06%, 12.08 ± 0.40%, and 16.23 ± 0.61% (d.b.) for kernels, and 12.51 ± 0.25%, 16.08 ± 0.14%, 20.07 ± 0.10%, and 23.86 ± 1.59% (d.b.) for shells, respectively.

2.2. Measurements of Moisture Content and True Density

The MC measurement was conducted following the AOAC Official Method 925.40 [30]. After separating shells of five stochastically chosen walnuts from the kernels, the two parts were separately ground by a blender (DE-300 g, 30–300 mesh, Zhejiang Hongjingtian Co., Ltd., Jinhua, China). Then, the powders of kernels and shells were placed in a vacuum oven (DZX-6020 B, Nanrong Lab equipment Inc., Shanghai, China) at a temperature of 105 °C and a pressure of 13.3 kPa and dried to a constant weight for about 5 h. To avoid the rapid absorption of moisture in the air by powders under the high temperature, a desiccator was applied to cool the samples to ambient temperature immediately after the drying. The MC of samples was expressed as the percentage of lost weight to the remaining one.

True density is a necessary parameter for preparation of walnut samples for dielectric property measurements. A liquid displacement method, of which the procedures were described in detail in Guo et al. (2008), was employed to determine the true density of walnut kernels and shells at different MCs [31]. To avoid being absorbed by samples, toluene (C_7H_8) was used as the displacement liquid because of the small surface tension. For each MC level, shells and kernels of five randomly chosen in-shell walnuts were separated and cut into small pieces. In each measurement, about 3 g samples were weighed and then placed into a 250 mL Lee's pycnometer with toluene, and the volumes occupied by samples were recorded. The true density was calculated by dividing the weight by the volume. Each test for measuring densities of both walnut shells and kernels was conducted in triplicate at each MC.

2.3. Measurements of Dielectric Properties

An open-ended coaxial probe system, consisting of an impedance analyzer (E4991 B-300, Keysight Technologies Co., Ltd., Palo Alto, CA, USA), a matched calibration kit (E4991 B-010), a high-temperature coaxial cable, the coaxial probe with dielectric probe kit (85070 E-020), a cylindrical customized sample test cell, an oil circulated bath (SST-20, Guanya Constant Temperature Cooling Technology Co., Ltd., Wuxi, China), and an auxiliary computer, was adopted to measure the dielectric properties of walnut kernels and shells within a frequency range between 10 and 3000 MHz. The detailed structure of the system was described by Li et al. (2017) [32].

Before the dielectric property determination, two calibrations were carried out. For the first calibration, the impedance analyzer was switched on for at least 20 min to warm up, and then calibrated using open, short, and 50 Ω load successively. Subsequently, the open-ended coaxial probe was attached to the system and calibrated by air, short circuit, and 25 °C deionized water in the order of the second calibration. To ensure the tight contact between the probe and samples to eliminate air gaps, kernels and shells of five stochastically selected walnuts were milled into powders, and about 12 g samples were put into the sample test cell (21 mm diameter × 50 mm height) and compressed into cylinders prior to experiments. Temperatures from 20 to 80 °C with an interval of 10 °C were adopted based on the maximal range commonly used for pasteurizing in-shell walnuts [11]. The temperature of samples was monitored by a thermocouple (HH-25 TC, Type-T, OMEGA Engineering Inc., Stamford, CT, USA) and maintained at selected levels by the oil bath. Each test for measuring dielectric properties of both walnut kernels and shells was performed in triplicate at each temperature and MC.

2.4. Measurements of Thermal Properties

A thermal property analyzer (KD2 Pro, Decagon Inc., Pullman, WA, USA) combining with an SH-1 sensor was used for determining thermal properties involving thermal conductivity and specific heat capacity of walnut kernels and shells. For each measurement, samples with different MCs were placed into the test cell and pressed into cylinders following the same aforementioned procedure. Subsequently, the Type-T thermocouple and the SH-1 sensor were inserted into the center of samples. The same temperature range of 20–80 °C with an interval of 10 °C as that for measuring DPs was employed. After each predetermined temperature was reached by the oil bath, the thermal property analyzer was turned on for measuring thermal properties. Each test for measuring thermal properties of both walnut kernels and shells was also conducted in triplicate at each temperature and MC.

2.5. Determination of Penetration Depth

Penetration depth (d_p, m) of MW and RF energy is defined as the distance travelled by the electromagnetic wave when the power declines to 1/e (e = 2.718) of the initial one at the entry surface of the material [33]. The penetration depth was determined by the following equation:

$$d_p = \frac{c}{2\pi f \sqrt{2\varepsilon' \left[\sqrt{1 + \left(\frac{\varepsilon''}{\varepsilon'}\right)^2} - 1 \right]}} \tag{2}$$

where c is the speed of light in free space (3×10^8 m/s) and f is the frequency (MHz) of electromagnetic wave.

2.6. Statistical Analysis

The experimental results were expressed as mean ± standard deviations over all three replicates. A significance test was performed using the statistical analysis software SPSS 23.0 version (SPSS Inc., Chicago, IL, USA). Differences ($p \leq 0.05$) between means were estimated by analysis of variance (ANOVA) and Tukey's post hoc test.

3. Results and Discussion

3.1. True Density

Table 1 lists true densities of walnut kernels and shells with different MCs. True density of kernels decreased from 1.019 to 1.003 g/cm^3 as the MC rose from 4.21% to 16.23% (d.b.). However, true density of shells increased from 0.988 to 1.067 g/cm^3 with increasing MC from 12.51% to 23.86% (d.b.). Similar results are also reported in the research on hazelnuts [34] and Turkish walnuts [35], which might be caused by greater increase proportions of volumes than those of weights for kernels with added moisture. Since the variation of the true density in both walnut kernels and shells was negligible, the effect of density on the dielectric properties was neglected in this study. As a result, the true densities of walnut kernels and shells used for determining thermal and dielectric properties were defined as their means of 1.010 and 1.030 g/cm^3, respectively.

Table 1. True densities (mean ± SD over three replicates) of walnut kernels and shells at different moisture contents.

	Moisture Content (%, d.b.)	True Density (g/cm^3)
Kernel	4.21	1.019 ± 0.006 a
	8.08	1.012 ± 0.004 a b
	12.08	1.007 ± 0.003 b
	16.23	1.003 ± 0.003 b
Shell	12.51	0.988 ± 0.023 c
	16.08	1.021 ± 0.015 b c
	20.07	1.045 ± 0.006 a b
	23.86	1.067 ± 0.007 a

Different lower-case letters indicate that means are significantly different at $p = 0.05$ among different moisture contents.

3.2. Frequency-Dependent Dielectric Properties

In general, dielectric constants and loss factors of walnut kernels and shells at seven temperatures and four respective MCs decreased with raised frequency from 10 to 3000 MHz (Figures 1–4). For kernels with a MC of 4.21% (d.b.), the dielectric constants were smaller than 4.0 while the loss factors were less than 1.2 at seven temperatures. The extremely small dielectric properties resulted mainly from low ionic conductivity and water relaxation caused by low MC [36]. In addition, both dielectric constants and loss factors increased slightly and reached a local maximum at the frequency of about 1600 MHz. The phenomenon was caused by the joint effects of ionic conductivity and free water relaxation whose contribution to dielectric losses reached its maximum at about 2000 MHz as the frequency increased from 10 to 3000 MHz [36]. Liu et al. (2020) and Li et al. (2017) found similar trends in the dielectric properties of honey [37] and almonds [32], respectively, and the different frequencies corresponding to the local maximum were due to the different chemical composition of various foods. In comparison, dielectric properties of shells with the lowest MC (12.51% d.b.) decreased steadily with no local maximum as the frequency increased. It was because the contribution of free water relaxation to the dielectric loss was much smaller than that of ionic conductivity at relatively high MCs. Particularly, standard deviations of dielectric properties of walnut kernels with an MC of 4.21% (d.b.) were relatively larger than those of kernels with larger MCs owing to the greater effect of systematic errors on experimental results.

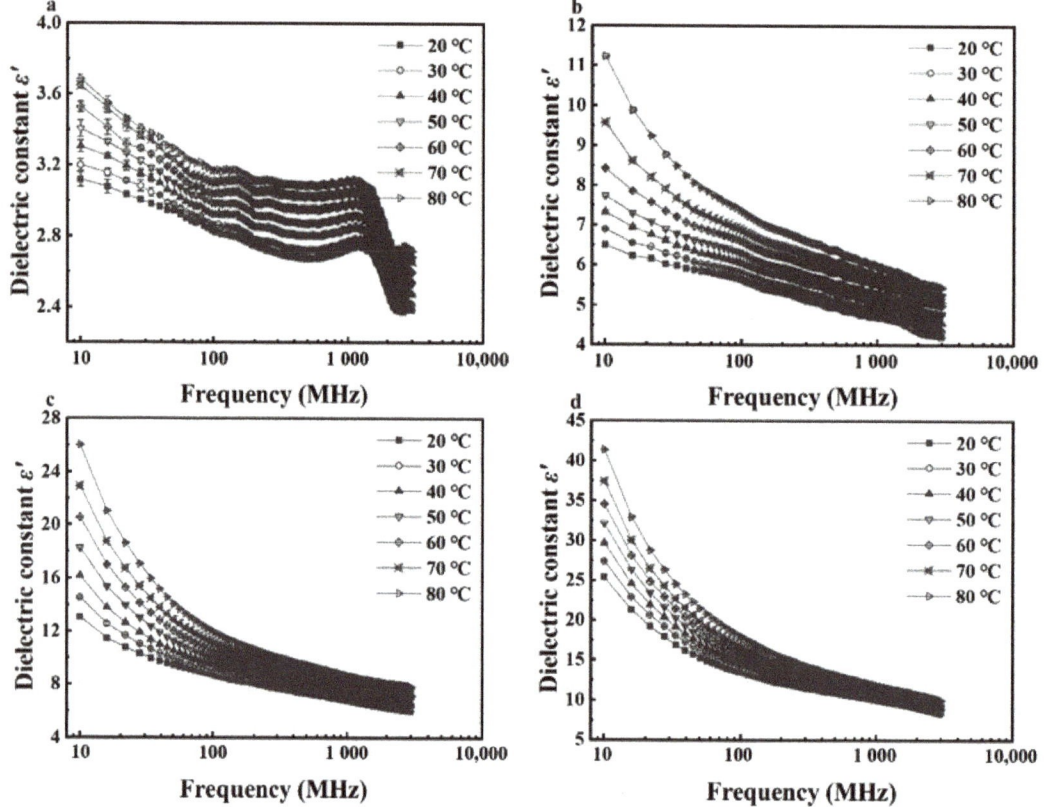

Figure 1. Frequency-dependent dielectric constant (ε') of walnut kernels at 7 temperatures with moisture contents of 4.21% (**a**), 8.08% (**b**), 12.08% (**c**), and 16.23% (**d**) (d.b.).

For kernels with MCs of 8.08%, 12.08%, and 16.23% (d.b.), and shells with MCs of 16.08%, 20.07%, and 23.86% (d.b.), their dielectric constants and loss factors decreased with increasing frequency with greater rates at lower frequencies (Figures 3 and 4). For example, as the frequency rose from 10 to 300 MHz, dielectric constant and loss factor of kernels with a MC of 16.23% (d.b.) descended from 25.41 and 42.18 to 11.55 and 3.87, respectively, at a temperature of 20 °C (Figures 1d and 2d), leading to decreasing amplitudes of 54.55% and 90.83%. However, as the frequency continued to increase from 300 to 3000 MHz, the dielectric constant and loss factor declined to 8.35 and 3.18, respectively, resulting in decreasing amplitudes of only 27.71% and 17.83%. The results were consistent with the theory in Feng et al. (2002) [36] and Gezahegn et al. (2021) [38]. To be specific, at frequencies between 10 and 300 MHz, the ionic conductivity dominated the dielectric dispersion and its contribution to DPs decreased rapidly with increasing frequency. As the frequency rose from 300 to 3000 MHz, the water relaxation was the dominant mechanism and varied slightly with increasing frequency. Those trends in dielectric properties were similar to those of tuna [18], edible fungi powder [39], and Camellia oleifera seed kernels [40].

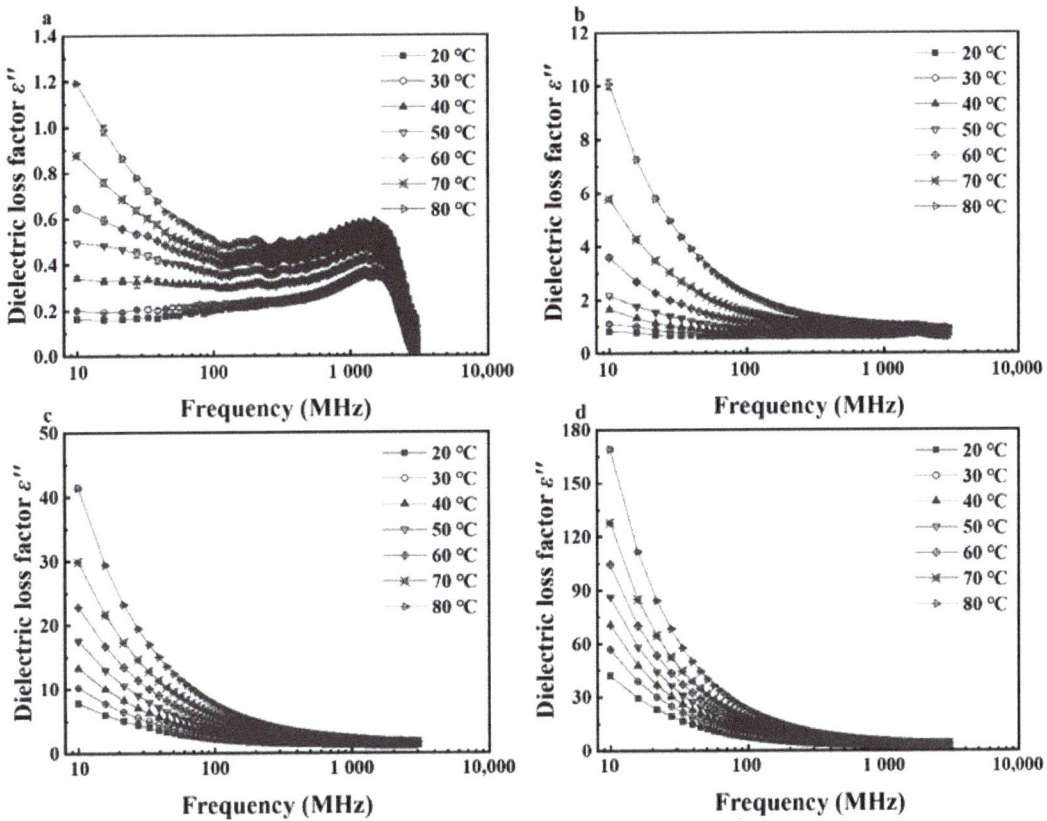

Figure 2. Frequency-dependent dielectric loss factor (ε'') of walnut kernels at 7 temperatures with moisture contents of 4.21% (**a**), 8.08% (**b**), 12.08% (**c**), and 16.23% (**d**) (d.b.).

3.3. Moisture and Temperature-Dependent Dielectric Properties

Effects of temperature (20–80 °C) and MC (4.21%–16.23% d.b. for kernels and 12.51%–23.86% d.b. for shells) on dielectric properties of walnut components at frequencies of 27, 40, 915, and 2450 MHz are shown in Figures 5–8. Generally, both dielectric constant and loss factor of walnut kernels and shells increased with increasing temperature and MC at a selected frequency. For instance, at a certain frequency of 27 MHz, dielectric constant ε' and dielectric loss factor ε'' rose from 10.03 and 18.83 to 50.87 and 45.41, respectively, with raised MC from 12.51% to 23.86% (d.b.) at a fixed temperature of 80 °C; when the MC was fixed at 23.86%, ε' and ε'' increased from 4.13 and 0.80 to 50.87 and 45.41, respectively, when the temperature increased from 20 to 80 °C. On the one hand, the enhancement of ionic conductivity by greater MC was the main cause of the increase in dielectric properties. To be specific, most water molecules in materials with low MCs were usually arranged in a monolayer form and bonded tightly with other polar molecules by hydrogen bonds, leading to inferior ionic conductivities. With the increase in the MC, the arrangement of water molecules transformed from the monolayer to a multilayer form and air voids in walnut kernels and shells were filled gradually, raising the ionic solubility and the water dipole mobility simultaneously; thus, improving the ionic conductivity [41]. On the other hand, as the temperature was elevated, the viscosity of materials was reduced, and the ionic migration and dipolar rotation were accelerated. Consequently, the dielectric constant and loss factor of walnut kernels and shells were enhanced by these two mechanisms. The results were found to be similar to those reported by Cao et al. (2018) for surimi [42]. Based on the aforementioned theory, during RF or MW drying, samples with larger MCs had greater dielectric constants and loss factors, making them tend to be heated rapidly, then the resulted high temperature leading to larger drying rates of samples. This phenomenon was commonly named as the "moisture levelling effect", which was conducive to improving the uniformity of moisture distribution in products and the product quality stability during storage. Moreover, the increasing rates of dielectric constant and loss factor with MC were greater at higher temperatures, which might be caused by the synergistic effect of temperature and MC and are also reported by Li et al. (2017) and Jiang et al. (2020) in the research on dielectric properties of almonds [32] and *Agaricus bisporus* slices [43]. The phenomenon suggested that applying relatively low temperature at the initial RF or MW pasteurization stage helped to avoid the "heating runaway" effect.

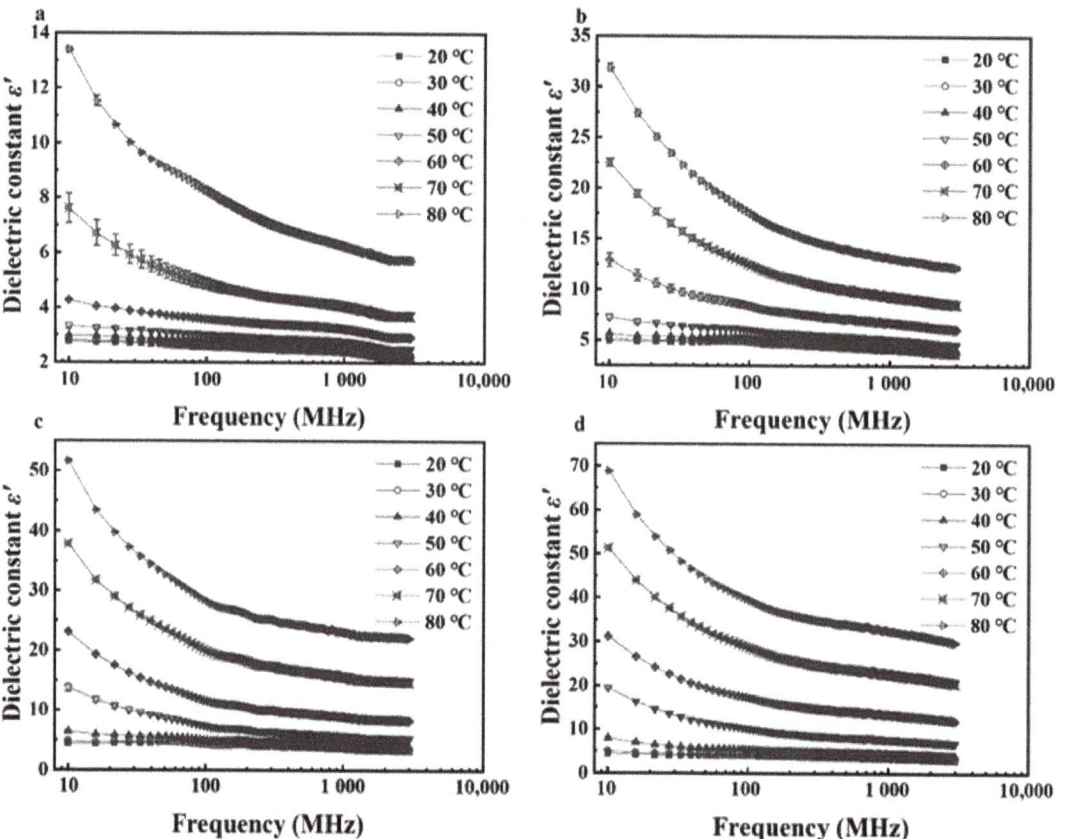

Figure 3. Frequency-dependent dielectric constant (ε') of walnut shells at 7 temperatures with moisture contents of 12.51% (**a**), 16.08% (**b**), 20.07% (**c**), and 23.86% (**d**) (d.b.).

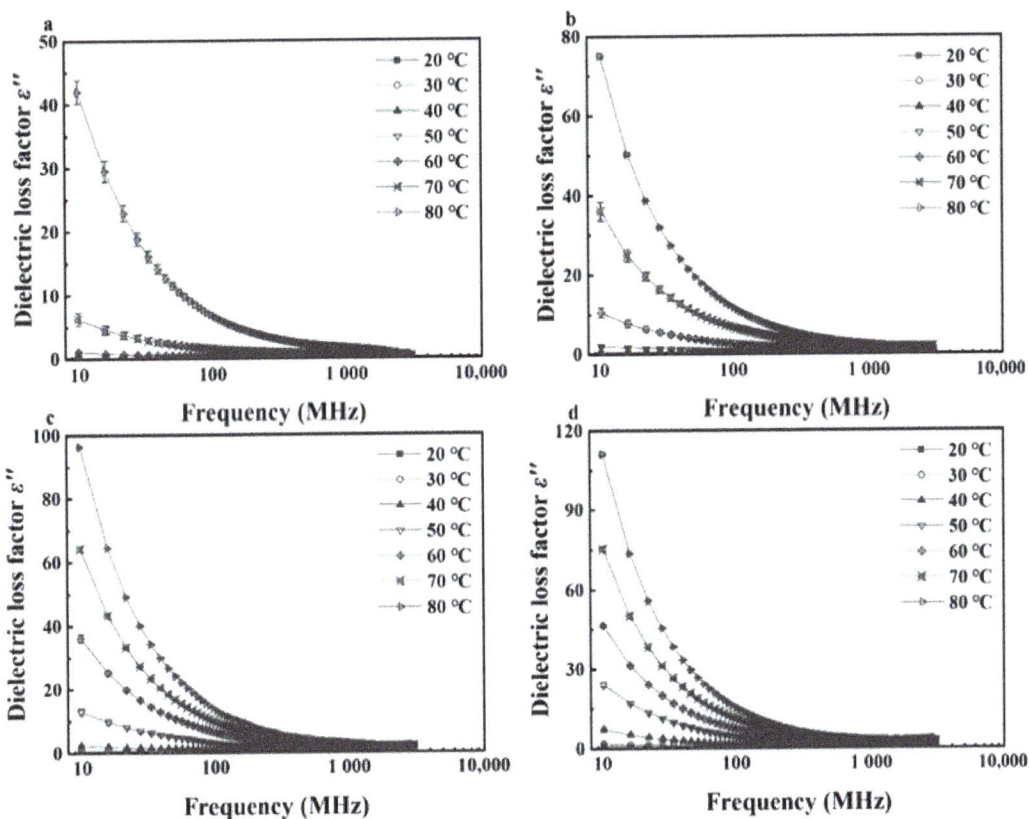

Figure 4. Frequency-dependent dielectric loss factor (ε'') of walnut shells at 7 temperatures with moisture contents of 12.51% (**a**), 16.08% (**b**), 20.07% (**c**), and 23.86% (**d**) (d.b.).

Figures 5–8 also show a trend that the variations in dielectric properties of kernels with MC from 4.21% to 16.23% (d.b.) were greater than those with temperatures from 20 to 80 °C whereas temperature had greater influence on dielectric properties of shells. For example, dielectric constant of walnut kernels at 20 and 80 °C dropped from 17.94 and 26.39 to 3.00 and 3.41, respectively, as the MC decreased from 16.23% to 4.21% (d.b.) under a frequency of 27 MHz, holding great decreasing amplitudes of 83.28% and 87.08%. By comparison, the ε' under MCs of 4.21% and 16.23% (d.b.) declined from 3.41 and 26.39 to 3.00 and 17.94 as the temperature decreased from 80 to 20 °C with decreasing amplitudes of only 12.02% and 32.02%. Yu et al. (2015) reported that dielectric properties of bulk canola seeds varied with temperature and MC following a similar trend [44]. Furthermore, Figures 5 and 7 demonstrate that dielectric loss factors of kernels were larger than those of shells with corresponding MCs at the same frequency and temperature, resulting in greater heating rates for kernels as compared with those of shells.

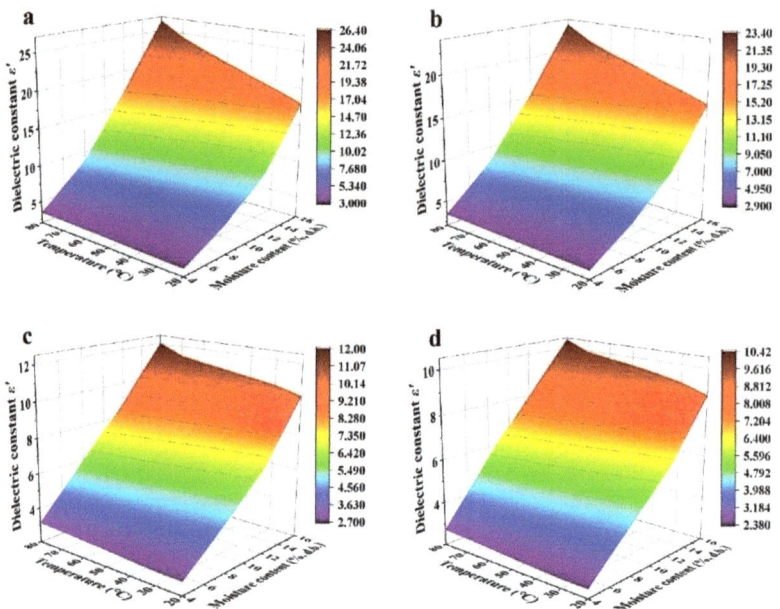

Figure 5. Moisture content and temperature-dependent dielectric constant (ε') of walnut kernels at frequencies of 27 (**a**), 40 (**b**), 915 (**c**), and 2450 (**d**) MHz.

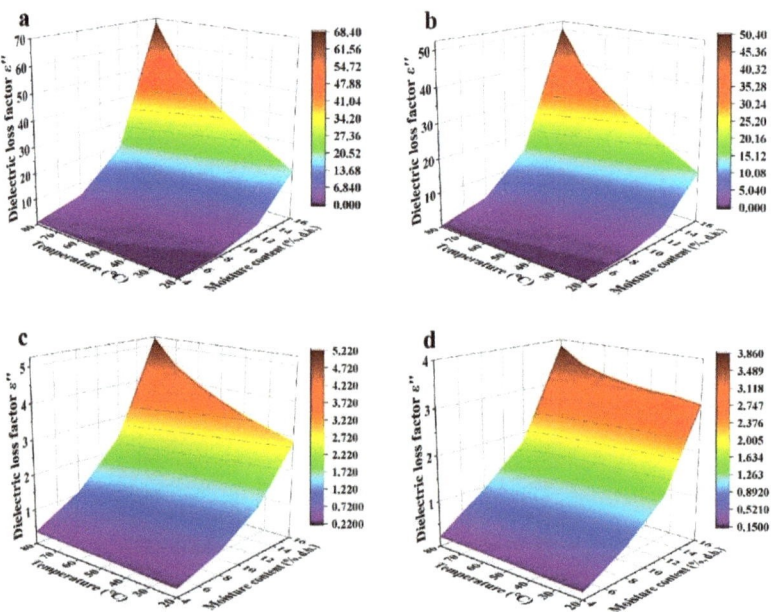

Figure 6. Moisture content and temperature-dependent dielectric loss factor (ε'') of walnut kernels at frequencies of 27 (**a**), 40 (**b**), 915 (**c**), and 2450 (**d**) MHz.

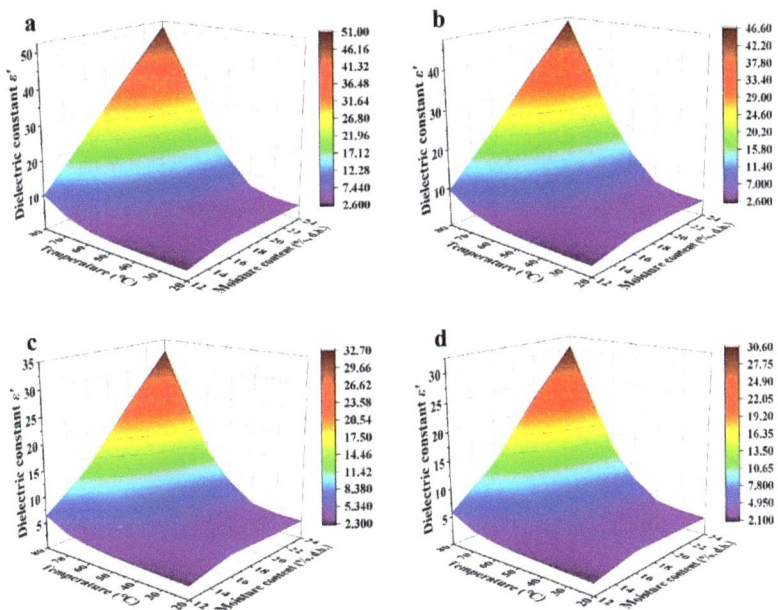

Figure 7. Moisture content and temperature-dependent dielectric constant (ε') of walnut shells at frequencies of 27 (**a**), 40 (**b**), 915 (**c**), and 2450 (**d**) MHz.

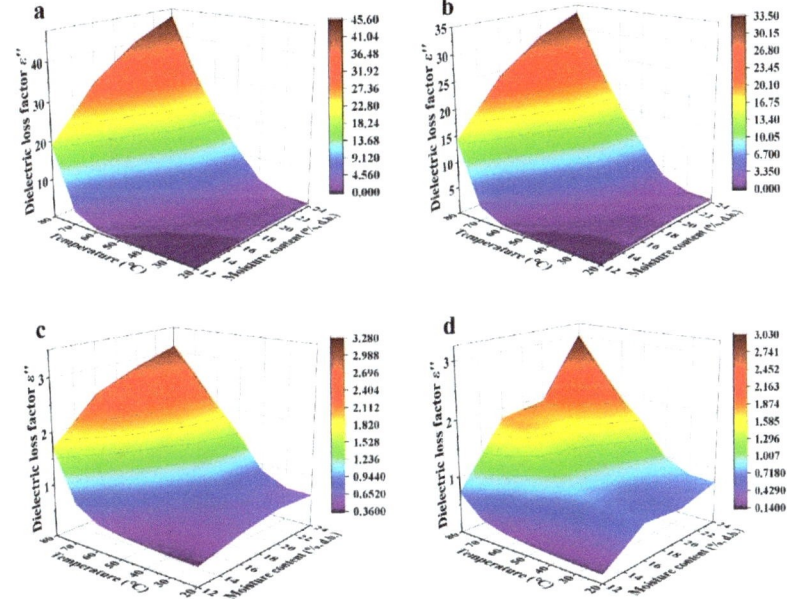

Figure 8. Moisture content and temperature-dependent dielectric loss factor (ε'') of walnut shells at frequencies of 27 (**a**), 40 (**b**), 915 (**c**), and 2450 (**d**) MHz.

3.4. Regression Models for Dielectric Properties

The established regression models for dielectric properties of walnut components as a function of temperature and MC at four frequencies of 27, 40, 915, and 2450 MHz are listed in Table 2. At frequencies in RF bands of 27 and 40 MHz, quadratic polynomial equations could predict moisture and temperature-dependent dielectric constant ε' of kernels and dielectric loss factor ε'' of shells with the best fit, while cubic polynomial models were the best fit for ε' of shells and ε'' of kernels. For frequencies of 915 and 2450 MHz in the MW band, cubic polynomial relationships were the best fit for dielectric properties of both walnut kernels and shells except for the ε'' of shells at 915 MHz whose optimal fitting model was a quadratic polynomial equation. According to results of analysis of variance (ANOVA), all established models could predict the measured dielectric properties of walnut components with a good fit at a significance level of 0.0001 ($p < 0.0001$) and their coefficients of determination (R^2) were all above 0.966. These results suggested that the developed mathematical models could predict the dielectric constants and loss factors of walnut kernels and shells with an adequate precision within a temperature range of 20–80 °C at MCs from 4.21% to 16.23% (d.b.) for kernels and 12.51% to 23.86% (d.b.) for shells and provide necessary data for further computer simulation.

Table 2. Regression equations of dielectric properties of walnut components at the 4 representative frequencies as a function of moisture content (M, %) and temperature (T, °C).

Frequency (MHz)	Material	Dielectric Properties	R^2
27	Kernel	$\varepsilon' = 4.24 - 8.78 \times 10^{-2} T - 0.37 M + 1.14 \times 10^{-2} TM + 4.68 \times 10^{-4} T^2 + 6.62 \times 10^{-2} M^2$	0.999
		$\varepsilon'' = -34.17 + 1.10 T + 9.48 M - 0.15 TM - 1.47 \times 10^{-2} T^2 - 0.95 M^2 + 6.58 \times 10^{-4} T^2 M + 7.20 \times 10^{-3} TM^2 + 7.51 \times 10^{-5} T^3 + 3.31 \times 10^{-2} M^3$	0.997
	Shell	$\varepsilon' = -46.15 + 0.61 T + 7.64 M - 7.18 \times 10^{-2} TM - 1.20 \times 10^{-2} T^2 - 0.34 M^2 + 1.08 \times 10^{-3} T^2 M + 6.95 \times 10^{-4} TM^2 + 1.64 \times 10^{-5} T^3 + 5.23 \times 10^{-3} M^3$	0.996
		$\varepsilon'' = 21.36 - 1.70 T + 0.31 M + 4.65 \times 10^{-2} TM + 1.38 \times 10^{-2} T^2 - 4.06 \times 10^{-2} M^2$	0.974
40	Kernel	$\varepsilon' = 3.52 - 7.32 \times 10^{-2} T - 0.17 M + 9.37 \times 10^{-3} TM + 3.98 \times 10^{-4} T^2 + 5.19 \times 10^{-2} M^2$	0.999
		$\varepsilon'' = -23.27 + 0.75 T + 6.51 M - 0.10 TM - 1.03 \times 10^{-2} T^2 - 0.66 M^2 + 4.63 \times 10^{-4} T^2 M + 4.95 \times 10^{-3} TM^2 + 5.31 \times 10^{-5} T^3 + 2.33 \times 10^{-2} M^3$	0.998
	Shell	$\varepsilon' = -43.54 + 0.72 T + 6.81 M - 7.53 \times 10^{-2} TM - 1.29 \times 10^{-2} T^2 - 0.29 M^2 + 1.03 \times 10^{-3} T^2 M + 7.79 \times 10^{-4} TM^2 + 2.35 \times 10^{-5} T^3 + 4.09 \times 10^{-3} M^3$	0.996
		$\varepsilon'' = 12.60 - 1.23 T + 0.55 M + 3.39 \times 10^{-2} TM + 1.01 \times 10^{-2} T^2 - 3.79 \times 10^{-2} M^2$	0.974
915	Kernel	$\varepsilon' = 0.19 + 7.10 \times 10^{-3} T + 0.59 M + 7.12 \times 10^{-2} TM - 6.19 \times 10^{-4} T^2 - 3.09 \times 10^{-2} M^2 + 1.00 \times 10^{-5} T^2 M - 3.12 \times 10^{-4} TM^2 + 3.93 \times 10^{-6} T^3 + 1.85 \times 10^{-3} M^3$	1.000
		$\varepsilon'' = -1.39 + 4.78 \times 10^{-2} T + 0.40 M - 5.72 \times 10^{-3} TM - 6.89 \times 10^{-4} T^2 - 3.26 \times 10^{-2} M^2 + 3.68 \times 10^{-5} T^2 M + 2.44 \times 10^{-4} TM^2 + 3.36 \times 10^{-6} T^3 + 1.39 \times 10^{-3} M^3$	0.999
	Shell	$\varepsilon' = -61.30 + 1.27 T + 8.64 M - 9.72 \times 10^{-2} TM - 1.73 \times 10^{-2} T^2 - 0.37 M^2 + 8.86 \times 10^{-4} T^2 M + 1.28 \times 10^{-3} TM^2 + 5.06 \times 10^{-5} T^3 + 5.61 \times 10^{-3} M^3$	0.997
		$\varepsilon'' = -0.96 - 7.15 \times 10^{-2} T + 0.25 M + 2.30 \times 10^{-2} TM + 6.37 \times 10^{-4} T^2 - 7.64 \times 10^{-3} M^2$	0.967
2450	Kernel	$\varepsilon' = 0.28 - 1.10 \times 10^{-2} T + 0.51 M + 8.95 \times 10^{-3} TM - 3.94 \times 10^{-4} T^2 - 2.43 \times 10^{-2} M^2 - 3.77 \times 10^{-6} T^2 M - 3.40 \times 10^{-4} TM^2 + 3.07 \times 10^{-6} T^3 + 1.38 \times 10^{-3} M^3$	1.000
		$\varepsilon'' = -1.72 + 2.90 \times 10^{-2} T + 0.55 M - 1.60 \times 10^{-3} TM - 5.17 \times 10^{-4} T^2 - 5.33 \times 10^{-2} M^2 + 2.09 \times 10^{-5} T^2 M + 2.01 \times 10^{-5} TM^2 + 2.61 \times 10^{-6} T^3 + 2.29 \times 10^{-3} M^3$	0.999
	Shell	$\varepsilon' = -49.63 + 1.20 T + 6.66 M - 8.69 \times 10^{-2} TM - 1.72 \times 10^{-2} T^2 - 0.27 M^2 + 8.49 \times 10^{-4} T^2 M + 1.04 \times 10^{-3} TM^2 + 5.25 \times 10^{-5} T^3 + 3.89 \times 10^{-3} M^3$	0.996
		$\varepsilon'' = -22.15 - 7.87 \times 10^{-2} T + 3.98 M + 1.48 \times 10^{-3} TM + 1.06 \times 10^{-3} T^2 - 0.22 M^2 + 1.32 \times 10^{-5} T^2 M - 5.48 \times 10^{-6} TM^2 - 6.34 \times 10^{-6} T^3 + 3.94 \times 10^{-3} M^3$	0.976

3.5. Penetration Depth

The penetration depths of electromagnetic waves into walnut kernels and shells calculated based on the measured dielectric properties at four frequencies, four temperatures, and different MCs are listed in Tables 3 and 4. The penetration depth was found to be smaller at greater frequencies, temperatures, and MCs. For example, the penetration depth d_p of the electromagnetic wave into walnut kernels with a MC of 4.21% (d.b.) decreased from 1845.75 to 421.55 cm as the temperature increased from 20 to 80 °C at a frequency of 27 MHz. The d_p dropped from 931.10 to 31.10 cm as the MC increased from 4.21% to 16.23% (d.b.) for kernels with a temperature of 40 °C at 27 MHz. When the frequency increased from 27 to 2450 MHz, the corresponding penetration depth in kernels with a

MC of 8.08% under 20 °C declined from 656.77 to 6.29 cm. It could be concluded that the penetration depth of the RF wave was much larger than that of the MW wave, enabling the RF technology to be more suitable for heating walnut samples with large volumes because of the better heating uniformity. The penetration depths in shells were greater than those in kernels with corresponding MCs at the same frequency and temperature, making the electromagnetic waves penetrate through shells and interact with kernels directly. Research for *Agaricus bisporus* slices [43] and almonds [32] also revealed similar results that penetration depth increased with reduced frequency, temperature, and MC.

Table 3. Penetration depth (cm) of electromagnetic waves into walnut kernels at different frequencies, moisture contents, and temperatures.

Moisture Content (%, d.b.)	Temperature (°C)	Penetration Depth (cm)			
		27 MHz	40 MHz	915 MHz	2450 MHz
4.21	20	1845.75 ± 108.31	1239.87 ± 73.91	27.59 ± 0.25	25.19 ± 0.04
	40	973.10 ± 73.58	645.63 ± 25.26	22.83 ± 0.01	16.43 ± 0.02
	60	604.89 ± 10.65	427.63 ± 2.83	18.89 ± 0.02	11.62 ± 0.01
	80	421.55 ± 5.41	325.36 ± 2.97	17.29 ± 0.02	10.37 ± 0.01
8.08	20	656.77 ± 44.57	424.37 ± 4.06	17.75 ± 0.07	6.29 ± 0.03
	40	435.42 ± 7.87	315.57 ± 6.41	16.39 ± 0.03	5.69 ± 0.00
	60	243.21 ± 5.65	188.68 ± 3.42	14.81 ± 0.05	5.38 ± 0.01
	80	109.35 ± 1.19	89.58 ± 1.22	11.94 ± 0.06	4.87 ± 0.01
12.08	20	131.12 ± 0.88	104.60 ± 0.41	10.03 ± 0.01	3.63 ± 0.00
	40	89.63 ± 0.29	72.14 ± 0.44	9.24 ± 0.00	3.57 ± 0.00
	60	62.18 ± 0.14	49.97 ± 0.15	8.03 ± 0.01	3.43 ± 0.00
	80	42.10 ± 0.21	33.91 ± 0.19	6.57 ± 0.02	3.12 ± 0.01
16.23	20	42.43 ± 0.91	35.61 ± 0.65	5.95 ± 0.05	1.93 ± 0.02
	40	31.10 ± 0.04	25.52 ± 0.01	5.23 ± 0.00	1.92 ± 0.00
	60	24.34 ± 0.04	19.91 ± 0.02	4.44 ± 0.00	1.85 ± 0.00
	80	18.26 ± 0.09	14.91 ± 0.07	3.55 ± 0.01	1.66 ± 0.00

Table 4. Penetration depth (cm) of electromagnetic waves into walnut shells at different frequencies, moisture contents, and temperatures.

Moisture Content (%, d.b.)	Temperature (°C)	Penetration Depth (cm)			
		27 MHz	40 MHz	915 MHz	2450 MHz
12.51	20	2648.13 ± 306.53	2015.44 ± 135.34	78.14 ± 9.05	18.97 ± 0.11
	40	2506.65 ± 194.41	1592.42 ± 99.18	73.97 ± 5.74	17.28 ± 0.08
	60	533.90 ± 45.83	427.54 ± 31.26	15.75 ± 1.35	13.75 ± 0.16
	80	37.25 ± 1.51	30.73 ± 1.22	1.10 ± 0.04	6.60 ± 0.07
16.08	20	983.50 ± 72.43	596.91 ± 19.48	19.22 ± 0.08	5.93 ± 0.01
	40	695.05 ± 53.40	481.09 ± 24.31	18.43 ± 0.13	5.54 ± 0.04
	60	103.81 ± 8.46	83.35 ± 6.83	11.53 ± 0.44	4.11 ± 0.10
	80	31.12 ± 0.33	25.74 ± 0.27	7.65 ± 0.03	3.78 ± 0.01
20.07	20	551.03 ± 3.04	344.42 ± 6.08	15.03 ± 0.01	6.40 ± 0.01
	40	261.02 ± 11.70	190.50 ± 10.86	13.96 ± 0.13	5.77 ± 0.05
	60	47.26 ± 0.88	38.38 ± 0.69	8.33 ± 0.05	4.00 ± 0.02
	80	29.92 ± 0.11	25.22 ± 0.09	8.86 ± 0.04	4.76 ± 0.01
23.86	20	453.58 ± 19.64	372.23 ± 5.47	15.87 ± 0.01	4.66 ± 0.00
	40	124.29 ± 7.13	101.60 ± 5.68	12.73 ± 0.27	3.99 ± 0.06
	60	45.68 ± 0.21	38.30 ± 0.22	9.57 ± 0.03	3.54 ± 0.01
	80	30.05 ± 0.27	25.75 ± 0.27	9.61 ± 0.09	3.57 ± 0.01

3.6. Moisture and Temperature-Dependent Thermal Properties

Specific heat capacity and thermal conductivity of walnut kernels and shells all increased with raised temperature and MC (Figure 9). Taking walnut shells as an example (Figure 9c,d), their specific heat capacity and thermal conductivity increased from

444.50 J/(kg·K) and 0.11 W/(m·K) to 1980.41 J/(kg·K) and 0.64 W/(m·K), respectively, with increased temperature from 20 to 80 °C and MC from 12.51% to 23.86% (d.b.). Oriola et al. (2021) and Aviara and Haque (2001) also found the similar trends in the thermal properties of Jack bean seeds [45] and sheanuts [46], respectively. Since the specific heat capacity C_p of water is a relatively high value of 4200 J/(kg·K), the C_p of materials was larger with higher MC. Moreover, the addition of moisture filled the voids in kernels and shells and increased the mobility of water, leading to the increase in the thermal conductivity. Since the thermal conductivity of shells was much larger than that of kernels for in-shell walnuts, addition of extra hot air is required to maintain sample temperatures during RF or MW pasteurization.

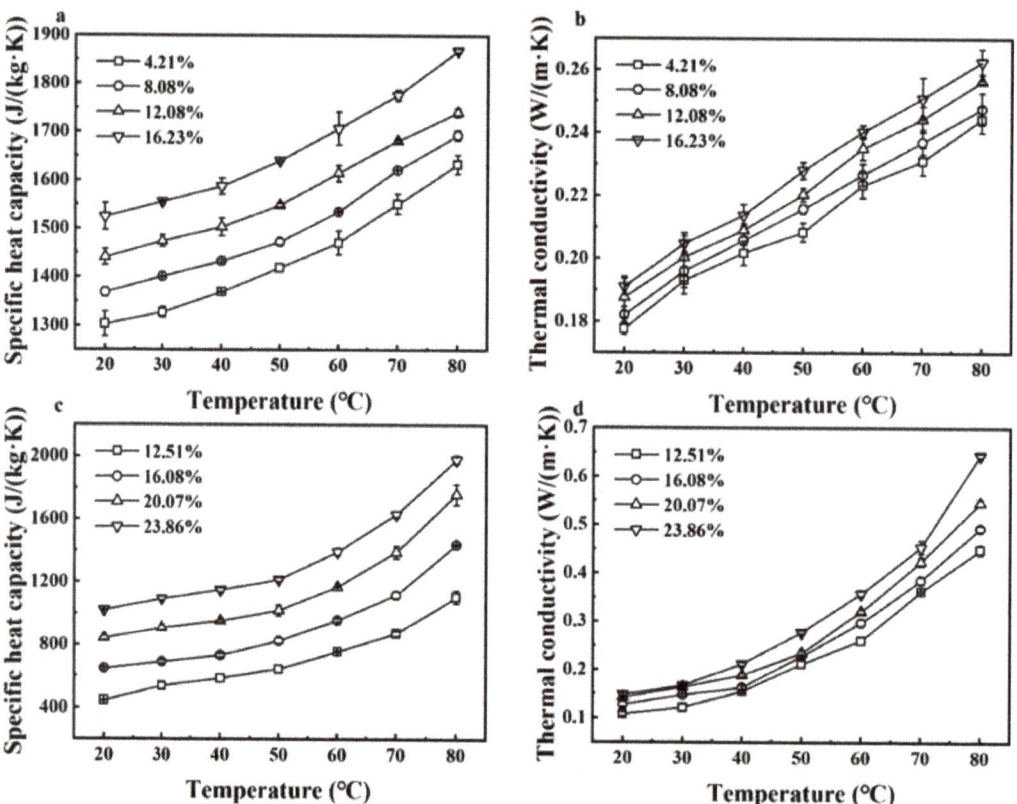

Figure 9. Moisture content and temperature-dependent specific heat capacity and thermal conductivity of walnut kernels (a,b) and shells (c,d).

The polynomial regression models describing the variation in thermal properties of walnut components with temperature and MC were developed and listed in Table 5. A two-factor interaction model and quadratic polynomial equation were the best fit for thermal conductivity and specific heat capacity of walnut kernels, respectively, and the cubic polynomial relationship was the best fit for the thermal properties of shells. Coefficients of determination (R^2) of established models were all greater than 0.995, demonstrating that those mathematical models provided a precise prediction for thermal properties of walnut components.

Table 5. Regression equations of thermal properties of walnut components as a function of moisture content (M, %) and temperature (T, °C).

Material		Thermal Properties	R^2
Kernel	Thermal conductivity	$k = 0.16 + 0.001\,T + 7.69 \times 10^{-4}\,M + 1.16 \times 10^{-5}\,TM$	0.996
	Specific heat capacity	$C_p = 1237.31 - 0.26\,T + 10.59\,M + 5.90 \times 10^{-3}\,TM + 5.63 \times 10^{-2}\,T^2 + 0.38\,M^2$	0.997
Shell	Thermal conductivity	$k = -0.41 + 8.77 \times 10^{-3}\,T + 6.73 \times 10^{-2}\,M - 8.54 \times 10^{-4}\,TM - 4.54 \times 10^{-5}\,T^2 - 2.68 \times 10^{-3}\,M^2 + 4.83 \times 10^{-6}\,T^2 M + 1.52 \times 10^{-5}\,TM^2 + 4.03 \times 10^{-7}\,T^3 + 3.79 \times 10^{-5}\,M^3$	0.996
	Specific heat capacity	$C_p = -216.30 + 31.28\,T - 4.50\,M - 0.11\,TM - 0.73\,T^2 + 2.97\,M^2 + 1.45 \times 10^{-2}\,T^2 M - 2.50 \times 10^{-2}\,TM^2 + 4.77 \times 10^{-3}\,T^3 - 3.64 \times 10^{-2}\,M^3$	0.998

4. Conclusions

Dielectric and thermal properties of both walnut kernels and shells were affected by temperature and MC, and dielectric properties decreased with increasing frequency. RF or MW heating rates of walnut kernels could be greater than those of shells because of the higher dielectric loss factors at the same temperature and MC. Penetration depth decreased with raised frequency, temperature, and MC, and great penetration depths in both kernels and shells at RF frequencies made the RF heating more effective for pasteurizing bulk and thick materials as compared with the MW method. Because of the lower dielectric loss factor and greater thermal conductivity of walnut shells than those of kernels, it is advisable to combine RF or MW heating with assisted hot air to maintain the temperature of shells. Both quadratic and cubic polynomial regression models provided the best fit for dielectric and thermal properties of walnut components. Consequently, the knowledge of dielectric and thermal properties of both walnut kernels and shells is conducive to developing rapid and effective RF or MW pasteurization technologies for in-shell walnuts at laboratory, pilot, and industrial scales, and provided necessary data for the further computer simulation.

Author Contributions: Y.M. conducted experiment, analyzed data, and wrote the first version of manuscript; Y.H. assisted with conducting the experiments; X.G. provided instruction for using the dielectric property measurement system; P.W. helped to analyze data; S.W. is the PI of the project, guided the experimental design and revised the manuscript. All authors have read and agreed to the published version of the manuscript.

Funding: This research was supported by research grants from General Program of National Natural Science Foundation of China (31772031).

Data Availability Statement: The datasets generated for this study are available on request to the corresponding author.

Acknowledgments: The authors would like to thank the instrument shared platform of College of Food Science and Engineering at NWAFU for providing the dielectric property measurement system.

Conflicts of Interest: The authors declare no conflict of interest.

References

1. FAOSTAT. Food and Agriculture Organization of the United States. 2022. Available online: https://www.fao.org/faostat/en/#data/QCL (accessed on 5 February 2022).
2. Davidson, G.R.; Frelka, J.C.; Yang, M.; Jones, T.M.; Harris, L.J. Prevalence of *Escherichia coli* O157:H7 and *Salmonella* on inshell California walnuts. *J. Food Prot.* **2015**, *78*, 1547–1553. [CrossRef] [PubMed]
3. Zhang, L.; Wang, S. Bacterial community diversity on in-shell walnut surfaces from six representative provinces in China. *Sci. Rep.* **2017**, *7*, 10054. [CrossRef] [PubMed]
4. Singh, P.K.; Shukla, A.N. Survey of mycoflora counts, aflatoxin production biochemical changes in walnut kernels. *J. Stored Prod. Res.* **2008**, *44*, 169–172. [CrossRef]
5. Frelka, J.C.; Harris, L.J. Evaluation of microbial loads and the effects of antimicrobial sprays in postharvest handling of California walnuts. *Food Microbiol.* **2015**, *48*, 133–142. [CrossRef]
6. Fan, X.; Annous, B.A.; Beaulieu, J.C.; Sites, J.E. Effect of hot water surface pasteurization of whole fruit on shelf life and quality of fresh-cut cantaloupe. *J. Food Sci.* **2008**, *73*, M91–M98. [CrossRef]

7. Chang, S.S.; Han, A.R.; Reyes-De-Corcuera, J.I.; Powers, J.R.; Kang, D.H. Evaluation of steam pasteurization in controlling *Salmonella* serotype Enteritidis on raw almond surfaces. *Lett. Appl. Microbiol.* **2010**, *50*, 393–398. [CrossRef]
8. Altemimi, A.; Aziz, S.N.; Al-Hilphy, A.R.S.; Lakhssassi, N.; Watson, D.G.; Ibrahim, S.A. Critical review of radio-frequency (RF) heating applications in food processing. *Food Qual. Saf.* **2019**, *3*, 81–91. [CrossRef]
9. Chen, L.; Jung, J.; Chaves, B.D.; Jones, D.; Subbiah, J. Challenges of dry hazelnut shell surface for radio frequency pasteurization of inshell hazelnuts. *Food Control* **2021**, *125*, 107948. [CrossRef]
10. Qu, Z.; Tang, Z.W.; Liu, F.; Sablani, S.S.; Ross, C.F.; Sankaran, S.; Tang, J.M. Quality of green beans (*Phaseolus vulgaris* L.) influenced by microwave and hot water pasteurization. *Food Control* **2021**, *124*, 107936. [CrossRef]
11. Zhang, L.; Lyng, J.G.; Xu, R.; Zhang, S.; Zhou, X.; Wang, S. Influence of radio frequency treatment on in-shell walnut quality and *Staphylococcus aureus* ATCC 25923 survival. *Food Control* **2019**, *102*, 197–205. [CrossRef]
12. Shirkolea, S.S.; Mujumdar, A.S.; Jayabalan, R.; Sutar, P.P. Dry pasteurization of paprika (*Capsicum annuum* L.) by short time intensive microwave-infrared radiation: Inactivation of *Salmonella* Typhimurium and *Aspergillus flavus* considering quality degradation kinetics. *Food Chem.* **2021**, *338*, 128012. [CrossRef] [PubMed]
13. Komarov, V.V. A review of radio frequency and microwave sustainability-oriented technologies. *Sustain. Mater. Technol.* **2021**, *28*, e00234. [CrossRef]
14. Jiang, H.; Liu, Z.; Wang, S. Microwave processing: Effects and impacts on food components. *Crit. Rev. Food Sci. Nutr.* **2018**, *58*, 2476–2489. [CrossRef] [PubMed]
15. Mukama, M.; Ambaw, A.; Opara, U.L. Thermophysical properties of fruit-a review with reference to postharvest handling. *J. Food Meas. Charact.* **2020**, *14*, 2917–2937. [CrossRef]
16. Sosa-Morales, M.E.; Valerio-Junco, L.; Lopez-Malo, A.; Garcia, H.S. Dielectric properties of foods: Reported data in the 21st Century and their potential applications. *LWT-Food Sci. Technol.* **2010**, *43*, 1169–1179. [CrossRef]
17. Bon, J.; Vaquiro, H.; Benedito, J.; Telis-Romero, J. Thermophysical properties of mango pulp (*Mangifera indica* L. cv. Tommy Atkins). *J. Food Eng.* **2010**, *97*, 563–568. [CrossRef]
18. Chen, Y.; He, J.; Li, F.; Tang, J.; Jiao, Y. Model food development for tuna (*Thunnus Obesus*) in radio frequency and microwave tempering using grass carp mince. *J. Food Eng.* **2021**, *292*, 110267. [CrossRef]
19. Evangelista, R.R.; Sanches, M.A.R.; de Castilhos, M.B.M.; Cantu-Lozano, D.; Telis-Romero, J. Determination of the rheological behavior and thermophysical properties of malbec grape juice concentrates (*Vitis vinifera*). *Food Res. Int.* **2020**, *137*, 109431. [CrossRef]
20. Taheri, S.; Brodie, G.; Jacob, M.V.; Antunes, E. Dielectric properties of chickpea, red and green lentil in the microwave frequency range as a function of temperature and moisture content. *J. Microw. Power. Electromagn. Energy* **2018**, *52*, 198–214. [CrossRef]
21. Zhu, X.; Guo, W.; Wang, S. Dielectric properties of ground hazelnuts at different frequencies, temperatures, and moisture contents. *Trans. ASABE* **2014**, *57*, 161–168.
22. Boldor, D.; Sanders, T.H.; Simunovic, J. Dielectric properties of in-shell and shelled peanuts at microwave frequencies. *Trans. ASAE* **2004**, *47*, 1159–1169. [CrossRef]
23. Huang, Z.; Zhang, B.; Marra, F.; Wang, S. Computational modelling of the impact of polystyrene containers on radio frequency heating uniformity improvement for dried soybeans. *Innov. Food Sci. Emerg. Technol.* **2016**, *33*, 365–380. [CrossRef]
24. Perussello, C.A.; Mariani, V.C.; Camargo do Amarante, A.C. Thermophysical properties of okara during drying. *Int. J. Food Prop.* **2014**, *17*, 891–907. [CrossRef]
25. Rusu, M.E.; Gheldiu, A.-M.; Mocan, A.; Vlase, L.; Popa, D.-S. Anti-aging potential of tree nuts with a focus on the phytochemical composition, molecular mechanisms and thermal stability of major bioactive compounds. *Food Funct.* **2018**, *9*, 2554–2575. [CrossRef] [PubMed]
26. Zheng, D.; Zhang, Y.; Guo, Y.; Yue, J. Isolation and Characterization of Nanocellulose with a Novel Shape from Walnut (*Juglans Regia* L.) Shell Agricultural Waste. *Polymers* **2019**, *11*, 1130. [CrossRef] [PubMed]
27. Wang, S.; Tang, J.; Johnson, J.A.; Mitcham, E.; Hansen, J.D.; Hallman, G.; Drake, S.R.; Wang, Y. Dielectric properties of fruits and insect pests as related to radio frequency and microwave treatments. *Biosyst. Eng.* **2003**, *85*, 201–212. [CrossRef]
28. Mao, Y.; Wang, P.; Wu, Y.; Hou, L.; Wang, S. Effects of various radio frequencies on combined drying and disinfestation treatments for in-shell walnuts. *LWT-Food Sci. Technol.* **2021**, *144*, 111246. [CrossRef]
29. Kader, A.A. Impact of nut postharvest handling, de-shelling, drying and storage on quality. In *Improving the Safety and Quality of Nuts*; Harris, L.J., Ed.; Woodhead Publishing Limited: Cambridge, MA, USA, 2013; Volume 250, pp. 22–34.
30. AOAC. *Official Methods of Analysis of the Association of Official Analytical Chemists*; AOAC: Rockville, MD, USA, 2002.
31. Guo, W.; Tiwari, G.; Tang, J.; Wang, S. Frequency, moisture and temperature-dependent dielectric properties of chickpea flour. *Biosyst. Eng.* **2008**, *101*, 217–224. [CrossRef]
32. Li, R.; Zhang, S.; Kou, X.; Ling, B.; Wang, S. Dielectric properties of almond kernels associated with radio frequency and microwave pasteurization. *Sci. Rep.* **2017**, *7*, 42452. [CrossRef]
33. Jiao, Y.; Tang, J.; Wang, Y.; Koral, T.L. Radio-frequency applications for food processing and safety. *Annu. Rev. Food Sci. Technol.* **2018**, *9*, 105–127. [CrossRef]
34. Aydin, C. Physical properties of hazel nuts. *Biosyst. Eng.* **2002**, *82*, 297–303. [CrossRef]
35. Altuntas, E.; Erkol, M. Physical properties of shelled and kernel walnuts as affected by the moisture content. *Czech J. Food Sci.* **2010**, *28*, 547–556. [CrossRef]

36. Feng, H.; Tang, J.; Cavalieri, R.P. Dielectric properties of dehydrated apples as affected by moisture and temperature. *Trans. ASAE* **2002**, *45*, 129–135. [CrossRef]
37. Liu, Y.; Yang, M.; Gao, Y.; Fan, X.; Zhao, K. Broadband dielectric properties of honey: Effects of temperature. *J. Food Sci. Technol.* **2020**, *57*, 1656–1660. [CrossRef]
38. Gezahegn, Y.A.; Tang, J.; Sablani, S.S.; Pedrow, P.D.; Hong, Y.K.; Lin, H.; Tang, Z. Dielectric properties of water relevant to microwave assisted thermal pasteurization and sterilization of packaged foods. *Innov. Food Sci. Emerg. Technol.* **2021**, *74*, 102837. [CrossRef]
39. Qi, S.; Han, J.; Lagnika, C.; Jiang, N.; Zhang, M. Dielectric properties of edible fungi powder related to radio-frequency and microwave drying. *Food Prod. Process. Nutr.* **2021**, *3*, 15. [CrossRef]
40. Xie, W.; Chen, P.; Wang, F.; Li, X.; Wei, S.; Jiang, Y.; Liu, Y.; Yang, D. Dielectric properties of Camellia oleifera seed kernels related to microwave and radio frequency drying. *Int. Food Res. J.* **2019**, *26*, 1577–1585.
41. Zhou, X.; Li, R.; Lyng, J.G.; Wang, S. Dielectric properties of kiwifruit associated with a combined radio frequency vacuum and osmotic drying. *J. Food Eng.* **2018**, *239*, 72–82. [CrossRef]
42. Cao, H.; Fan, D.; Jiao, X.; Huang, J.; Zhao, J.; Yan, B.; Zhou, W.; Zhang, W.; Zhang, H. Heating surimi products using microwave combined with steam methods: Study on energy saving and quality. *Innov. Food Sci. Emerg. Technol.* **2018**, *47*, 231–240. [CrossRef]
43. Jiang, N.; Liu, C.; Li, D.; Lagnika, C.; Zhang, Z.; Huang, J.; Liu, C.; Zhang, M.; Yu, Z. Dielectric properties of *Agaricus bisporus* slices relevant to drying with microwave energy. *Int. J. Food Prop.* **2020**, *23*, 354–367. [CrossRef]
44. Yu, D.U.; Shrestha, B.L.; Baik, O.D. Radio frequency dielectric properties of bulk canola seeds under different temperatures, moisture contents, and frequencies for feasibility of radio frequency disinfestation. *Int. J. Food Prop.* **2015**, *18*, 2746–2763. [CrossRef]
45. Oriola, K.O.; Hussein, J.B.; Oke, M.O.; Ajetunmobi, A. Description and evaluation of physical and moisture-dependent thermal properties of jack bean seeds (*Canavalia ensiformis*). *J. Food Process. Pres.* **2021**, *45*, e15166. [CrossRef]
46. Aviara, N.A.; Haque, M.A. Moisture dependence of thermal properties of sheanut kernel. *J. Food Eng.* **2001**, *47*, 109–113. [CrossRef]

Article

Effect of Load Spatial Configuration on the Heating of Chicken Meat Assisted by Radio Frequency at 40.68 MHz

Sandro M. Goñi [1,2], Matteo d'Amore [3], Marta Della Valle [4], Daniela F. Olivera [1,5], Viviana O. Salvadori [1,2] and Francesco Marra [4,*]

1. Centro de Investigación y Desarrollo en Criotecnología de Alimentos (CIDCA), Conicet La Plata-Universidad Nacional de La Plata-Comisión de Investigaciones Científicas, 47 y 116, La Plata 1900, Argentina; smgoni@quimica.unlp.edu.ar (S.M.G.); danielaolivera@conicet.gov.ar (D.F.O.); vosalvad@ing.unlp.edu.ar (V.O.S.)
2. Facultad de Ingeniería, Universidad Nacional de La Plata, 1 y 47, La Plata 1900, Argentina
3. Dipartimento di Farmacia, Università degli Studi di Salerno, 84084 Fisciano, Italy; mdamore@unisa.it
4. Dipartimento di Ingegneria Industriale, Università degli Studi di Salerno, 84084 Fisciano, Italy; m.dellavalle1@studenti.unisa.it
5. Facultad de Ciencias Veterinarias, Universidad Nacional de La Plata, 60 y 118, La Plata 1900, Argentina
* Correspondence: fmarra@unisa.it

Abstract: Food heating assisted by radio frequencies has been industrially applied to post-harvest treatment of grains, legumes and various kind of nuts, to tempering and thawing of meat and fish products and to post-baking of biscuits. The design of food processes based on the application of radiofrequencies was often based on rules of thumb, so much so that their intensification could lead significant improvements. One of the subjects under consideration is the shape of the food items that may influence their heating assisted by radiofrequency. In this work, a joint experimental and numerical study on the effects of the spatial configuration of a food sample (chicken meat shaped as a parallelepiped) on the heating pattern in a custom RF oven (40.68 MHz, 50 Ohm, 10 cm electrodes gap, 300 W) is presented. Minced chicken breast samples were shaped as cubes ($4 \times 4 \times 4$ cm^3) to be organized in different loads and spatial configurations (horizontal or vertical arrays of 2 to 16 cubes). The samples were heated at two radiofrequency operative power levels (225 W and 300 W). Heating rate, temperature uniformity and heating efficiency were determined during each run. A digital twin of the experimental system and process was developed by building and numerically solving a 3D transient mathematical model, taking into account electromagnetic field distribution in air and samples and heat transfer in the food samples. Once validated, the digital tool was used to analyze the heating behavior of the samples, focusing on the most efficient configurations. Both experiments and simulations showed that, given a fixed gap between the electrodes (10 cm), the vertically oriented samples exhibited a larger heating efficiency with respect to the horizontally oriented ones, pointing out that the gap between the top electrode and the samples plays a major role in the heating efficiency. The efficiency was larger (double or even more; >40% vs. 10–15%) in thicker samples (built with two layers of cubes), closer to the top electrode, independently from nominal power. Nevertheless, temperature uniformity in vertical configurations was poorer (6–7 °C) than in horizontal ones (3 °C).

Keywords: radiofrequency; heating; energy efficiency; digital twin

Citation: Goñi, S.M.; d'Amore, M.; Della Valle, M.; Olivera, D.F.; Salvadori, V.O.; Marra, F. Effect of Load Spatial Configuration on the Heating of Chicken Meat Assisted by Radio Frequency at 40.68 MHz. *Foods* **2022**, *11*, 1096. https://doi.org/10.3390/foods11081096

Academic Editors: Shaojin Wang and Rui Li

Received: 16 March 2022
Accepted: 5 April 2022
Published: 11 April 2022

Publisher's Note: MDPI stays neutral with regard to jurisdictional claims in published maps and institutional affiliations.

Copyright: © 2022 by the authors. Licensee MDPI, Basel, Switzerland. This article is an open access article distributed under the terms and conditions of the Creative Commons Attribution (CC BY) license (https://creativecommons.org/licenses/by/4.0/).

1. Introduction

In dielectric heating (which can be referred to both radio frequency—RF—and microwaves—MW) heat is volumetrically generated inside a sample (often simply called load) due to the interaction of the electromagnetic field the load is placed in, with the ionic and polar molecules contained in the sample. In traditional heating, heat is transferred from the surrounding medium to the sample by convection and conduction mechanisms, the latter being particularly slow in foods due to their low thermal conductivity. Dielectric heating provides higher heating rates and reduces processing times with respect to processes controlled by heat conduction or convection [1].

Guo et al. [2] reviewed the recent literature on RF heating applied to fresh food processing. The benefits of RF, basically due to rapid heating, deep thermal penetration and possibility of quality control, were investigated in different processes, such as cooking [3,4], post-harvest treatment of agriculture commodities [5,6], pasteurization [7,8], drying [9], tempering [10] and thawing [11–13].

Dielectric heating in the RF range is generally performed at three defined frequencies, i.e., 13.56 MHz, 27.12 MHz and 40.68 MHz, even if most of the studies reported in the scientific literature refer to RF-assisted processes at 27.12 MHz. In RF cooking of different meats, Laycock et al. [14] investigated the effect of RF on quality, heating rate and temperature profiles of three types of meat products (ground, comminuted and muscle) cooked in a 1.5 kW, 27.12 MHz RF heater. Authors concluded that ground and comminuted products seem to be promising for RF cooking, as cooking times decreased by more than 80%. Kirmaci and Singh [15] compared RF with water bath (WB) cooking of fresh and marinated chicken breast meat, using a 6 kW, 27.12 MHz RF oven. RF cooking resulted in a higher heating rate, achieving cooking times that were 42% shorter. Similar quality parameters were measured in both procedures, although RF cooked meat had lower redness. Muñoz et al. [16] analyzed a two-step cooking process for pork hams, which included a RF tunnel and a steam oven, and compared the results with hams completely processed in the steam oven. Authors pointed out that cooking time decreased by 50% using RF, with no significant differences observed in the terms of the sensory quality of the final products. Wang et al. [17] applied RF energy at 27.12 MHz to pork tenderloin, demonstrating that RF heating improved water retention of pork myofibrillar protein gel and it had the potential to improve meat quality and could greatly reduce the processing time, furthermore facilitating the formation of a stable and orderly gel network structure.

Although RF heating is claimed to be a method to obtain uniform temperature distribution in the heated product, a common issue of RF heating is the lack of temperature uniformity, which produces zones or spots of high and low temperatures inside the heated food. The temperature uniformity depends on the way electromagnetic energy is spatially absorbed by food, which has, in turn, a complex dependence on several factors [18]. A paramount factor affecting heating uniformity is the deflection of the electric field because of the sample, which concentrates the net electric field in certain regions that are heated quicker [19,20]. Other factors can be grouped into three classes: food characteristics, such as dielectric and thermal properties, geometry, size and mass; RF system features such as power, electrodes shape and size, gap between electrodes; and factors related to the RF application, such as food orientation on the cavity, position of the sample between the electrodes, manipulation of electrode positions (for adjustable gap devices), food area related to electrode area, and the use of special food containers [18]. Different strategies to improve heating uniformity have been extensively reviewed by [18]. Among others, Romano and Marra [21] used modeling and simulation to analyze the effects of regular geometries (a cube, cylinder and sphere of equal volume) and orientation on temperature uniformity during heating at 27.12 MHz, 400 W maximum power, using properties of luncheon meat batter. The cubic shape presented the best temperature uniformity and the highest heating rate, followed by the vertical cylinder. The sphere and horizontal cylinder had lower heating rates and worse temperature uniformity. Recently, Li et al. [22] discussed a strategy for improving the uniformity of radio frequency tempering for frozen beef with cuboid and step shapes; Bedane et al. [23] experimentally investigated the same aspects using regular geometries of equal volume (a cube, cylinder and sphere) of tylose with different salt concentrations. A 27.12 MHz, 10 cm electrode gap, 600 W maximum power RF cavity was employed, and the samples were heated for 6 min. The temperature was measured at different locations. The vertical cylinder and cube presented the best temperature uniformity, while the sphere and horizontal cylinder had the worst one.

Both studies clearly demonstrate the effect of geometry and orientation on RF heating performance. Tiwari et al. [20] simulated the heating of dry foods in a 27.12 MHz, 12 kW RF system. Several simulations varying the geometry (cuboids, cylinders and ellipsoids), the size and position of cuboid-shaped samples, the gap between electrodes, etc., were performed. A power uniformity index (PUI) was easily obtained from the numerical results. The best (lowest) PUI was obtained for ellipsoids in the middle of the oven, followed by the cylinder and cuboid. Theoretical predictions show that for cuboids the uniformity can be improved increasing the sample size. Additionally, for different cuboid sizes, different vertical positions led to different optimal PUI locations. Later, Tiwari et al. [24] successfully validated the previous model through experiments using hard red spring wheat flour. In cooking of fresh and marinated chicken breast, Kirmaci and Singh [15] obtained a better temperature uniformity with water bath cooking (0.9 °C) with respect to RF (5.3 °C), even if in the water bath a longer time was required. Uyar et al. [3] used a simulation to analyze the effect of volume of cubic meat samples on heating rate and efficiency at 27.12 MHz. Two cases were considered: a fixed electrode gap and a fixed electrode-sample distance (air gap). Results showed the great impact of the sample load and air gap on the heating rate and efficiency. For a fixed electrode gap, the heating rate and efficiency improved with the sample volume increase. For a fixed air gap, the heating rate was higher for small loads, but the efficiency was higher for large loads. Information about temperature uniformity was not reported. In a further study, Uyar et al. [4] simulated the effect of the projected area during heating of meat samples on several uniformity indexes, such as PUI. It was found that configurations leading to a higher temperature increase also have a less uniform temperature distribution. So, this opposite behavior between desirable objectives led to the need for an increase in the studies in the design of the RF cavity [18].

In summary, the influence of geometric characteristics on power absorption uniformity is proven. Notwithstanding, there is a lack of studies oriented to the heating (cooking) of fresh products. Besides, the applicability of different RF frequencies, such as 40.68 MHz, which is more appropriate for heating small volume pieces due to its smaller penetration depth, has not been sufficiently investigated [2]. Therefore, the objective of this work was to study the effect of load spatial configuration on heating rate, temperature uniformity and energy efficiency during RF heating of chicken meat at 40.68 MHz.

2. Materials and Methods

2.1. Food Samples

Minced lean chicken breast bought at a local butcher shop in Salerno, Italy, was used in the experimental heating tests. Sixteen plastic cubic containers ($4 \times 4 \times 4$ cm^3) were fabricated with an advanced desktop 3D printer, Replicator 2X Experimental 3D (MakerBot Industries, Brooklyn, NY, USA), with dual extrusion to print with filaments. Acrylonitrile butadiene styrene (ABS), a thermoplastic polymer characterized by a high impact resistance and toughness, and electrical properties that are constant over the RF range [25] were used to print the containers.

Each mold was filled with 60 g of minced chicken breast, paying attention to filling all the mold's volume, and then wrapped with plastic film as shown in Figure 1a. Before each heating test, the samples were equilibrated in a refrigerator at 277 K.

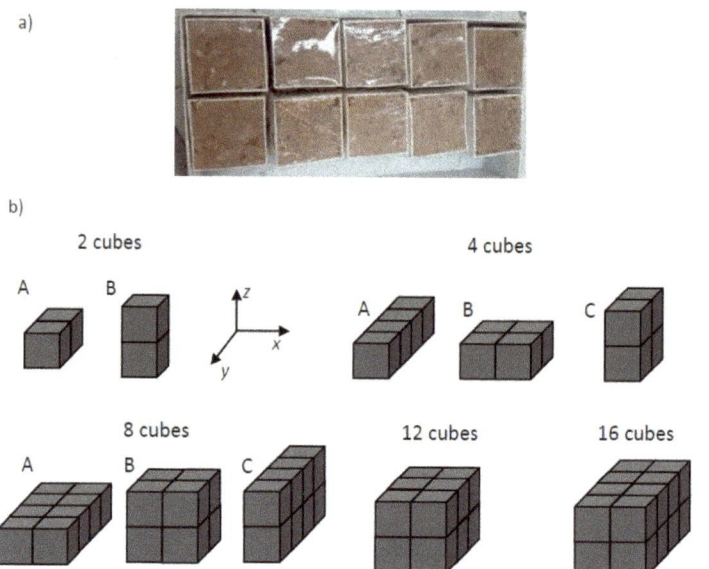

Figure 1. (a) Minced breast meat into the cubic containers, sealed with plastic film; (b) spatial configuration of the different samples used in the heating tests.

2.2. RF Oven

An experimental prototype of RF oven (40.68 MHz, 50 Ohm, 300 W maximum power) was used to perform the heating tests. The RF oven consisted of an electrically insulated chamber 275 mm width, 304 mm depth, 170 mm height, with a front door of 260 mm × 95 mm, a RF generator, a RF amplifier and a RF applicator (280 mm × 240 mm fixed parallel rectangular electrodes, 10 mm gap). The RF power is controlled by a matching system provided by the RF cavity manufacturer, which adjusts the impedance and the resistivity of the whole circuit to make it resonate at the operating frequency. The whole system is remotely controlled by an Android app [26]. Many details of the RF prototype cannot be revealed due to an ongoing non-disclosure agreement with the manufacturer.

2.3. Heating Tests

Two power levels were used during the heating experiments: the maximum power of the prototype, i.e., 300 W, and 225 W. Complete details of the experiments are given in Table 1: spatial configuration, sample size, area projected over the electrodes, sample mass and heating times. All experiments were run in five replicates.

Ten sample configurations (formed by sets of 2 to 16 meat cubes) were tested in each set. The cubic shape is used as a reference geometry as this geometry exhibits a more uniform heating rate and power absorption [21,23]. The cubes were arranged as samples of different sizes, all of them exposing a planar surface to the electrodes in the RF cavity. Figure 1b shows the geometric configuration of the samples as they were placed inside the cavity. The area projected over the electrodes is the sample area crossed by the electric field formed between the two electrodes [4].

Experimental heating times (also reported in Table 1) were defined to avoid reaching the saturation temperature, i.e., 100 °C, at atmospheric pressure. This condition also allows us to ignore the heat losses due to moisture transfer.

Table 1. Spatial configuration and main characteristics of samples employed in each experiment.

Sample Code	Spatial Configuration (N Cubes per Width) × (N of Cubes per Length) × (N of Cubes along the Height)	Sample Size (Width × Length × Height) (cm)	Projected Area (cm^2)	Mass (kg)	Heating Time t_h (min)
2A	1 × 2 × 1	4 × 8 × 4	32	0.12	2
2B	1 × 1 × 2	4 × 4 × 8	16	0.12	2
4A	1 × 4 × 1	4 × 16 × 4	64	0.24	4
4B	2 × 2 × 1	8 × 8 × 4	64	0.24	4
4C	1 × 2 × 2	4 × 8 × 8	32	0.24	4
8A	2 × 4 × 1	8 × 16 × 4	128	0.48	8
8B	2 × 2 × 2	8 × 8 × 8	64	0.48	8
8C	1 × 4 × 2	4 × 16 × 8	64	0.48	8
12	2 × 3 × 2	8 × 12 × 8	96	0.72	12
16	2 × 4 × 2	8 × 16 × 8	128	0.96	16

Optical fibers (TS2, Optocon, Weidmann Technologies Deutschland GMBH, Dresden, Germany) were used to measure the sample temperature, every 10 s, at characteristic points: the center and one or more corners of each cube; thus providing several transient values of temperature profiles for each sample.

The experiments allowed us to evaluate the influence of the sample mass and spatial configuration on the heating performance of the oven, measured in terms of heating rate (referred to the heating time), temperature uniformity and energy efficiency.

The heating rate HR (in K s^{-1}) was defined in as:

$$HR = \frac{T - T_0}{t_h} \quad (1)$$

where T is the sample temperature (K), T_0 is the initial sample temperature (K) and t_h is the heating time (s). HR was evaluated for a single point of the cube, for a cube (averaging the HR for all measured points), or for a sample (averaging HR for all cubes constituting the sample).

The experimental average temperature increase $T_{Inc,Exp}$ (Equation (2)) was calculated as the difference between the average final temperature $T_{Ave,Exp}$ (Equation (3)) and the initial temperature T_0.

$$T_{Inc,Exp} = T_{Ave,Exp} - T_0 \quad (2)$$

$$T_{Ave,Exp} = \frac{\sum_{i=1}^{N} T_i}{N} \quad (3)$$

where N is the number of points where the temperature has been measured.

Additionally, the experimental temperature uniformity $T_{U,Exp}$ was assessed as the absolute temperature deviation of the temperature of each point T_i from the average temperature:

$$T_{U,Exp} = \frac{\sum_{i=1}^{N} |T_i - T_{Ave,Exp}|}{N} \quad (4)$$

Energy efficiency measures the quantity of energy absorbed by the sample, with reference to the energy supplied by the RF oven, i.e., the nominal value. According to the experimental value of final average temperature $T_{Ave,Exp}$, the averaged absorbed energy was calculated as:

$$Q_{Exp} = m\, C_P\, (T_{Ave,Exp} - T_0) \quad (5)$$

where m is the sample mass (kg), C_P is the specific heat capacity (J kg^{-1} K^{-1}), while experimental energy efficiency $\eta_{Exp}(\%)$ was defined as:

$$\eta_{Exp}(\%) = 100 \frac{Q_{Exp}}{(t_h \ NP)} \qquad (6)$$

where t_h is the experimental heating time (s), and NP is the nominal power (W), then the term $(t_h \ NP)$ (in J) is the nominal energy provided by the RF oven.

2.4. RF Heating Model

Dielectric heating is a complex multi-physics problem where heat transfer and electric field displacement inside and around the food sample must be considered simultaneously, and the two phenomena must be described and solved together. The phenomena are strictly interconnected: as a matter of fact, while the electric field distribution is affected by the sample dielectric properties (which, in turn, are strongly dependent on the temperature and the food composition), the heat balance within the food sample depends on the electromagnetic heat source, which in turn depends on the local modulus of the electric field and on food dielectric properties. In order to solve the coupled system, a specialized frequency-transient solver in COMSOL™ Multiphysics was employed (COMSOL AB, Sweden [27]). In the built model, the COMSOL AC/DC module (which provides mathematical description for the electromagnetic field in the RF range), and the transient heat transfer module were used.

A tridimensional scheme of the system was built (Figure 2). The distance between the sample and the bottom electrode was 0.015 m; the distance between the sample and the top electrode depends on sample configuration. Samples sizes and configurations detailed in Table 1 were simulated.

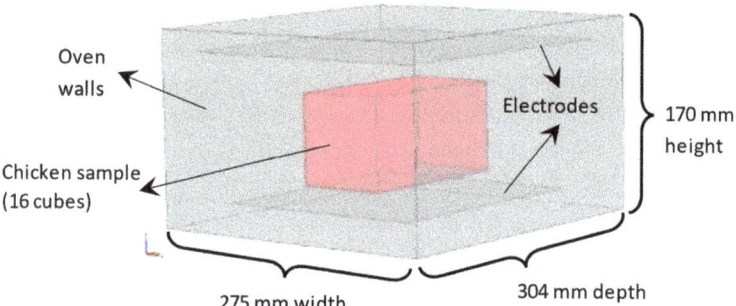

Figure 2. Scheme of the RF oven and a food sample used in simulations.

2.4.1. Governing Equations

Heat transfer balance was solved only in the food sample, imposing boundary conditions at the interfaces between sample and surrounding environment, while the electromagnetic field distribution was solved in both air and food domains.

The transient heat transfer inside the food sample was described according to Equation (7):

$$\rho C_P \frac{\partial T}{\partial t} = \nabla \cdot (k \nabla T) + \dot{Q} \qquad (7)$$

where ρ is the density (kg m^{-3}), C_P the specific heat capacity (J kg^{-1} K^{-1}) and k the thermal conductivity (W m^{-1} K^{-1}) of the food sample. The electromagnetic heat source $\dot{Q}$ (W m^{-3}) was defined as (Huang et al., 2018):

$$\dot{Q} = 2\pi \omega \varepsilon_0 \varepsilon'' |E|^2 \qquad (8)$$

where ω is the electric field frequency (Hz), ε_0 is the permittivity of free space (8.85 × 10^{-12} Fm^{-1}), ε'' is the dielectric loss factor of the sample and $|E|$ (V m^{-1}) is the modulus of the electric field vector.

Initial uniform condition for temperature was considered (277 K, experimental data), while convective heat transfer boundary condition between air and sample surface was modeled using a heat transfer coefficient ($h = 10$ W m^{-2} K^{-1}, $T_{air} = 293$ K):

$$-k\nabla T = h(T - T_{air}) \qquad (9)$$

The RF wavelength (about 7.37 m) at the employed frequency was larger than the electrode gap, so the Maxwell's equations reduce to Equation (10), which was used to model the displacement of the electric field [3,19,24,28]:

$$\nabla \cdot \varepsilon E = 0 \qquad (10)$$

where ε is the relative complex permittivity of the material (F m^{-1}), which depends on the dielectric constant ε' and the dielectric loss factor ε'' of the material ($\varepsilon = \varepsilon' - j\varepsilon''$). It should be noted that although the electric field model is steady, it is solved at each time step of the heat transfer transient simulation.

Boundary conditions to determine the electric field distribution inside the RF oven were set up as:

- Bottom electrode was maintained at the ground condition ($V = 0$);
- Top electrode was maintained at a constant potential V_0 according with the applied output power (300 or 225 W) with a frequency of 40.68 MHz;
- Oven walls were electrically insulated, $\nabla \cdot E = 0$.

The chicken breast thermophysical properties were estimated from its composition and temperature T, according to the relationships proposed by [29]. In general, the lipid content of chicken breast meat is lower than 1.2% [30]; therefore, for the purpose of this work, the composition was simplified as 75% water and 25% protein. Then, the resulting thermophysical properties were expressed as:

$$\rho = -3.087 \; 10^{-3} T^2 + 1.597 \; T + 839.5 \qquad (11)$$

$$C_P = 3.602 \; 10^{-3} T^2 - 1.721 \; T + 3834 \qquad (12)$$

$$k = -5.579 \; 10^{-6} \; T^2 + 4.471 \; 10^{-3} T - 0.3319 \qquad (13)$$

As was described in Equation (8), the dielectric properties characterize the absorption of incident RF power. In this work, chicken dielectric properties were estimated from experimental data measured by [31], who measured the dielectric constant ε' and the dielectric loss factor ε'' of uncooked chicken breast meat at 51 different frequencies, in the range from 10 MHz to 1.8 GHz. Additionally, the influence of temperature on dielectric properties was recorded, with measures in the range from 5 to 85 °C.

To adequately represent the dielectric properties of chicken breast meat in the numerical model, the values reported by [31] at 40 MHz and different temperatures were fitted. Average dielectric properties of pectoralis minor and major muscles were employed in the fitting procedure. Figure 3 shows the dependence of both dielectric properties on temperature T. The dielectric constant monotonically increases with temperature from 278 to 358 K (5 to 85 °C). This behavior is also observed at other frequencies. Instead, the loss factor increases until it reaches a maximum, which depends on the frequency; at 40 MHz the maximum is observed near to 348 K. Equations (14) and (15) represent the fitted equations used in the model:

$$\varepsilon' = 2.865 \; 10^{-3} T^2 - 1.4469 \; T + 265.23 \qquad (14)$$

$$\varepsilon'' = -1.4042 \; 10^{-3} T^3 + 1.3107 \; T^2 - 400.7 \; T + 40499 \qquad (15)$$

For air, the dielectric constant ε' is 1, whereas dielectric loss factor ε'' is 0.

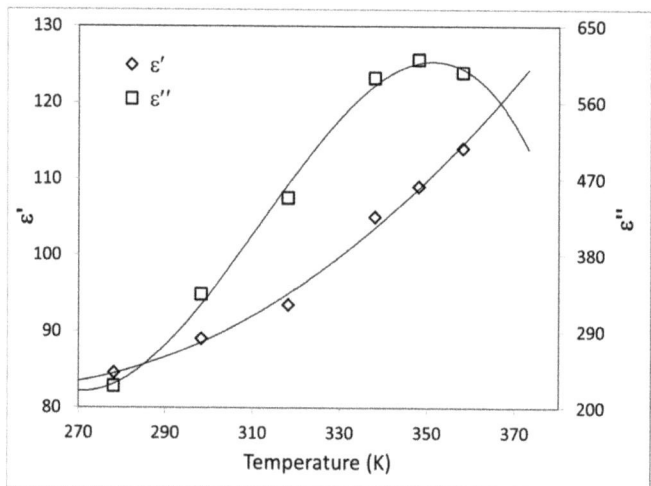

Figure 3. Estimated dielectric properties of chicken breast. Dielectric constant ε' ($\diamond$) and loss factor ε'' ($\square$) vs. temperature, at 40.68 MHz. The lines are the fitting of Equations (14) and (15).

2.4.2. Numerical Solution

We used a custom computer equipped with 8 CPUs Intel I7-3820 @ 3600 GHz FSB (front side bus, Intel, Santa Clara, CA, USA); motherboard MSI X79A-GD65 (8D) certified for military use according to the standard MIL-STD-810G, 1.600 MHz Socket (Micro-Start Int'l Co., Ltd., New Taipei City, Taiwan); Socket R (LGA 2011, Intel, Santa Clara, CA, USA) 4 cores; equipped with a RAM of 64 Gb DDR3 1600 MHz. The workstation runs under Windows 7 Professional (Microsoft, Redmond, WA, USA) operating system at Department of Industrial Engineering, University of Salerno, Italy. Time-dependent equations were discretized using the 2nd-order Backward Differentiation Formula (BDF) method, with maximum time steps of 5 s; MUMPS solver was used to solve the resulting equations with relative and absolute tolerances of 0.01 and 0.1, respectively.

Simulated Variables

From the mathematical model, different features can be determined: among others, average temperature, average temperature increase, temperature uniformity, absorbed energy and energy efficiency.

Volume average temperature at any time was calculated as:

$$T_{Ave,Sim}(t) = \frac{1}{V_{sample}} \int_V T dV \qquad (16)$$

where V_{sample} (m^3) is the sample volume. Average temperature increase was evaluated at any time during heating as the volumetric deviation from initial temperature T_0:

$$T_{Inc,Sim}(t) = \frac{1}{V_{sample}} \int_V (T - T_0) dV \qquad (17)$$

Temperature uniformity was defined as the absolute volumetric deviation from average temperature $T_{Ave,sim}$:

$$T_{U,Sim}(t) = \frac{1}{V_{sample}} \int_V |T - T_{Ave,Sim}| dV \qquad (18)$$

To estimate the efficiency of the process using the numerical model, the energy absorbed by the sample during RF heating was calculated as:

$$Q_{Sim} = \int_0^{t_h} \int_V \rho C_p \frac{\partial T}{\partial t} dV dt \qquad (19)$$

The simulated energy efficiency $\eta_{Sim}(\%)$ was defined in Equation (20) as:

$$\eta_{Sim}(\%) = 100 \frac{Q_{Sim}}{t_h \, NP} \qquad (20)$$

Thus, the performance of the RF oven was completely characterized through the temperature uniformity and the energy efficiency. These parameters were defined analogously to the experimental variables, for comparison and validation purposes. In this sense, accuracy of the mathematical model was assessed using the percentage average absolute relative deviation (AARD) and average absolute deviation (AAD) between experimental and simulated values of average temperature (K) profiles:

$$AARD(\%) = \frac{100}{n} \sum_{i=1}^{n} \left| \frac{T_{Ave,Exp,i} - T_{Ave,Sim,i}}{T_{Ave,Exp,i}} \right| \qquad (21)$$

$$AAD(K) = \frac{1}{n} \sum_{i=1}^{n} \left| T_{Ave,Exp,i} - T_{Ave,Sim,i} \right| \qquad (22)$$

Mesh Independence Analysis

Mesh convergence studies were performed to ensure that the results were independent of mesh resolution. With this aim, the influence of meshing on the electric field and heat transfer solution was analyzed using the biggest sample (16 cubes). Eight default mesh configurations, from extra fine to extremely coarse, were tested.

The volume average temperature increase (Equation (17)) and the volume average electromagnetic heat source (Equation (23)) were calculated as a function of time and used to evaluate the mesh influence.

$$\dot{Q}_{Ave}(t) = \frac{1}{V_{sample}} \int_V \dot{Q}(t) dV \qquad (23)$$

Figure 4 shows the influence of the mesh's refinement on the average temperature increase and the average heat source versus time. Thus, the mesh size was determined based on these mesh analyses, seeking acceptable differences between successive calculations [32].

Table 2 reports the final average values of the mentioned quantities and the required simulations times (related to the hardware employed in this study) for the different mesh sizes. From the results of mesh analysis, the "Extra Fine" mesh was selected and employed in all cases. Different mesh analyses can be also performed, for instance setting personalized maximum and minimum finite element sizes or using different element sizes for air and chicken (due to differences in the dielectric properties of both materials).

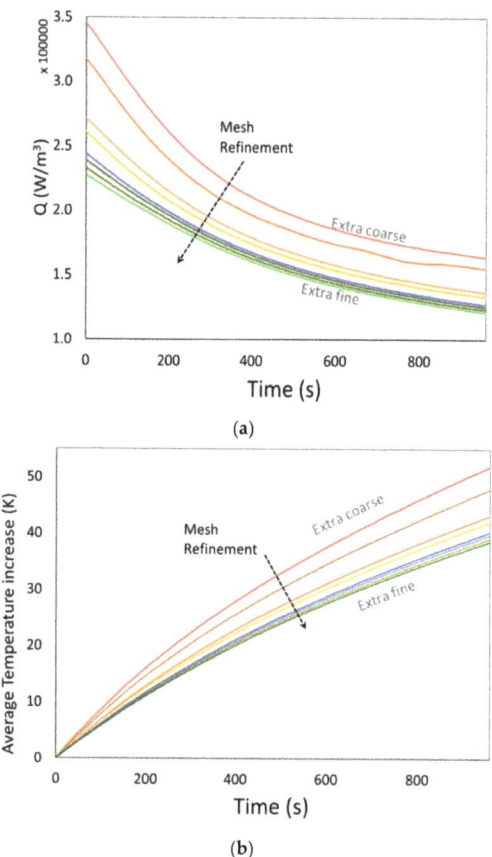

Figure 4. Effect of mesh refinement on simulated values, for the sample with 16 cubes, 300 W RF nominal power. (**a**) volume average electromagnetic heat source (Equation (23)) vs. time; (**b**) volume average temperature increase (Equation (17)) vs. time.

Table 2. Effect of meshing refinement on temperature increase and electromagnetic heat source for sample 16 (16 cubes) for 300 W RF nominal power.

Mesh Refinement	N° of Elements	Final Average Values		Simulation Time (s)
		$T_{Inc,sim}$ (K)	$\dot{Q}_{Ave}$ (W m^{-3})	
Extremely coarse	240	51.9	163,930	23
Extra coarse	597	47.9	155,420	24
Coarser	1442	43.2	137,150	33
Coarse	2521	42.1	133,630	43
Normal	7912	40.3	127,470	108
Fine	14258	39.8	125,470	179
Finer	48326	39.2	123,390	593
Extra fine	187631	38.7	121,560	2735

3. Results and Discussion

3.1. Heating Rate. Effect of Sample Mass and Spatial Configuration

The behavior of the tested sample arrays was analyzed employing the heating rate defined in Equation (1), *T* being the temperature measured at the end of heating.

The average values of *HR* for each spatial configuration, obtained in the tests with 225 and 300 W of nominal power, are presented in Table 3.

Comparing the heating rates at different nominal power for the same configuration, it is possible to appreciate that an increase in the nominal power corresponded to an increase in the heating rate, except for the configuration 4B, which did not increase its heating rates. At a given power, it was observed that—at a given sample mass—all the configurations with the shorter possible distance from the top electrode and surface exhibited a higher heating rate. Configurations characterized by the same mass (and then by the same number of cubes) were compared: configuration 2A was compared with configuration 2B; configuration 4A with 4B and 4C; configurations 8A with 8B and 8C.

When comparing samples made by two cubes (configurations 2A and 2B), in configuration 2B (which was vertically oriented) heating was observed to be three times faster than in the 2A one. The different orientation corresponded to a closer distance between the top electrode and the upper free surface of the sample, though the configuration 2B had a smaller projected area with respect to 2A. Uyar et al. [3] reported that a wider sample projected area together with a lower electrode gap provide higher heating rates. In the case analyzed in this work, it appeared evident that the distance between the top surface and the top electrode was the key parameter to distinguish the two cases: in configuration 2A, the sample had a greater projected area and the major gap between the top surface and the electrode, while in configuration 2B the sample had half the projected area but it was closer to the top electrode.

Table 3. Average experimental heating rate calculated according to Equation (1), for two RF nominal powers.

Sample Code	Heating Rate HR (K s^{-1})	
	225 W	300 W
2A	0.045	0.052
2B	0.133	0.217
4A	0.029	0.041
4B	0.031	0.034
4C	0.101	0.126
8A	0.018	0.025
8B	0.057	0.073
8C	0.051	0.073
12	0.036	0.050
16	0.032	0.049

In samples formed by four cubes, the highest heating rate was observed in scheme 4C (three times higher than the heating rate of samples 4A and 4B for both power levels), which again presented two rows of cubes in a vertical direction, this sample being the closest to the top electrode. Although samples 4A–4B had a greater projected area than 4C, the effect of the lower gap between sample and electrode was more relevant and dominant over the projected area.

Regarding the configuration with eight cubes, for both power levels the highest heating rate was found in the sample 8B and 8C, with two rows in a vertical direction. Samples 8B and 8C had the same projected area and the same gap between the surface and the electrode. As they were closer to the top electrode than 8A, their heating rates triplicated the heating rate of the latter.

As the distance between the top and bottom electrode was fixed, in configurations with more than one layer of cubes the gap between the sample surface and the top electrode was smaller. Therefore, these experimental results confirmed previous theoretical results: the larger the air gap between sample and top electrode, the slower the heating rate (Uyar et al., 2014 [3]).

3.2. Model Validation

Experimental and simulated average temperature profiles, $T_{Inc,Exp}$ and $T_{Inc,Sim}$, are shown in Figures 5 and 6. Additionally, the accuracy of the mathematical model is evaluated by AARD and AAD, these data are summarized in Table 4.

As can be seen, the model predictions were in good agreement with experimental values; the predicted values were generally slightly lower than the experimental ones. The higher differences can be observed for sample 8B at 225 W, and 16 cubes for both powers. The numerical results were slightly lower than the experimental ones in all the considered cases, and this can be attributed to the expressions used to represent the dielectric property of the sample, taken from [31], since differences in dielectric constant and loss factor will affect the determination of EM field and the calculation of the heat generated by the interactions of the EM field with the food material, and thus the evolution of the temperature in the space and during the time.

Table 4. Prediction accuracy of the mathematical model.

Sample Code	225 W Power		300 W Power	
	AARD (%)	AAD (K)	AARD (%)	AAD (K)
2A	0.25	0.70	0.05	0.15
2B	0.14	0.41	0.21	0.63
4A	0.06	0.18	0.19	0.53
4B	0.33	0.94	0.14	0.40
4C	0.35	1.03	0.17	0.51
8A	0.04	0.11	0.09	0.27
8B	0.67	1.98	0.11	0.33
8C	0.23	0.66	0.25	0.74
12	0.51	1.50	0.34	1.03
16	0.53	1.58	0.54	1.66
Average	0.31	0.91	0.21	0.63

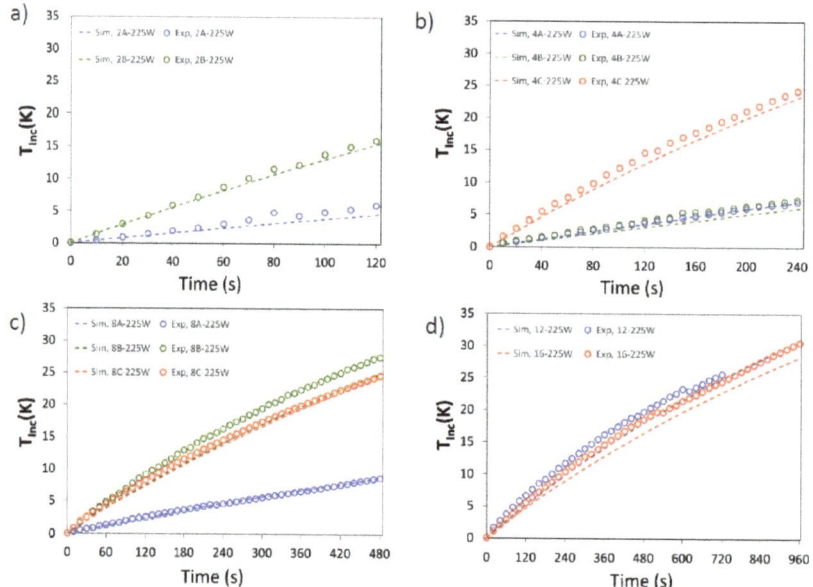

Figure 5. Experimental (symbols) and simulated (dashed lines) average temperature increase (T_{Inc}, K) for 225 W. Samples: (**a**) 2 cubes; (**b**) 4 cubes; (**c**) 8 cubes; (**d**) 12 and 16 cubes.

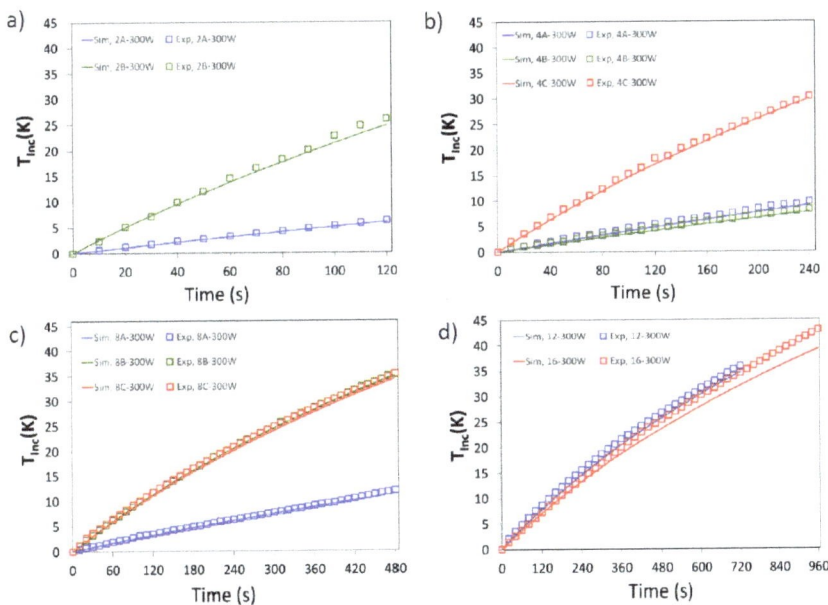

Figure 6. Experimental (symbols) and simulated (lines) average temperature increase (T_{Inc}, K) for 300 W. Samples: (**a**) 2 cubes; (**b**) 4 cubes; (**c**) 8 cubes; (**d**) 12 and 16 cubes.

To complete the analysis of temperature evolution, Table 5 detailed the final temperature increase, $T_{Inc,Exp}$ and $T_{Inc,Sim}$, for both powers. Samples with the same size (and mass) presented different thermal histories and, in consequence, different average final temperatures.

Table 5. Experimental and simulated values of average temperature increase ($T-T_0$), calculated at the final heating time, for two RF nominal powers.

Sample Code	$T_{Inc,exp}$ (K) 225 W	$T_{Inc,sim}$ (K) 225 W	$T_{Inc,exp}$ (K) 300 W	$T_{Inc,sim}$ (K) 300 W
2A	5.99	4.49	6.22	5.96
2B	16.01	15.35	26.02	24.78
4A	6.96	6.86	9.75	9.02
4B	7.36	5.91	8.18	7.76
4C	24.17	23.10	30.21	29.78
8A	8.72	8.68	11.95	11.28
8B	27.46	24.87	35.16	34.95
8C	24.58	24.37	35.11	34.24
12	25.60	24.80	35.81	34.63
16	30.52	28.23	42.99	39.16

The spatial configuration of the samples affected the evolution of temperature profile inside them; in those close to the top electrode, the electric field deflection is more pronounced [19,20]), increasing the heating rate and the average final temperature. Although this effect is desirable, an additional consequence is that the electric field is concentrated in the corners and edges of the sample, leading to uneven power absorption and poor temperature uniformity. Table 6 details the experimental and simulated temperature uniformity. These results indicate that samples with higher heating rates and temperature increases present worse temperature uniformity.

Table 6. Experimental and simulated values of average temperature uniformity, Equations (4) and (18) respectively, for two RF nominal powers.

Sample Code	$T_{U,exp}$ (K) 225 W	$T_{U,sim}$ (K) 225 W	$T_{U,exp}$ (K) 300 W	$T_{U,sim}$ (K) 300W
2A	1.5	1.9	2.0	2.1
2B	3.7	3.5	5.7	4.9
4A	3.5	2.5	3.5	2.8
4B	3.4	2.7	3.9	3.0
4C	5.5	5.4	6.3	6.2
8A	4.1	3.6	4.5	4.0
8B	6.4	5.8	7.6	7.5
8C	6.2	5.9	7.3	7.3
12	6.6	6.0	7.9	7.6
16	6.9	6.6	7.9	8.3

As was shown, the distance between the sample and electrodes influences the thermal response of the sample processed in a RF oven. In the experiments, the distance to the bottom electrode remains constant for the complete set of sample configurations tested. On the contrary, when a second layer of cubes is employed, the distance to the top electrode diminishes significantly. This effect was analyzed through numerical analysis. Figures 7 and 8 show the electric behavior and the sample temperature profile in the middle of y-dimension after 1 min and 4 min RF heating. As can be seen, the electric field largely varies with position and time. At the sample center, the initial norm of the electric field was 1065.7 V m^{-1}, while at 4 min it was 595.95 V m^{-1} (results not shown). Electric field variations, together with dielectric properties variations, affect the heat absorbed by the sample.

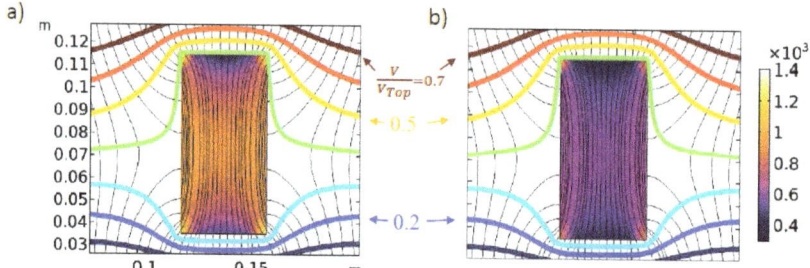

Figure 7. Simulated electric field values during heating of sample 4C, at middle y-dimension, 300 W RF nominal power, after (**a**) 1 min and (**b**) 4 min heating. Colored surface: electric field norm (V m^{-1}); colored level curves: ratio of electric potential to maximum electric potential; black lines correspond to electric field lines.

After 1 min heating (Figure 7a), the simulated volume average electromagnetic heat source (Equation (23)) was 524 kW m^{-3}, whereas after 4 min heating (Figure 7b), it was 369 kW m^{-3}. This variation represents a 29.6% reduction in the average heat source. Figure 8 shows the temperature profile in the sample, which follows the electric field distribution depicted in Figure 7, with higher temperature values in the corners and the sample center. The simulated electric field deformations due to the sample are like the ones reported by [6,20], confirming that the uneven distribution of the electric field is responsible of the lack of uniformity in the sample temperature. In this sense, Tiwari et al. [20] found that power uniformity could be improved when the sample projected area is similar to the electrode size, since less electric field distortion is verified.

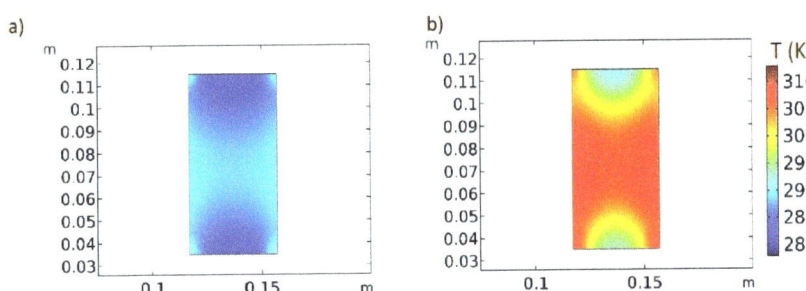

Figure 8. Simulated temperature distribution (K) during heating of sample 4C, at middle y-dimension, 300 W RF nominal power, after (**a**) 1 min and (**b**) 4 min heating.

3.3. Energy Efficiency

In order to analyze the heating efficiency of the RF oven tested in this work, Table 7 details experimental and simulated values of energy consumption and efficiency, for the whole set of studied cases. The experimental values of energy consumption (i.e., energy absorbed by the sample) were calculated as the numerator of Equation (5), and the simulated ones by Equation (18). Considering the nominal energy provided in each condition, efficiency was evaluated by using Equations (5) and (19) for experimental and simulated values, respectively. Despite the differences between experimental and simulated values, both sets are highly correlated ($r \cong 0.9991$).

Table 7. Experimental and simulated values of absorbed energy and efficiency, for two RF nominal powers.

Sample Code	Nominal Supplied Energy, $t_h \times NP$ (kJ)	Absorbed Energy (kJ)		Efficiency (%)	
		Q_{Exp}	Q_{Sim}	η_{Exp}	η_{Sim}
225 W power					
2A	27	2.8	2.2	10.3%	8.1%
2B	27	7.5	7.5	27.6%	27.6%
4A	54	6.5	6.7	12.0%	12.4%
4B	54	6.9	5.7	12.7%	10.6%
4C	54	22.5	22.4	41.7%	41.5%
8A	108	16.3	16.9	15.1%	15.6%
8B	108	51.2	48.3	47.4%	44.7%
8C	108	45.8	47.3	42.4%	43.8%
12	162	71.6	72.2	44.2%	44.6%
16	216	113.8	109.6	52.7%	50.7%
300 W power					
2A	36	2.9	2.9	8.1%	8.1%
2B	36	12.1	12	33.7%	33.3%
4A	72	9.1	8.8	12.6%	12.2%
4B	72	7.6	7.5	10.6%	10.4%
4C	72	28.2	28.9	39.1%	40.1%
8A	144	22.3	21.9	15.5%	15.2%
8B	144	65.6	67.8	45.5%	47.1%
8C	144	65.5	66.4	45.5%	46.1%
12	216	100.2	100.8	46.4%	46.7%
16	288	160.3	151.9	55.7%	52.7%

The results show that efficiency is higher when the spatial configuration has more than one layer, with a smaller gap between the sample and the top electrode (2B, 4C, 8B, 8C, 12, 16), in coincidence with the higher temperature increase and worse temperature uniformity. Notwithstanding, the bigger samples (8B, 8C, 12, 16) the higher the efficiencies, denoting a better exploitation of the provided energy.

The tendencies of experimental and simulated efficiency values presented in this work agree with simulated values obtained in a similar system [3], despite the differences in the operating characteristics (27.12 vs. 40.68 MHz) of the cavity and sample configurations. The bigger the load and the smaller the gap between the sample and the top electrode, the higher the heating efficiency. In Uyar et al. [3], the predicted heating efficiency was near zero for a hypothetical small sample that was 1% of the maximum simulated sample size (0.398 L). Efficiency increased as the sample size increased, reaching values near 40–50% for the maximum simulated sample size. Additionally, for the same mass, the spatial configuration with the smallest gap between the top electrode and the sample had the highest efficiencies.

In general, the energy efficiency observed in bigger samples is higher than the efficiency observed in traditional convective ovens: efficiency from 7.7 to 18.3% was measured in meat cooking [33] and from 6 to 13% in bakery products [34,35].

4. Conclusions

In the RF heating of chicken meat at 40.68 MHz, the geometrical factors, such as the spatial configuration of the sample and the gap between top electrode and samples, play a major role in heating rate, temperature uniformity, power absorption and energy efficiency. A decrease in the electrode-sample gap improves heating rates and energy efficiency, but a less uniform temperature distribution is attained. Therefore, the possibility of changing the electrode gap is a qualifying point in the design of RF assisted heating systems.

Numerical results agree with experimental measurements and show the same general tendencies.

Both measured and simulated energy efficiencies were close to 50% for the maximum loads, substantially higher than efficiencies obtained in traditional ovens. Notwithstanding, the lack of temperature uniformity is still an issue that deserves a major research effort. The developed mathematical model, validated by comparing the average temperature evolution in the samples, can be extended to different food geometries and RF systems, so that the presented methodology and the digital tool developed in this work can be used in RF heating system design.

Author Contributions: Conceptualization, S.M.G., V.O.S. and F.M.; Methodology, S.M.G. and F.M.; Software, S.M.G. and F.M.; Validation, S.M.G., F.M., M.D.V. and M.d.; Formal Analysis, S.M.G., V.O.S. and F.M.; Investigation, S.M.G., D.F.O. and F.M.; Resources, F.M., M.d., M.D.V.; Data Curation, F.M., M.d.; Writing—Original Draft Preparation, S.M.G., V.O.S., D.F.O. and F.M.; Writing—Review & Editing, S.M.G., V.O.S., D.F.O., M.d. and F.M.; Visualization, S.M.G.; Supervision, F.M.; Project Administration, F.M.; Funding Acquisition, V.O.S. and F.M. All authors have read and agreed to the published version of the manuscript.

Funding: This work was supported by the VII Executive Program of Scientific and Technological Cooperation between MAECI-Italy and MINCYT-Argentina 2017–2019, Project IT/AR-17-MO2.

Data Availability Statement: The data presented in this study are available in this article.

Acknowledgments: The authors would like to thank to Consejo Nacional de Investigaciones Científicas y Técnicas (Argentina), Universidad Nacional de La Plata (Argentina) and Università degli Studi Si Salerno (Italy).

Conflicts of Interest: The authors declare no conflict of interest.

References

1. Marra, F.; Zhang, L.; Lyng, J. Radio frequency treatment of foods: Review of recent advances. *J. Food Eng.* **2009**, *91*, 497–508. [CrossRef]
2. Guo, C.; Mujumdar, A.; Zhang, M. New development in radio frequency heating for fresh food processing: A review. *Food Eng. Rev.* **2019**, *11*, 29–43. [CrossRef]

3. Uyar, R.; Erdogdu, F.; Marra, F. Effect of load volume on power absorption and temperature evolution during radio-frequency heating of meat cubes: A computational study. *Food Bioprod. Process.* **2014**, *92*, 243–251. [CrossRef]
4. Uyar, R.; Erdogdu, F.; Sarghini, F.; Marra, F. Computer simulation of radio-frequency heating applied to block-shaped foods: Analysis on the role of geometrical parameters. *Food Bioprod. Process.* **2016**, *98*, 310–319. [CrossRef]
5. Hou, L.; Johnson, J.A.; Wang, S. Radio frequency heating for postharvest control of pests in agricultural products: A review. *Postharvest Biol. Technol.* **2016**, *113*, 106–118. [CrossRef]
6. Huang, Z.; Marra, F.; Wang, S. A novel strategy for improving radio frequency heating uniformity of dry food products using computational modeling. *Innov. Food Sci. Emerg. Technol.* **2016**, *34*, 100–111. [CrossRef]
7. Xu, J.; Zhang, M.; Bhandari, B.; Kachele, R. ZnO nanoparticles combined radio frequency heating: A novel method to control microorganism and improve product quality of prepared carrots. *Innov. Food Sci. Emerg. Technol.* **2017**, *44*, 46–53. [CrossRef]
8. Zhu, J.; Zhang, D.; Zhou, X.; Cui, Y.; Jiao, S.; Shi, X. Development of a pasteurization method based on radiofrequency heating to ensure microbiological safety of liquid egg. *Food Control.* **2019**, *123*, 107035. [CrossRef]
9. Zhou, X.; Wang, S. Recent developments in radio frequency drying of food and agricultural products: A review. *Dry. Technol.* **2019**, *37*, 271–286. [CrossRef]
10. Palazoğlu, T.K.; Miran, W. Experimental investigation of the effect of conveyor movement and sample's vertical position on radio frequency tempering of frozen beef. *J. Food Eng.* **2018**, *219*, 71–80. [CrossRef]
11. Bedane, T.F.; Chen, L.; Marra, F.; Wang, S. Experimental study of radio frequency (RF) thawing of foods with movement on conveyor belt. *J. Food Eng.* **2017**, *201*, 17–25. [CrossRef]
12. Bedane, T.F.; Altin, O.; Erol, B.; Marra, F.; Erdogdu, F. Thawing of frozen food products in a staggered through-field electrode radio frequency system: A case study for frozen chicken breast meat with effects on drip loss and texture. *Innov. Food Sci. Emerg. Technol.* **2018**, *50*, 139–147. [CrossRef]
13. Llave, Y.; Liu, S.; Fukuoka, M.; Sakai, N. Computer simulation of radiofrequency defrosting of frozen foods. *J. Food Eng.* **2015**, *152*, 32–42. [CrossRef]
14. Laycock, L.; Piyasena, P.; Mittal, G.S. Radio frequency cooking of ground, comminuted and muscle meat products. *Meat Sci.* **2003**, *65*, 959–965. [CrossRef]
15. Kirmaci, B.; Singh, R.K. Quality of chicken breast meat cooked in a pilot-scale radio frequency oven. *Innov. Food Sci. Emerg. Technol.* **2012**, *14*, 77–84. [CrossRef]
16. Muñoz, I.; Serra, X.; Guàrdia, M.D.; Fartdinov, D.; Arnau, J.; Picouet, P.A.; Gou, P. Radio frequency cooking of pork hams followed with conventional steam cooking. *LWT* **2020**, *123*, 109104. [CrossRef]
17. Wang, X.; Wang, L.; Yang, K.; Wu, D.; Ma, J.; Wang, S.; Zhang, Y.; Sun, W. Radio frequency heating improves water retention of pork myofibrillar protein gel: An analysis from water distribution and structure. *Food Chem.* **2021**, *350*, 129265. [CrossRef]
18. Huang, Z.; Marra, F.; Subbiah, J.; Wang, S. Computer simulation for improving radio frequency (RF) heating uniformity of food products: A review. *Crit. Rev. Food Sci. Nutr.* **2018**, *58*, 1033–1057. [CrossRef]
19. Marra, F.; Lyng, J.; Romano, V.; McKenna, B. Radio-frequency heating of foodstuff: Solution and validation of a mathematical model. *J. Food Eng.* **2007**, *79*, 998–1006. [CrossRef]
20. Tiwari, G.; Wang, S.; Tang, J.; Birla, S.L. Analysis of radio frequency (RF) power distribution in dry food materials. *J. Food Eng.* **2011**, *104*, 548–556. [CrossRef]
21. Romano, V.; Marra, F. A numerical analysis of radio frequency heating of regular shaped foodstuff. *J. Food Eng.* **2008**, *84*, 449–457. [CrossRef]
22. Li, F.; Zhu, Y.L.; Li, S.; Wang, P.Z.; Zhang, R.Y.; Tang, J.M.; Koral, T.; Jiao, Y. A strategy for improving the uniformity of radio frequency tempering for frozen beef with cuboid and step shapes. *Food Contr.* **2021**, *123*, 107719. [CrossRef]
23. Bedane, T.F.; Erdogdu, F.; Lyng, J.G.; Marra, F. Effects of geometry and orientation of food products on heating uniformity during radio frequency heating. *Food Bioprod. Processing* **2021**, *125*, 149–160. [CrossRef]
24. Tiwari, G.; Wang, S.; Tang, J.; Birla, S.L. Computer simulation model development and validation for radio frequency (RF) heating of dry food materials. *J. Food Eng.* **2011**, *105*, 48–55. [CrossRef]
25. Fahmy, T. Dielectric Relaxation Spectroscopy of Poly (Vinyl Chloride-co-Vinyl Acetate-co-2-Hydroxypropyl Acrylate)/Poly (Acrylonitrile-Butadiene-Styrene) Polymer Blend. *Polym.-Plast. Technol. Eng.* **2007**, *46*, 7–18. [CrossRef]
26. Della Valle, M. Analysis of Chicken Meat Heating: Product Characterization and RF Assisted Processing. Master's Thesis, University of Salerno, Fisciano, Italy, 2018.
27. COMSOL Multiphysics®. COMSOL AB: Stockholm, Sweden. 2021. Available online: www.comsol.com (accessed on 1 November 2021).
28. Chen, L.; Wang, K.; Li, W.; Wang, S. A strategy to simulate radio frequency heating under mixing conditions. *Comput. Electron. Agr.* **2015**, *118*, 100–110. [CrossRef]
29. Choi, Y.; Okos, M.R. Effects of temperature and composition on the thermal properties of foods. In *Food Engineering and Process Applications*; Elsevier Applied Science Publishers: London, UK, 1986; Volume 1, pp. 93–101.
30. Chmiel, M.; Roszko, M.; Adamczak, L.; Florowski, T.; Pietrzak, D. Influence of storage and packaging method on chicken breast meat chemical composition and fat oxidation. *Poult. Sci.* **2019**, *98*, 2679–2690. [CrossRef]
31. Zhuang, H.; Nelson, S.O.; Trabelsi, S.; Savage, E.M. Dielectric properties of uncooked chicken breast muscles from ten to one thousand eight hundred Megahertz. *Poult. Sci.* **2007**, *86*, 2433–2440. [CrossRef]

32. Chen, J.; Lau, S.K.; Chen, L.; Wang, K.; Li, W.; Wang, S.; Subbiah, J. Modeling radio frequency heating of food moving on a conveyor belt. *Food Bioprod. Process.* **2017**, *102*, 307–319. [CrossRef]
33. Goñi, S.M.; Salvadori, V.O. Energy consumption estimation during oven cooking of food. In *Energy Consumption: Impacts of Human Activity, Current and Future Challenges, Environmental and Ecological Effects*; Reiter, S., Ed.; Nova Publishers: New York, NY, USA, 2014; pp. 99–116.
34. Paton, J.; Khatir, Z.; Thompson, H.; Kapur, N.; Toropov, V. Thermal energy management in the bread baking industry using a system modelling approach. *Appl. Therm. Eng.* **2013**, *53*, 340–347. [CrossRef]
35. Ureta, M.M.; Goñi, S.M.; Salvadori, V.O.; Olivera, D.F. Energy requirements during sponge cake baking: Experimental and simulated approach. *Appl. Therm. Eng.* **2017**, *115*, 637–643. [CrossRef]

Article

Evaluation of Pilot-Scale Radio Frequency Heating Uniformity for Beef Sausage Pasteurization Process

Ke Wang [1], Lisong Huang [2], Yangting Xu [1], Baozhong Cui [1], Yanan Sun [1], Chuanyang Ran [1], Hongfei Fu [1], Xiangwei Chen [1], Yequn Wang [1] and Yunyang Wang [1,*]

1 College of Food Science and Engineering, Northwest A&F University, Yangling, Xianyang 712100, China; wangkejxsj@nwafu.edu.cn (K.W.); 2017013554@nwafu.edu.cn (Y.X.); 2015014870@nwafu.edu.cn (B.C.); synsyn@nwafu.edu.cn (Y.S.); 2019055072@nwafu.edu.cn (C.R.); fuhongfei@nwsuaf.edu.cn (H.F.); chenxiangwei@nwsuaf.edu.cn (X.C.); yequnw@163.com (Y.W.)
2 College of Food Science and Engineering, NanJing University of Finance &Economics, Nanjing 210023, China; 1120211124@stu.edu.nufe.cn
* Correspondence: wyy10421@nwafu.edu.cn; Tel.: +86-135-7241-2298

Abstract: Radio frequency (RF) heating has the advantages of a much faster heating rate as well as the great potential for sterilization of food compared to traditional thermal sterilization. A new kettle was designed for sterilization experiments applying RF energy (27.12 MHz, 6 kW). In this research, beef sausages were pasteurized by RF heating alone, the dielectric properties (DPs) of which were determined, and heating uniformity and heating rate were evaluated under different conditions. The results indicate that the DPs of samples were significantly influenced ($p < 0.01$) by the temperature and frequency. The electrode gap, sample height and NaCl content had significant effects ($p < 0.01$) on the heating uniformity when using RF energy alone. The best heating uniformity was obtained under an electrode gap of 180 mm, a sample height of 80 mm and NaCl content of 3%. The cold points and hot spots were located at the edge of the upper section and geometric center of the sample, respectively. This study reveals the great potential in solid food for pasteurization using RF energy alone. Future studies should focus on sterilization applying RF energy and SW simultaneously using the newly designed kettle.

Keywords: RF sterilization; beef sausage; dielectric properties; heating uniformity

Citation: Wang, K.; Huang, L.; Xu, Y.; Cui, B.; Sun, Y.; Ran, C.; Fu, H.; Chen, X.; Wang, Y.; Wang, Y. Evaluation of Pilot-Scale Radio Frequency Heating Uniformity for Beef Sausage Pasteurization Process. *Foods* **2022**, *11*, 1317. https://doi.org/10.3390/foods11091317

Academic Editors: Olivier Rouaud, Pierre Sylvain Mirade and Hyun-Gyun Yuk

Received: 8 March 2022
Accepted: 28 April 2022
Published: 30 April 2022

Publisher's Note: MDPI stays neutral with regard to jurisdictional claims in published maps and institutional affiliations.

Copyright: © 2022 by the authors. Licensee MDPI, Basel, Switzerland. This article is an open access article distributed under the terms and conditions of the Creative Commons Attribution (CC BY) license (https://creativecommons.org/licenses/by/4.0/).

1. Introduction

Nowadays, people's increasing awareness of the convenience and health aspects of ready-to-eat foods has stimulated the demand for more organic and natural foods, thus promoting the development of a sausage food free of artificial additives [1,2]. However, a majority of the manufactured beef products continue to be periodically contaminated with micro-organisms like *Escherichia coli* (*E. coli*) and *Salmonella* [3,4].

The traditional process of sterilization often has a harmful impact on product quality as the result of the slow heating rate and non-uniform heating [5–8]. Some novel sterilization technologies have proven to be an effective replacement for conventional methods, including infrared, ultrasonic, ohmic heating, microwave (MW) and radio frequency (RF) technologies [9–13].

RF heating processing, an emerging technology for pasteurization, has been carried out as an effective method for microbial inactivation in different food materials [14]. Its function depends on the heat energy transformed by electric power as a consequence of charged ions or polar molecules migrating and rotating under radiofrequency electromagnetic fields (3000 Hz–300 MHz) [15,16]. Therefore, RF heating technology has the advantages of higher penetration depth and faster heating rate, resulting in much less processing time to complete the pasteurization process compared to conventional methods [17,18].

DPs composed of dielectric constant (ε') and dielectric loss factor (ε'') are important factors affecting heating uniformity and heating rate [19]. To achieve a better understanding of the RF heating of beef sausage, it is essential to measure and evaluate the DPs of beef sausage. In previous research, the effects of temperature (T) and frequency (f) on dielectric properties of different kinds of meat and meat products have been analyzed, including chicken meat [20,21] and meat batters [22]. In addition, research on dielectric properties involving beef meat blends [23] and beef biceps femoris muscle [24], whose DPs are similar, have been reported. However, the DPs of beef sausage have been rarely reported, and further study of DPs is needed for pilot-scale RF heating.

Non-uniform heating is an important factor restricting the large-scale application of RF pasteurization. Therefore, it is necessary to study the uniformity of RF heating. Several experimental investigations on heating uniformity have been reported using RF energy. Li et al. [25] evaluated the effects on frozen beef of the heating uniformity in RF systems, observing that the RF tempering rate with immersion in glycerol solution was much higher than the traditional refrigeration tempering method. Dong et al. [26] revealed that added water and fat had negative effects on the RF heating uniformity, whereas high salt content had a positive influence on the RF heating uniformity. In addition, the uniformity index (λ) has been applied to analyze the heating uniformity of food, and numerous studies have been carried out on RF treatments, including those involving walnut kernels [27] and peanuts [28]. However, there is limited literature on heating uniformity in meat products during RF heating in terms of the uniformity index (λ).

The objectives of this research were (1) to study the DPs of beef sausage as influenced by temperature and frequency and analyze the influence of DPs on temperature at RF frequency compared to MW frequency; (2) to assess the effects of electrode gap, sample height and NaCl content in terms of the heating profile and temperature distribution of beef sausage by RF heating; (3) to evaluate the heating uniformity on the basis of the uniformity index (λ) when subjected to RF energy.

2. Materials and Methods

2.1. Beef Sausage Manufacturing

Qin-chuan beef (hind leg), as the main raw material of beef sausage, was obtained from a local store. Sausages were prepared with fresh beef (250 ± 1 g), ginger powder (5.0 ± 0.1 g), green Chinese onion (5.0 ± 0.1 g) and carrot powder (5.0 ± 0.1 g), minced through a high-speed blender (SP903, Supor Co., Ltd., Zhejiang, China), with the following additives: wheat starch (30.0 ± 1 g), egg (60.0 g ± 1 g) and soy sauce (5.0 ± 0.1 g). After packaging and steaming, the sample was cut into a cylindrical shape. For RF heating experiments, the beef sausages were prepared with dimensions (diameter × height) of 26 × 75 mm, 26 × 80 mm and 26 × 85 mm with mean ± SD weights of 38.8 ± 0.4 g, 40.5 ± 0.5 g and 41.8 ± 0.5 g, respectively. For DPs measurement, the samples were prepared with dimensions of 21.5 mm × 45 mm and weighed 25.5 ± 0.3 g. The samples were prepared with different NaCl content (3.0%, 2.0% and 1.0%). All batches of sausages were manufactured at a temperature of 25 °C within 1 h and then packaged and stored in a refrigerator at 4 °C. To keep the same initial temperature, the samples were transferred to an incubator at 25 ± 1 °C for 1–2 h before the experiments.

2.2. Physiochemical Composition of Beef Sausage

The moisture of beef sausage was measured according to the AOAC (1985) [29] standard methods 985.14. The total nitrogen content of beef sausage was measured using the Kjeldahl method following the method number 992.15 (AOAC, 1993) [30]. Crude protein was estimated by multiplying total nitrogen content by a factor of 6.25. Moreover, fat content of the sample was determined by following the method of Soxhlet extraction according to AOAC 991.36 [31]. The ash and starch of the sample were analyzed according to AOAC 923.03 [32] and AOAC 958.06 [33], respectively. The composition of the sample is summarized in Table 1. Each experiment was replicated three times.

Table 1. The composition of beef sausage.

Moisture Content (g kg^{-1} w.b.)	Ash (g kg^{-1})	Protein (g kg^{-1})	Carbohydrate (g kg^{-1})	Fat (g kg^{-1})
633.5	22.5	171.1	122.5	38.9

2.3. Determination of DPs

The open-ended coaxial probe method is commonly used in the field of food research for measuring the DPs of different kinds of foods [34]. The method described above was chosen to measure the DPs of beef sausage in this study. The measuring system mainly includes an E4991B-300 model impedance analyzer (Keysight Technologies Inc., Santa Rosa, CA, USA), a calibration kit (E4991B-010), a SST-20 oil circulated oil bath (Wuxi Guanya Refrigeration Technology Co., Ltd., Jiangsu, China) and a high-temperature coaxial cable and dielectric probe (85070E-20). A more detailed description can be found in Cui et al. [14].

A 30–45 min warm-up and calibration of the impedance analyzer were performed prior to each test. The E4991B calibration kit was used to calibrate the impedance analyzer, with an Open, Short, and 50 Ω resistance in order. The coaxial probe was calibrated by air at first, then Short, and finally deionized water (25 °C) [17]. After calibration, the sample was put into a cylindrical sample holder with dimensions of 21.5 mm × 45 mm, which was surrounded by a circulating temperature-controlled oil bath. A pre-calibrated type-T thermocouple temperature sensor was applied to test the central temperature of the sample. The sample was kept in close contact with the dielectric probe and sealed in the test cell. The temperature of the three samples was controlled from 25 to 90 °C at an interval of 10 °C (25, 30, 40, 50, 60, 70, 80 and 90 °C), with frequency range between 1 MHz and 3000 MHz. This method was similar to studies on lean beef meats [35] and meat lasagna [36]. Each treatment was replicated three times.

2.4. RF Heating System

A 6 KW, 27.12 MHz pilot-scale free-running oscillator RF heating system (GJG-2.1-10A-JY; Hebei Huashijiyuan High Frequency Equipment Co., Ltd., Shijiazhuang, China) was used in this research. This system was composed of an RF feeder, four inductors and two electrode plates. Different electrode gaps between 100–300 mm could be obtained by adjusting the top electrode plate. Detailed descriptions of the RF system can be found in Cui et al. [14]. The RF heating system was equipped with two parts of a sterilization kettle and a fluorescence optical fiber temperature measurement system (Figure 1). A customized sterilization kettle was designed, consisting of two lid plates (Ø 280 mm × 20 mm), a hollow cylindrical vessel (Ø 280 mm × 110 mm) made of PTFE, and a flange (Figure 2). For sealing, the top and bottom lid, which were both made of aluminum alloy, were fixed with to vessel by screws. The real-time changes in the temperature of samples was tested through a fluorescence-based optical fiber temperature measure system (HQ-FTS-D1F00, Xi'an Heqi Opto-Electronic Technology Co., Ltd., Xi'an, China) in the processing of RF heating. The sterilization kettle provided a relatively thermal insulation environment for materials, and a sterilization kettle using RF energy alone was selected in this study.

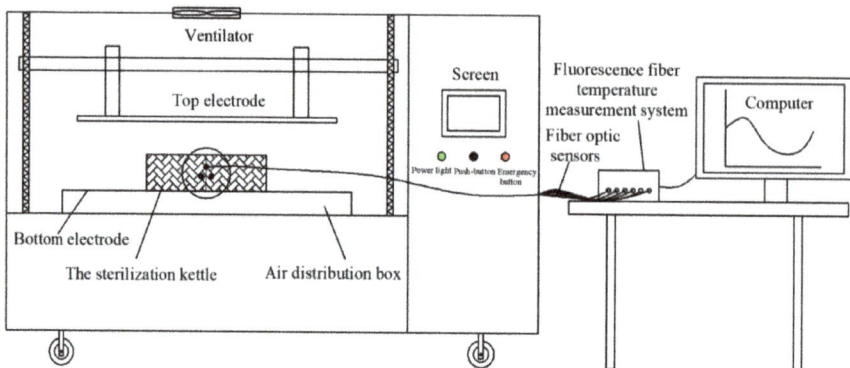

Figure 1. Simplified schematic diagram of sterilization kettle in the RF heating system.

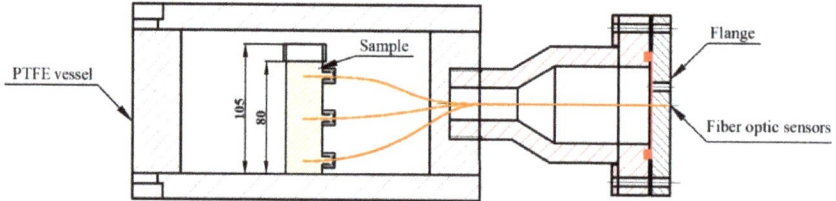

Figure 2. Detailed cross-section view of the sterilization kettle with beef sausage placed in the center and fiber optic sensor inserted in the sample.

2.5. Evaluating the Heating Uniformity and Heating Rate

Prior to starting this experiment, the cylinder-shape sample (Ø 26 mm × 80 mm) was filled in a flat-bottom tube made of PTFE, and the tube filled with sample was placed on the geometric center of the bottom lid in the sterilization kettle (Figure 2). Three fiber optic sensors were inserted at positions 1, 2 and 3 in the sample to test the internal temperature during RF heating processing (Figure 3a). Then, the top lid was tightened with bolts. The sterilization kettle was placed on the bottom electrode in the RF cavity (Figure 1).

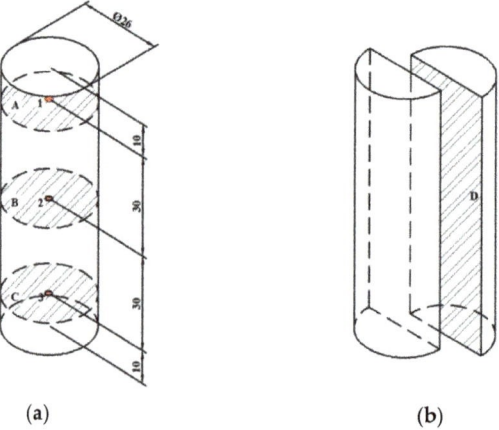

Figure 3. The location of fiber optic sensors (1, 2, 3) and four layers (upper, A; middle, B; lower, C; longitudinal, D) in the beef sausage sample for temperature distribution measurement (all measured in millimeters). (**a**) Detailed diagram of sample. (**b**) Longitudinal section of sample.

After RF heating experiments, the sample temperature distribution was determined instantly using the infrared thermography camera (FLIR A300, FLIR Systems Inc., Wilsonville, OR, USA). Individual surface temperature data points of 2500–10,000 for each thermal image were recorded and collected using the software (BM_IR V7.4).

2.5.1. Electrode Gap

Three electrode gaps of 175, 180 and 185 mm were selected according to the arrangement of the experiments. Then, the heating system of the RF was opened. The time-temperature profile was recorded to find the point with the slowest heating rate, which was considered to be the cold point. The RF heating system was turned off when the temperature at the cold spot reached 70 °C [37–39] for pasteurization. After RF heating was turned off, the sample was removed from the tube carefully and quickly. The sample was photographed using the infrared thermography camera to obtain thermal images of upper (A), middle (B) and lower layers (C), as well as the longitudinal section (D) (Figure 3). The whole process lasted less than 30 s from turning off the RF system to the end of photographing.

2.5.2. Sample Height

Samples with different heights (75, 80 and 85 mm) were selected in the processing of RF heating. The electrode gap of 120 mm was used based on previous experiments. The PTFE tube filled with the sample was placed on the geometric center of the bottom lid in the sterilization kettle (Figure 2). Three fiber optic sensors were inserted at positions 1, 2 and 3 of the beef sausage to test the internal temperature during RF heating, as illustrated in Figure 3a. The RF heating system was turned off when the temperature at the cold spot reached 70 °C [37]. The temperatures of the upper, middle and lower layer and the longitudinal section were analyzed and mapped as mentioned previously.

2.5.3. NaCl Content

Based on previous experiments, the determined electrode gap and sample height were 180 mm and 80 mm, respectively. Sausage samples were prepared with different NaCl content (3.0%, 2.0% and 1.0%) for RF heating. The sample and the fiber optic sensors were placed as described in Sections 2.5.1 and 2.5.2. At the slowest heating point, which was considered to be the cold point reaching the target temperature (70 °C) for pasteurization, the RF system was turned off. When RF heating was stopped, the temperatures of the upper, middle and lower layers, as well as the longitudinal section, were analyzed using the same method as described above.

2.6. Assessment of the Heating Uniformity

The heating uniformity was evaluated by the thermal images captured by the infrared camera, and the heating uniformity index λ was calculated using Equation (1) [13,16,40]

$$\lambda = \frac{\sqrt{\sigma^2 - \sigma_0^2}}{\mu - \mu_0} \quad (1)$$

where μ_0 and μ are the initial and final sample temperatures, and σ and σ_0 are the final and initial SDs of sample temperatures in the RF process, respectively.

2.7. Statistical Analysis

Three trials were conducted to obtain mean values and standard error for statistical analysis. The contour plots of data were created by Origin 2020 (Origin Lab Corp., Northampton, MA, USA) to analyze the temperature distribution of different layers of the sample, and the average temperature as well as the SDs of each layer were obtained using Microsoft Excel (Microsoft Office, Redmond, WA, USA). Microsoft Excel was also used to analyze the variance (ANOVA), and the Tukey test with a confidence level of 99% was performed in SPSS version 16.0 (SPSS, Chicago, IL, USA).

3. Results and Discussion

3.1. DPs of Beef Sausage

Figure 4 indicates the DPs of beef sausages as a function of frequency (f) at different temperatures. Both dielectric constant (ε') and loss factor (ε'') values decreased as the frequency increased in the samples. Similar trends were observed in the dielectric properties of noodles, beef meatballs and sauce [36]. At a frequency lower than 300 MHz, ε' significant increased ($p < 0.01$) at temperatures ranging from 25 to 90 °C, while it increased slightly at 300 MHz or higher as the test temperature increased. The effects of frequency on ε'' showed a different trend with ε'. ε'' decreased with the decrease of temperature among the range of frequencies from 13.56 MHz to 2450 MHz. This was probably because of free water dispersion and ionic conduction, which govern the change of ε'' [41]. At a frequency higher than 2450 MHz, both ε' and ε'' varied within a narrow range.

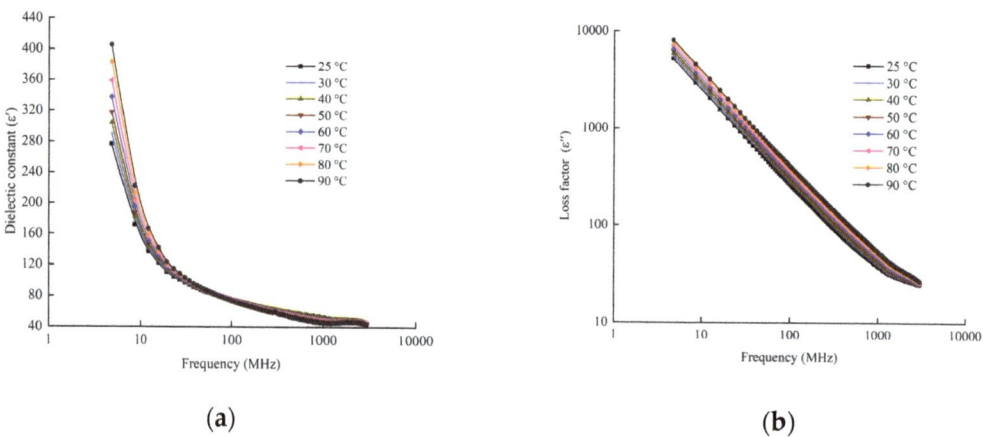

Figure 4. DPs related with frequency of beef sausage at different temperatures (25, 30, 40, 50, 60, 70, 80 and 90 °C): (**a**) dielectric constant (ε') as dependent variable; (**b**) dielectric loss factor (ε'') as dependent variable.

The measured DPs of the sample as a function of temperature are presented in Figure 5. The dielectric constant at three RF frequencies (13.56, 27.12 and 40.68 MHz) are higher than those at microwave frequencies (915 and 2450 MHz), and this increased with an increasing temperature as compared with no significant changes at microwave frequencies. Similar results have also been reported in meat batters [22]. The denaturation of protein at high temperature will cause the release and shrinkage of water, which is considered the cause of the significant changes in the dielectric properties of beef sausage. The loss factor of beef sausages at RF frequencies is much higher than that at microwave frequencies, and this increased with an increasing temperature both in RF and MW. Similar tendencies were also seen in previous research [21,22]. Providing a uniform and stable electric field is significantly important for pilot-scale RF pasteurization at 27.12 MHz [42].

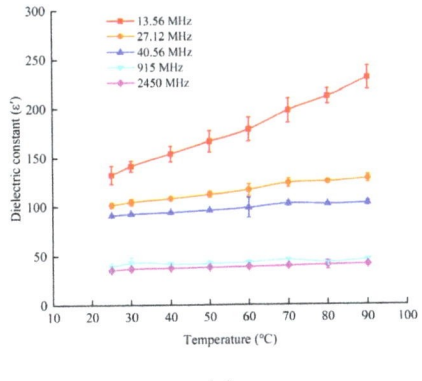
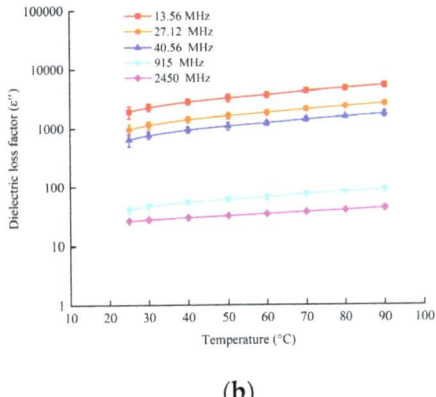

Figure 5. Temperature-dependent DPs of samples at five selected frequencies: (**a**) dielectric constant (ε') as dependent variable; (**b**) dielectric loss factor as dependent variable. RF band included 13.56, 27.12 and 40.68 MHz; microwave frequency included 915 and 2450 MHz.

Two predictive models of ε' and ε'' were created as a function of temperature (T) and frequency (f) by nonlinear regression analysis in beef sausage samples in Equations (2) and (3).

Both temperature (T) and frequency (f) had significant effects on (ε') and (ε''), and the R^2 values were 0.941 and 0.951, respectively ($p < 0.001$).

$$\varepsilon' = 183.937 - 3.129f + 0.160T + 0.004f^2 + 0.006T^2 - 0.01fT - 1.316 \times 10^{-6}f^3 - 1.323 \times 10^{-5}T^3 + 4.171 \times 10^{-7}Tf^2 - 1.758 \times 10^{-6}fT^2 \tag{2}$$

$$\varepsilon'' = 2925.437 - 90.397f + 28.767T + 0.131f^2 + 0.024T^2 - 0.043fT - 3.84 \times 10^{-5}f^3 + 3.797 \times 10^{-4}T^3 + 1.345 \times 10^{-5}Tf^2 - 1.837 \times 10^{-5}fT^2 \tag{3}$$

The regression models are suited for predicting the DPs of beef sausage under a temperature of 25 to 90 °C and at the frequency of 13.56 to 2450 MHz. Further research should focus on the DPs of the different moisture content and containers of beef sausage, using computer simulation system to improve RF heating uniformity.

The DPs of samples with different NaCl contents were measured at 27.12 MHz and are shown in Table 2. The results revealed that ε' increased with the increase of NaCl content, and a similar trend was obtained with ε''.

Table 2. The DPs of samples with different NaCl contents (1%, 2% and 3%) at an RF frequency of 27.12 MHz and a temperature of 25 °C.

Frequency (MHz)	NaCl Content (%)	Dielectric Properties	
		$\varepsilon' \pm$ SD	$\varepsilon'' \pm$ SD
27.12	1	96.69 ± 2.28	676.52 ± 14.57
	2	98.95 ± 2.01	806.92 ± 16.99
	3	101.01 ± 2.83	937.28 ± 21.76

3.2. Determination of Cold Spot of RF Heating

Testing the cold spots during the RF heating process is crucial for pasteurization and affects the processing technology, safety, quality, and cost of food materials [43–45]. According to the description above, three electrode gaps (175, 180 and 185 mm), sample heights (75, 80 and 85 mm), and NaCl contents (1%, 2%, and 3%) were selected. The results of the temperature distribution are shown in Table 3. Within the same period of time, the temperature of the upper layer in Figure 3a was much lower that of the middle layer and lower layer when subjected to RF heating. The table also showed that the temperature of the middle part was slightly higher than that of the lower part, which meant the heating hot spot was generally near the geometric center. A similar phenomenon was reported for potato cuboids [16], which was attributed to the electric field behavior and heat loss at the sample edge. Therefore, the sample position 1 was selected as the cold spot for subsequent study.

Table 3. The temperature of different layers of samples with different NaCl contents (1, 2 and 3%) under different sample heights (75, 80 and 85 mm) at different electrode gaps (175, 180 and 185 mm) when the temperature of the cold spot reached 70 °C and the temperature was initialized as 25 °C.

Electrode Gap (mm)	Sample Height (mm)	NaCl (%)	Layer		
			Upper	Middle	Lower
175	80	3	70.7 ± 0.7	79.9 ± 1.1	69.0 ± 0.9
180	80	3	70.3 ± 0.4	93.2 ± 1.6	99.4 ± 1.1
185	80	3	70.4 ± 0.6	83.9 ± 2.2	87 ± 2.3
180	75	3	70.4 ± 0.6	90.6 ± 0.7	78.3 ± 0.6
180	80	3	70.3 ± 0.4	93.2 ± 1.6	99.4 ± 1.1
180	85	3	70.0 ± 1.2	96.8 ± 0.9	94.7 ± 1.3
180	80	1	70.2 ± 0.5	91.3 ± 0.8	81.3 ± 0.9
180	80	2	70.1 ± 0.7	95.5 ± 1.2	102.4 ± 1.4
180	80	3	70.3 ± 0.4	93.2 ± 1.6	99.4 ± 1.1

3.3. Evaluating the Heating Uniformity and Heating Rate in the Different Conditions

3.3.1. Electrode Gap

Figure 6 reveals the heating profile of time–temperature and contour plots of samples during the RF treatment with different electrode gaps. The RF heating rates increased as the electrode gap decreased (Figure 6a), and the maximum heating rate was obtained under an electrode gap of 175 mm. This phenomenon may be due to the distance between the sample and upper electrode plate. The sample could absorb more electric energy when the distance was closer [46]. Similar trends were reported for frozen chicken breast meat [47] and frozen tilapia fillets [48]. The detailed temperature distribution of the beef sausage at the upper, middle, lower and longitudinal sections is shown in Figure 6b when the RF system was shutting down. The results in this figure indicated that the cold spots were at the edges, while the hot spots were at the center of the beef sausage, and the temperature close to the top was much lower than temperature near the middle and bottom, which was consistent with the results in Table 3. It also can be seen from the contour plots that the temperature distribution was more uniform at an electrode gap of 180 mm as compared to the others. The temperatures of upper parts were below 70 °C because of the heat dissipation when the beef sausage samples were removed from the sterilization kettle.

Table 4 shows the heating uniformity index (λ) values of beef sausage samples in the upper, middle and lower layers and the longitudinal section at different NaCl (1, 2 and 3%) content with different sample heights (75, 80 and 85 mm) under different electrode gaps (175, 180 and 185 mm) at the initial temperature of 25 °C when applied to RF heating. The results showed that there were no significant effects ($p > 0.01$) on the electrode gaps and layers, and the lowest uniformity index was obtained under the electrode gap of 180 mm. High heating rates corresponding to high throughputs may have caused the non-heating uniformity due to rapid and runaway heating [49,50]. Thus, the electrode gap of 180 mm was used for additional research.

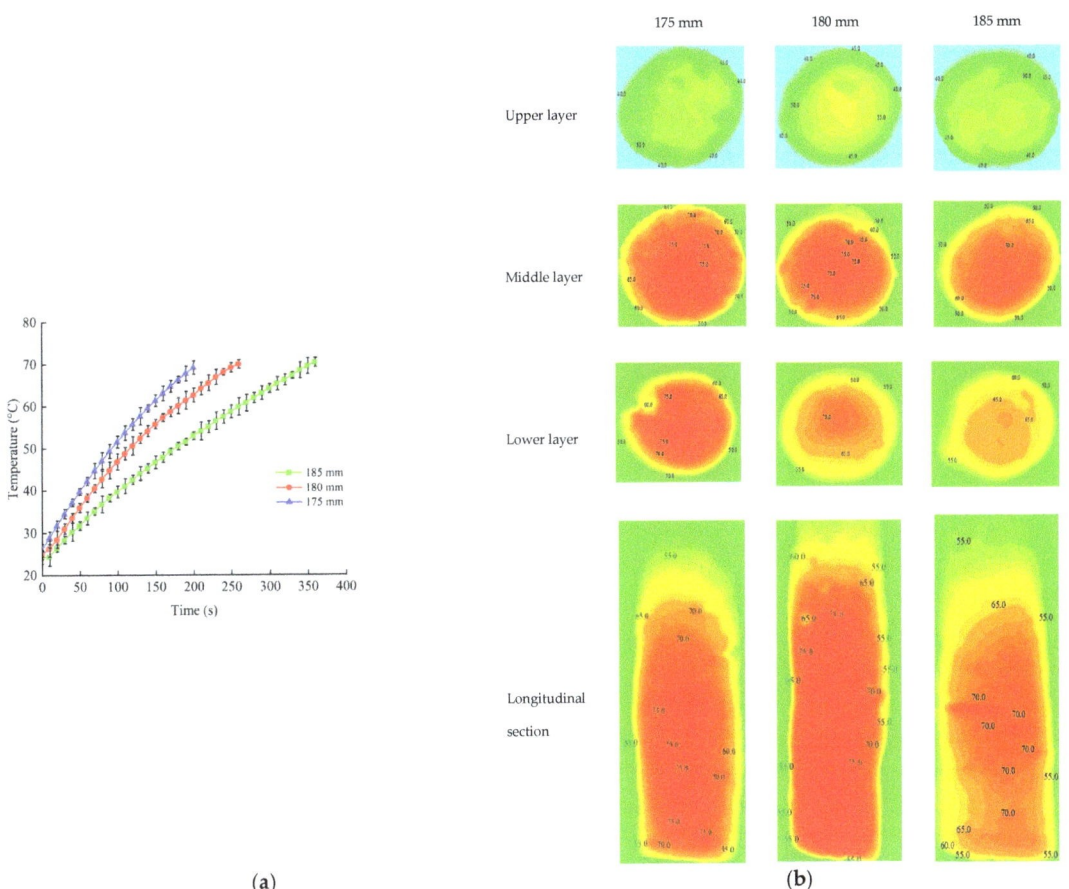

Figure 6. Time–temperature heating curves of beef sausages (**a**) and the contour plots of temperature distribution (**b**) at different electrode gaps (175, 180 and 185 mm) at a sample height of 80 mm and when the temperature was initialized at 25 °C.

Table 4. Beef sausage λ value of different layers at different NaCl contents (1, 2 and 3%) with different sample heights (75, 80 and 85 mm) under different electrode gaps (175, 180 and 185 mm) at the initial temperature of 25 °C.

Electrode Gap (mm)	Sample Height (mm)	NaCl (%)	λ			
			Upper	Middle	Lower	Longitudinal
175	80	3	0.341 ± 0.0025 [a]	0.343 ± 0.0044 [a]	0.464 ± 0.0054 [d]	0.455 ± 0.0039 [cd]
180	80	3	0.335 ± 0.0016 [a]	0.337 ± 0.0051 [a]	0.406 ± 0.0026 [b]	0.441 ± 0.0015 [c]
185	80	3	0.404 ± 0.0054 [b]	0.341 ± 0.0036 [a]	0.414 ± 0.0046 [b]	0.453 ± 0.0047 [cd]
180	75	3	0.344 ± 0.0012 [bc]	0.331 ± 0.0034 [a]	0.414 ± 0.0033 [d]	0.404 ± 0.0046 [d]
180	80	3	0.335 ± 0.0016 [a]	0.337 ± 0.0051 [a]	0.406 ± 0.0026 [b]	0.441 ± 0.0015 [c]
180	85	3	0.351 ± 0.0041 [c]	0.348 ± 0.0022 [c]	0.438 ± 0.0013 [e]	0.449 ± 0.0012 [f]
180	80	1	0.366 ± 0.0011 [b]	0.344 ± 0.0022 [a]	0.445 ± 0.0031 [e]	0.475 ± 0.0016 [f]
180	80	2	0.378 ± 0.0034 [c]	0.359 ± 0.0041 [b]	0.438 ± 0.0039 [e]	0.467 ± 0.0021 [f]
180	80	3	0.335 ± 0.0016 [a]	0.337 ± 0.0017 [a]	0.406 ± 0.0021 [d]	0.441 ± 0.0023 [e]

Different lowercase letters indicate a significant difference ($p < 0.01$).

3.3.2. Sample Height

Figure 7 shows the profiles of the time–temperature and contour distribution of samples when subjected to RF energy with three different sample heights (75 mm, 80 mm and 85 mm). The RF heating rates increased with an increasing sample height when the electrode gap was 180 mm and NaCl content was 3%. Li et al. (2018) obtained similar trends of the heating rate increasing as the food thickness increases from 4 cm and 5 cm to 6 cm of frozen beef samples with various thicknesses under RF treatment [46]. The contour plots in Figure 7b show the temperature distribution of beef sausage with different sample heights after RF heating. It can be observed in the contour plots that temperature distribution was more uniform when the sample height was 80 mm. At the same time, the upper layer was more uniform according to the results of temperature distribution in general. The highest temperatures were obtained at the middle layer compared to those of upper and lower layers. Table 4 compares the heating uniformity index (λ) of three sample heights after RF heating when all beef sausage sample temperatures of cold spots reached 70 °C. The lowest λ of the sample was obtained with a height of 80 mm. This was because the deflection of the electric field increased the electric field intensity at the geometric center of the sample [51]. A sample of 80 mm had the fastest heating rate and the lowest heating uniformity index when all beef sausage sample temperatures of cold spots reached 70 °C. Based on the above results, a fixed sample height of 80 mm was selected for further research.

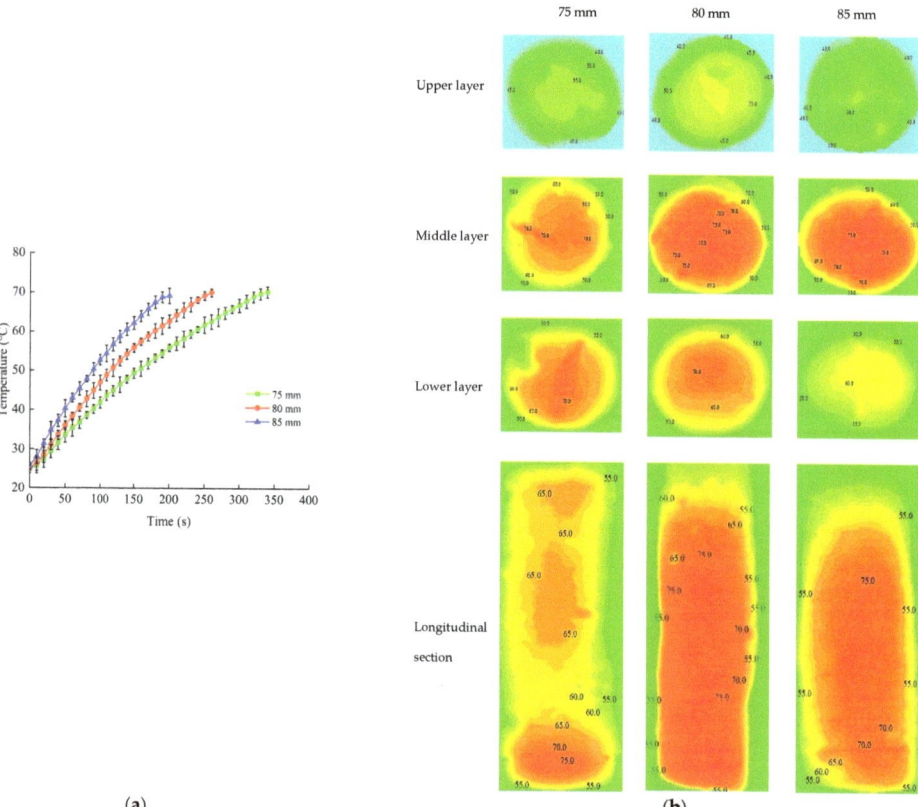

Figure 7. Time–temperature heating curves of beef sausages (**a**) and the contour plots of temperature distribution (**b**) at different sample heights (75, 80 and 85 mm) under an electrode gap of 180 mm and when the temperature was initialized at 25 °C.

3.3.3. NaCl Content

Figure 8 demonstrates the time–temperature profile and contour plots of temperature distribution of samples with different NaCl contents by RF heating. Figure 8a shows that the slowest heating rate was achieved when the NaCl content was 1%. The heating rate increased with the increase of NaCl content at first. However heating rates slowed down when NaCl content was more than 2%, which showed a similar trend to the study of Jeong et al. [52]. This phenomenon may be due to the effect of salty shield, which leads to the slowdown of the heating rate [53]. It revealed that the temperature distribution was more uniform when NaCl contents were 1% and 3%, as seen in Figure 8b. The minimum λ value was obtained with the NaCl content of 3%, as seen in Table 4. Dong et al. (2021) also reported that RF heating uniformity might be improved with the high salt concentrations of ground beef [26]. In addition, the sample with a higher salt content could provide a better value of sensory evaluation and preservation characteristics. Based on the above results, a sample with 3% NaCl content had the fastest heating rate and the lowest heating uniformity index. Therefore, the sample with the NaCl content of 3% was selected for future studies.

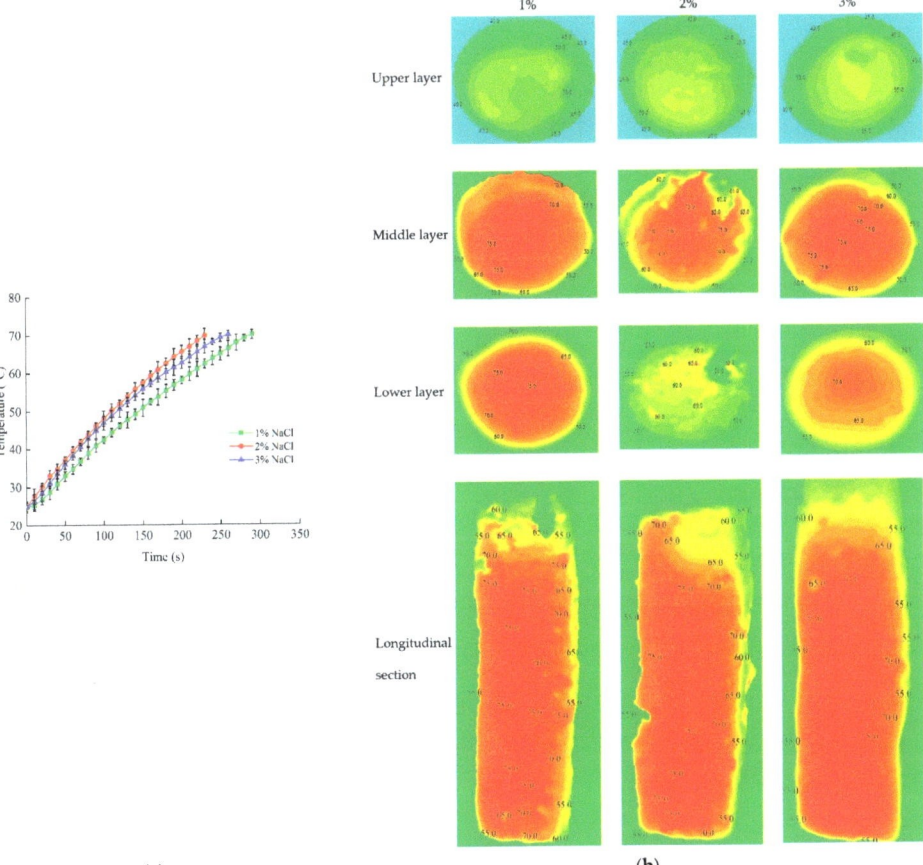

Figure 8. Time–temperature heating curves of samples (**a**) and the contour plots of temperature distribution (**b**) with different NaCl contents (1%, 2% and 3%) at an electrode gap of 180 mm, a sample height of 80 mm and when the temperature was initialized at 25 °C.

4. Conclusions

A customized sterilization kettle was designed for RF pasteurization of beef sausage. The sterilization kettle was well applied in an RF heating system according to efficiency and stability tests. The temperature distribution indicated that the cold points were at the upper layer and edge of the sample, whereas the hot spots were at the geometric center. Temperature and frequency had significant influence ($p < 0.01$) on the DPs of the beef sausage sample. The best heating uniformity was obtained at an electrode gap of 180 mm, sample height of 80 mm and NaCl content of 3%. Overall, this research enriches the information on the heating uniformity of foods for pasteurization under RF systems. Additional studies should focus on the quality effects and pasteurization kinetics of food. Moreover, the sterilization system described above can also realize RF combined with super high temperature water (RFSW) sterilization for subsequent tests in the future.

Author Contributions: Conceptualization, Y.W. (Yunyang Wang) and K.W.; methodology, K.W., C.R. and Y.X.; software, B.C., Y.S., L.H. and K.W.; validation, K.W. and Y.X.; formal analysis, Y.W. (Yunyang Wang), Y.S., C.R. and K.W.; investigation, Y.W. (Yunyang Wang) and K.W.; resources, Y.W. (Yequn Wang), X.C., H.F. and C.R.; data curation, B.C., K.W. and L.H.; writing—original draft preparation, K.W.; writing—review and editing, Y.W. (Yunyang Wang) and K.W.; visualization, K.W. and B.C.; supervision, Y.W. (Yunyang Wang); project administration, Y.W. (Yunyang Wang); funding acquisition, Y.W. (Yunyang Wang). All authors have read and agreed to the published version of the manuscript.

Funding: This research was funded by the general program (Grant No. 31371854) of the National Nature Science Foundation of China, the Key Research Project of Shaanxi Province (2017ZDXM-SF-104).

Institutional Review Board Statement: Not applicable.

Data Availability Statement: Data on this study are available in the article.

Acknowledgments: The authors would like to thank the instrument sharing platform of the College of Food Science & Engineering (Northwest A & F University, China).

Conflicts of Interest: The authors declare no conflict of interest.

References

1. Xiong, Y.; Zhang, P.; Warner, R.D.; Hossain, M.N.; Leonard, W.; Fang, Z. Effect of sorghum bran incorporation on the physicochemical and microbial properties of beef sausage during cold storage. *Food Control.* **2022**, *132*, 108544. [CrossRef]
2. Hashemi, A.; Jafarpour, A. Rheological and microstructural properties of beef sausage batter formulated with fish fillet mince. *J. Food Sci. Technol.* **2016**, *53*, 601–610. [CrossRef] [PubMed]
3. Gill, C.O.; Badoni, M. Microbiological and organoleptic qualities of vacuum-packaged ground beef prepared from pasteurized manufacturing beef. *Int. J. Food Microbiol.* **2002**, *74*, 111–118. [CrossRef]
4. Scanga, J.A.; Grona, A.D.; Belk, K.E.; Sofos, J.N.; Bellinger, G.R.; Smith, G.C. Microbiological contamination of raw beef trimmings and ground beef. *Meat Sci.* **2000**, *56*, 145–152. [CrossRef]
5. Enns, D.K.; Crandall, P.G.; O'Bryan, C.A.; Griffis, C.L.; Martin, E.M. A 2-step cooking method of searing and hot water pasteurization to maximize the safety of refrigerated, vacuum packaged, chicken breast meat. *J. Food Sci.* **2007**, *72*, M113–M119. [CrossRef]
6. Huang, L. Computer simulation of heat transfer during in-package pasteurization of beef frankfurters by hot water immersion. *J. Food Eng.* **2007**, *80*, 839–849. [CrossRef]
7. Gill, C.O.; Badoni, M. The effects of hot water pasteurizing treatments on the appearances of pork and beef. *Meat Sci.* **1997**, *46*, 77–87. [CrossRef]
8. Badoni, C. The effects of hot water pasteurizing treatments on the microbiological conditions and appearances of pig and sheep carcasses. *Food Res. Int.* **1998**. [CrossRef]
9. Tang, J.; Hong, Y.-K.; Inanoglu, S.; Liu, F. Microwave pasteurization for ready-to-eat meals. *Curr. Opin. Food Sci.* **2018**, *23*, 133–141. [CrossRef]
10. Cho, W.-I.; Yi, J.Y.; Chung, M.-S. Pasteurization of fermented red pepper paste by ohmic heating. *Innov. Food Sci. Emerg. Technol.* **2016**, *34*, 180–186. [CrossRef]
11. Monteiro, S.; Silva, E.K.; Alvarenga, V.O.; Moraes, J.; Freitas, M.Q.; Silva, M.C.; Raices, R.S.L.; Sant'Ana, A.S.; Meireles, M.A.A.; Cruz, A.G. Effects of ultrasound energy density on the non-thermal pasteurization of chocolate milk beverage. *Ultrason Sonochem.* **2018**, *42*, 1–10. [CrossRef] [PubMed]
12. Bingol, G.; Yang, J.; Brandl, M.T.; Pan, Z.; Wang, H.; McHugh, T.H. Infrared pasteurization of raw almonds. *J. Food Eng.* **2011**, *104*, 387–393. [CrossRef]

13. Li, Y.; Zhang, Y.; Lei, Y.; Fu, H.; Chen, X.; Wang, Y. Pilot-scale radio frequency pasteurisation of chili powder: Heating uniformity and heating model. *J. Sci. Food Agric.* **2016**, *96*, 3853–3859. [CrossRef] [PubMed]
14. Cui, B.; Fan, R.; Ran, C.; Yao, Y.; Wang, K.; Wang, Y.; Fu, H.; Chen, X.; Wang, Y. Improving radio frequency heating uniformity using a novel rotator for microorganism control and its effect on physiochemical properties of raisins. *Innov. Food Sci. Emerg. Technol.* **2021**, *67*, 102564. [CrossRef]
15. Zhang, Z.; Yao, Y.; Shi, Q.; Zhao, J.; Fu, H.; Wang, Y. Effects of radio-frequency-assisted blanching on the polyphenol oxidase, microstructure, physical characteristics, and starch content of potato. *Lwt-Food Sci. Technol.* **2020**, *125*, 109357. [CrossRef]
16. Zhang, Z.; Guo, C.; Gao, T.; Fu, H.; Chen, Q.; Wang, Y. Pilot-scale radiofrequency blanching of potato cuboids: Heating uniformity. *J. Sci. Food Agric.* **2018**, *98*, 312–320. [CrossRef]
17. Li, Y.; Zhou, L.; Chen, J.; Subbiah, J.; Chen, X.; Fu, H.; Wang, Y. Dielectric properties of chili powder in the development of radio frequency and microwave pasteurisation. *Int. J. Food Prop.* **2018**, *20*, S3373–S3384. [CrossRef]
18. Basaran-Akgul, N.; Basaran, P.; Rasco, B.A. Effect of temperature (−5 to 130 °C) and fiber direction on the dielectric properties of beef Semitendinosus at radio frequency and microwave frequencies. *J. Food Sci.* **2008**, *73*, E243–E249. [CrossRef]
19. Ozturk, S.; Kong, F.; Singh, R.K.; Kuzy, J.D.; Li, C.; Trabelsi, S. Dielectric properties, heating rate, and heating uniformity of various seasoning spices and their mixtures with radio frequency heating. *J. Food Eng.* **2018**, *228*, 128–141. [CrossRef]
20. Traffano-Schiffo, M.V.; Castro-Giraldez, M.; Colom, R.J.; Talens, P.; Fito, P.J. New methodology to analyze the dielectric properties in radiofrequency and microwave ranges in chicken meat during postmortem time. *J. Food Eng.* **2021**, *292*, 110350. [CrossRef]
21. Trabelsi, S. Measuring changes in radio-frequency dielectric properties of chicken meat during storage. *J. Food Meas. Charact.* **2018**, *12*, 683–690. [CrossRef]
22. Zhang, L.; Lyng, J.G.; Brunton, N.; Morgan, D.; McKenna, B. Dielectric and thermophysical properties of meat batters over a temperature range of 5–85 °C. *Meat Sci.* **2004**, *68*, 173–184. [CrossRef] [PubMed]
23. Farag, K.W.; Lyng, J.G.; Morgan, D.J.; Cronin, D.A. Dielectric and thermophysical properties of different beef meat blends over a temperature range of −18 to +10 °C. *Meat Sci.* **2008**, *79*, 740–747. [CrossRef] [PubMed]
24. Brunton, N.P.; Lyng, J.G.; Zhang, L.; Jacquier, J.C. The use of dielectric properties and other physical analyses for assessing protein denaturation in beef biceps femoris muscle during cooking from 5 to 85°C. *Meat Sci.* **2006**, *72*, 236–244. [CrossRef]
25. Li, F.; Zhu, Y.; Li, S.; Wang, P.; Zhang, R.; Tang, J.; Koral, T.; Jiao, Y. A strategy for improving the uniformity of radio frequency tempering for frozen beef with cuboid and step shapes. *Food Control.* **2021**, *123*, 107719. [CrossRef]
26. Dong, J.; Kou, X.; Liu, L.; Hou, L.; Li, R.; Wang, S. Effect of water, fat, and salt contents on heating uniformity and color of ground beef subjected to radio frequency thawing process. *Innov. Food Sci. Emerg. Technol.* **2021**, *68*, 102604. [CrossRef]
27. Zuo, Y.; Zhou, B.; Wang, S.; Hou, L. Heating uniformity in radio frequency treated walnut kernels with different size and density. *Innov. Food Sci. Emerg. Technol.* **2022**, *75*, 102899. [CrossRef]
28. Zhang, S.; Ramaswamy, H.; Wang, S. Computer simulation modelling, evaluation and optimisation of radio frequency (RF) heating uniformity for peanut pasteurisation process. *Biosyst. Eng.* **2019**, *184*, 101–110. [CrossRef]
29. American Oil Chemists Society (AOCS). *AOAC Official Method 985.14: Moisture in Meat and Poultry Products*; AOCS Press: Champaign, IL, USA, 1985.
30. American Oil Chemists Society (AOCS). *AOAC Official Method 992.15: Crude Protein in Meat and Meat Products Including Pet Foods*; AOCS Press: Champaign, IL, USA, 1993.
31. American Oil Chemists Society (AOCS). *AOAC Official Method 991.36: Fat (Crude) in Meat and Meat Products*; AOCS Press: Champaign, IL, USA, 2000.
32. American Oil Chemists Society (AOCS). *AOAC Official Method 923.03:Ash of Flour*; AOCS Press: Champaign, IL, USA, 2000.
33. American Oil Chemists Society (AOCS). *AOAC Official Method 958.06: Starch in Meat*; AOCS Press: Champaign, IL, USA, 2000.
34. Herve, A.G.; Tang, J.; Luedecke, L.; Feng, H. Dielectric properties of cottage cheese and surface treatment using microwaves. *J. Food Eng.* **1998**, *37*, 389–410. [CrossRef]
35. Farag, K.W.; Duggan, E.; Morgan, D.J.; Cronin, D.A.; Lyng, J.G. A comparison of conventional and radio frequency defrosting of lean beef meats: Effects on water binding characteristics. *Meat Sci.* **2009**, *83*, 278–284. [CrossRef]
36. Wang, Y.; Luechapattanaporn, K.; Wang, Y.; Tang, J. Radio-frequency heating of heterogeneous food—Meat lasagna. *J. Food Eng.* **2012**, *108*, 183–193. [CrossRef]
37. Inmanee, P.; Kamonpatana, P.; Pirak, T. Ohmic heating effects on Listeria monocytogenes inactivation, and chemical, physical, and sensory characteristic alterations for vacuum packaged sausage during post pasteurization. *LWT* **2019**, *108*, 183–189. [CrossRef]
38. Cichoski, A.J.; Rampelotto, C.; Silva, M.S.; de Moura, H.C.; Terra, N.N.; Wagner, R.; de Menezes, C.R.; Flores, E.M.M.; Barin, J.S. Ultrasound-assisted post-packaging pasteurization of sausages. *Innov. Food Sci. Emerg. Technol.* **2015**, *30*, 132–137. [CrossRef]
39. Roering, A.M.; Wierzba, R.K.; Ihnot, A.M.; Luchansky, J.B. Pasteurization of vacuum-sealed packages of summer sausage inoculated with listeria monocytogenes. *J. Food Saf.* **2010**, *18*, 49–56. [CrossRef]
40. Jiao, Y.; Shi, H.; Tang, J.; Li, F.; Wang, S. Improvement of radio frequency (RF) heating uniformity on low moisture foods with Polyetherimide (PEI) blocks. *Food Res. Int.* **2015**, *74*, 106–114. [CrossRef]
41. Guo, W.; Tiwari, G.; Tang, J.; Wang, S. Frequency, moisture and temperature-dependent dielectric properties of chickpea flour. *Biosyst. Eng.* **2008**, *101*, 217–224. [CrossRef]
42. Wang, Y.; Wig, T.D.; Tang, J.; Hallberg, L.M. Dielectric properties of foods relevant to RF and microwave pasteurization and sterilization. *J. Food Eng.* **2003**, *57*, 257–268. [CrossRef]

43. Lin, Y.; Subbiah, J.; Chen, L.; Verma, T.; Liu, Y. Validation of radio frequency assisted traditional thermal processing for pasteurization of powdered infant formula milk. *Food Control.* **2020**, *109*, 106897. [CrossRef]
44. Bornhorst, E.R.; Tang, J.; Sablani, S.S.; Barbosa-Cánovas, G.V. Thermal pasteurization process evaluation using mashed potato model food with Maillard reaction products. *LWT—Food Sci. Technol.* **2017**, *82*, 454–463. [CrossRef]
45. Peng, J.; Tang, J.; Luan, D.; Liu, F.; Tang, Z.; Li, F.; Zhang, W. Microwave pasteurization of pre-packaged carrots. *J. Food Eng.* **2017**, *202*, 56–64. [CrossRef]
46. Li, Y.; Li, F.; Tang, J.; Zhang, R.; Wang, Y.; Koral, T.; Jiao, Y. Radio frequency tempering uniformity investigation of frozen beef with various shapes and sizes. *Innov. Food Sci. Emerg. Technol.* **2018**, *48*, 42–55. [CrossRef]
47. Bedane, T.F.; Altin, O.; Erol, B.; Marra, F.; Erdogdu, F. Thawing of frozen food products in a staggered through-field electrode radio frequency system: A case study for frozen chicken breast meat with effects on drip loss and texture. *Innov. Food Sci. Emerg. Technol.* **2018**, *50*, 139–147. [CrossRef]
48. Zhang, Y.; Li, S.; Jin, S.; Li, F.; Tang, J.; Jiao, Y. Radio frequency tempering multiple layers of frozen tilapia fillets: The temperature distribution, energy consumption, and quality. *Innovative Food Sci. Emerg. Technol.* **2021**, *68*, 102603. [CrossRef]
49. Ozturk, S.; Kong, F.; Singh, R.K.; Kuzy, J.D.; Li, C. Radio frequency heating of corn flour: Heating rate and uniformity. *Innov. Food Sci. Emerg. Technol.* **2017**, *44*, 191–201. [CrossRef]
50. Xie, Y.; Zhang, Y.; Xie, Y.; Li, X.; Liu, Y.; Gao, Z. Radio frequency treatment accelerates drying rates and improves vigor of corn seeds. *Food Chem.* **2020**, *319*, 126597. [CrossRef]
51. Huang, Z.; Marra, F.; Wang, S. A novel strategy for improving radio frequency heating uniformity of dry food products using computational modeling. *Innov. Food Sci. Emerg. Technol.* **2016**, *34*, 100–111. [CrossRef]
52. Jeong, J.Y.; Lee, E.S.; Choi, J.H.; Lee, J.Y.; Kim, J.M.; Min, S.G.; Chae, Y.C.; Kim, C.J. Variability in temperature distribution and cooking properties of ground pork patties containing different fat level and with/without salt cooked by microwave energy. *Meat Sci.* **2007**, *75*, 415–422. [CrossRef]
53. Chan, C.-C.; Chen, Y.-C. Demulsification of W/O emulsions by microwave radiation. *Sep. Sci. Technol.* **2002**, *37*, 3407–3420. [CrossRef]

Article

Effect of Large-Scale Paddy Rice Drying Process Using Hot Air Combined with Radio Frequency Heating on Milling and Cooking Qualities of Milled Rice

Karn Chitsuthipakorn [1] and Sa-nguansak Thanapornpoonpong [2,*]

[1] Postharvest Technology Research Center, Faculty of Agriculture, Chiang Mai University, Chiang Mai 50200, Thailand; karn_c@cmu.ac.th

[2] Department of Plant and Soil Sciences, Faculty of Agriculture, Chiang Mai University, Chiang Mai 50200, Thailand

* Correspondence: sa-nguansak.t@cmu.ac.th; Tel.: +66-84-6525-196

Abstract: The objectives of the study on a continuous flow hot air dryer combined with radio frequency heating at different temperatures (HA/RF) (38 °C, 42 °C, 46 °C, and 50 °C) in a large-scale process compared with conventional continuous flow hot air drying (HA) were (1) to investigate the drying characteristics, drying kinetics, and milling quality of the process and (2) to observe the cooking quality and compare the sensory differences of the cooked rice after treatment. The drying characteristics and moisture diffusivity showed that the higher the radio frequency (RF) heating temperature, the shorter the drying time. The specific energy consumption and energy cost decreased when the RF heating temperature increased. The optimal condition in terms of fissure percentage was HA/RF42. In addition, there were no significant differences in head rice yield and white rice color determination, amylose content, texture profiles, and pasting properties in all HA/RF treatments. In the triangle test, it was found that at least 6% of the population could perceive a difference between HA and HA/RF50. In conclusion, this study proposes the further development of the HA/RF drying process at low-temperature profiles and shows the great potential of RF technology for commercial drying in rice industry.

Keywords: paddy; rice; hot air drying; radio frequency heating; milling quality; cooking quality

1. Introduction

Paddy rice cultivation plays an important role in the Thai economy. About 30 million tons of paddy rice are harvested annually. It is usually harvested with a combine harvester at a moisture content of 18–28% (w.b.) and immediately dried to below 14% for safe storage. The most common large-scale mechanical drying systems used in Thailand are continuous flow dryers and recirculating batch dryers, as they offer advantages in terms of cost and drying speed. Paddy is dried with hot air (HA) at a temperature of (50 °C–80 °C) to achieve the shortest drying time because, the shorter the drying time, the more the business turnover can be raised.

Radio frequency (RF) heating is a new type of processing technology that uses electromagnetic waves in the frequency range of 1–300 MHz [1]. It has recently been used for paddy drying [2], accelerated aging [3–5], paddy seed treatment [2], Aspergillus flavus control [6], and insect control in milled rice [7–10]. RF heating, as a type of dielectric heating method, involves the generation of heat as a result of the interaction between an electromagnetic field and polarized molecules in the crops, which are either bipolar or ionic [11]. In hot air drying combined with the RF heating application, the paddy is placed between two electrodes, causing the polar molecules in the paddy to rotate and generate heat energy throughout the paddy volume; it also penetrates the products and generates heat uniformly and instantaneously in the paddy kernel without regard to convective

or conductive media. The heat generated promotes evaporation of moisture from the paddy surface and stimulates moisture migration from the inside to the outside. The most successful past applications have often combined two or more technologies (RF–HA, etc.). Most notably, the use of hot air (HA) in combination with RF heating to dry paddy has been investigated to develop a commercial dryer for the rice processing industry with a lower carbon footprint in the future.

Based on the vertical operating prototype that supported an RF heating system by Vearasilp et al. [10] and the earlier study by Chitsuthipakorn and Thanapornpoonpong [12], the electrical voltage control of the RF heating machine was set manually, which proved to be a disadvantage for the authors. There were two different ways to solve this problem: (1) installing an automatic system with a vibration feeder to control the paddy release rate from the RF heating machine to control the paddy temperature in the RF heating chamber, and (2) installing an automatic system to control the electrical voltage to control the RF heating energy for the paddy. After many experiments, it was found that installing an automatic system to control the electrical voltage resulted in a constant paddy temperature in the RF heating chamber. Therefore, the authors decided to conduct the study with a lower temperature in a narrower range. In addition, the lowest temperature of the study was 38 °C because it was close to the ambient temperature, which was about 30–35 °C. Likewise, the optimal condition in the previous study was at the lowest RF heating temperature (HA/RF45) in the experimental design, and it may be too early to conclude that HA/RF45 is the appropriate condition for this improved system. The prototype machine in this study was upgraded with respect to the RF heating chamber by replacing a Polypropylene (PP) chamber with a Polytetrafluoroethylene (PTFE) chamber because it is nonstick, nonwetting, the United States Food and Drug Administration (FDA) compliant, and resistant to high temperatures, and by using an automatic RF heating temperature control system. The specific objectives of the study of the hot air dryer combined with RF heating in a large-scale process compared to conventional hot air drying were (1) to investigate the drying characteristics, drying kinetics, and milling quality of the process and (2) to observe the cooking quality and compare the sensory differences of the cooked rice after treatment.

2. Materials and Methods

2.1. Materials

The study used a randomized complete block design (RCBD) with three replicates. A continuous flow hot air dryer was used in combination with different radio frequency (RF) heating temperatures measured at paddy in the RF heating chamber (38 °C, 42 °C, 46 °C, and 50 °C). A total of 9000 kg of freshly harvested paddy rice (variety RD 41) was purchased from local farmers in Nakhon Sawan province, Thailand, in November 2021. Initial moisture content averaged 25–26% (w.b.). It was stored in PP-woven bags separated for each replicate (450 kg paddy/bag). To avoid quality deterioration, each replicate was treated within 4 days after harvest. After treatment, 25 kg paddy samples from each replicate were stored for quality determination.

2.2. Hot Air Dryer and RF Heating Systems

Figure 1a shows the industrial-scale continuous flow hot air dryer with an RF heating machine that was used in this study. It mainly consists of a recirculating hot air dryer (QS -500, Quaser Engineering, Pathum Thani, Thailand) capable of drying up to 450 kg of paddy at a time, a 15 kW, 27.12 MHz RF heating machine (BiO-Q model S-1, Yont Phol Dee, Nakhon Sawan, Thailand), and a bucket elevator. The electrode gap of RF heating was 170 mm, with the length and width of the electrode plate to be 450 mm and 300 mm. The RF energy was regulated to produce the desired paddy temperatures by automatically increasing or decreasing the electrical voltage on the control panel of the machine. The voltage supplied to the RF heating machine was adjusted by a variable autotransformer (Variac) to stabilize the RF heating temperature for paddy. The maximum electrical input voltage was 380 V, and the voltage used in HA/RF38, HA/RF42, HA/RF46, and HA/RF50 was 90–100 V, 100–140 V, 120–180 V, and 150–220 V, respectively. Air is circulated from a blower (2 HP/2850 rpm/28 CMM/80 mmAq) at the rear of the dryer through the heating chamber at the front, which contains three rows of 1000-watt infrared heaters. The hot air temperature was regulated at the beginning and automatically controlled by a sensor and control device during drying. The hot air from the drying section circulated through the paddy, with evaporated water being extracted by a blower and released to the environment. As shown in Figure 1b, the dryer is equipped with five temperature sensors (PT100, Primus Thai, Bangkok, Thailand), two relative humidity sensors (RHM, Primus Thai, Bangkok, Thailand), and one humidity and temperature data logger (EL-USB-2-LCD, Lascar Electronics Inc., Erie, PA, USA). Fresh paddy was dumped into the intake pit and transported to a hot air dryer by a bucket elevator. To maintain the quality of the milled rice, the hot air temperature was manually adjusted with a temperature controller and divided into four different levels: 80 °C at a moisture content of 25–26%, 70 °C at a moisture content of 20–25%, 60 °C at a moisture content of 17–20%, and 50 °C at a moisture content of 13–17%. Hot air was blown through the paddy in the drying chamber while the paddy was continuously conveyed from the hopper into the RF heating machine. As the paddy moved through the RF heating chamber, it was heated to the desired temperature (38 °C, 42 °C, 46 °C, and 50 °C) using the required RF energy. A temperature sensor 6 (PT100, Primus Thai, Bangkok, Thailand) was placed in the RF heating chamber to monitor the temperature of the paddy after heating. After the paddy was discharged from the RF heating machine into the bucket elevator, it was conveyed through a vibratory feeder controlled by a fixed input voltage electrical control. The average flow rate of the recirculation was 494 kg of paddy/hour. This process was repeated continuously and not in batches until the moisture content of the paddy fell below 14% (w.b.).

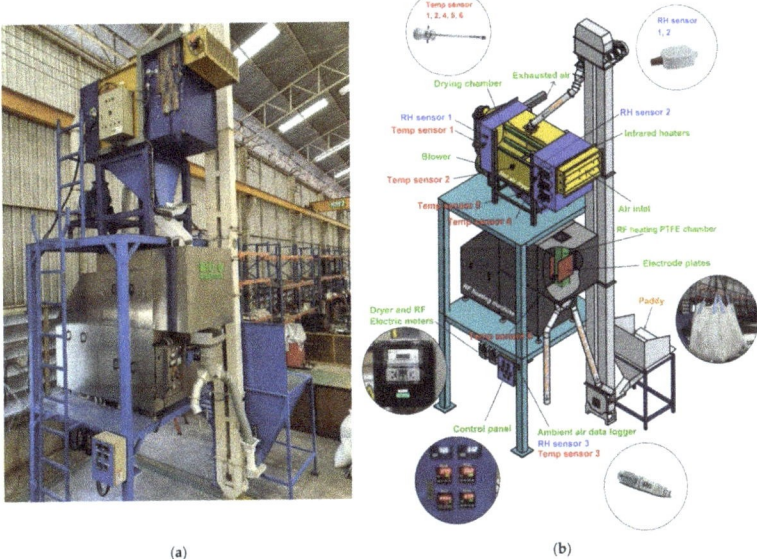

Figure 1. The industrial-scale continuous-flow hot-air dryer with RF heating machine: (**a**) prototype machine (**b**) schematic diagram.

2.3. Moisture Content Determination

Moisture content was measured in triplicate every 30 min throughout the drying process using a rapid capacitance-type moisture meter (Granomat, Pfeuffer GmbH, Kitzingen, Germany) to observe the value of moisture content during the drying process, which was reported as mean and standard deviation.

2.4. Drying Efficiency

2.4.1. Drying Kinetics

Because the initial moisture content of paddy was not identical for all treatments, the moisture content data were converted using Equation (1) to obtain the dimensionless moisture ratio. Over time, the equilibrium moisture content (M_e) becomes insignificant and relatively small compared with the moisture content of paddy at each time point (M_t) and the initial moisture content of paddy (M_0) [13,14], as prolonged exposure of the grain to infrared radiation eventually causes the material to burn [15]. The moisture ratio (MR) can be simply expressed as follows:

$$MR = \frac{M_t - M_e}{M_0 - M_e} \quad (1)$$

The drying kinetics of this study were described using the Lewis model and the Henderson–Pabis model, as shown in Equations (2) and (3). These models are based on Newton's law of cooling and Fick's second law of diffusion, respectively.

$$MR_{LW} = e^{-kt} \quad (2)$$

$$MR_{HP} = a\, e^{-kt} \quad (3)$$

where MR_{LW} denotes moisture ratio using the Lewis model, MR_{HP} denotes moisture ratio using the Henderson–Pabis model, a denotes model constant, k denotes rate constant, and t is drying time in minutes. The rate constant (k) represents drying performance, with a high value representing a fast-drying rate [16].

2.4.2. Effective Moisture Diffusivity

Using Equations (4)–(6), the effective moisture diffusivity was derived to characterize moisture movement in paddy during the drying process [17].

$$MR = \frac{6}{\pi^2} \exp\left(-\frac{\pi^2 D_{eff} t}{R_s^2}\right) \quad (4)$$

where D_{eff} is the effective moisture diffusivity (m^2/s), t is the drying time in minutes, and R_s^2 is the equivalent radius of the sphere (m). Equation (4) evaluated numerically for Fourier number (F_o) given by ($D_{eff} t / R_s^2$) for diffusion and can be rewritten as:

$$MR = \frac{6}{\pi^2} \exp\left(-\pi^2 F_o\right) \quad (5)$$

$$F_o = -0.1013 \ln(MR) - 0.0504 \quad (6)$$

$$D_{eff} = \left(\frac{F_o}{t/R_s^2}\right) \quad (7)$$

where D_{eff} was estimated by substituting the positive values of F_o and the drying time (t), along with the equivalent radius of paddy grain in Equation (7). The paddy grain is approximated as isotropic spheres of 3 mm diameter [17].

2.4.3. Energy Consumption

A specific energy consumption (SEC) of the drying was determined using Equation (8) [18]:

$$\text{Specific energy consumption (SEC)} = \frac{E_{total}}{m_{eva}} \quad (8)$$

where SEC is the specific energy consumption (MJ/kg-H$_2$O), E_{total} is the total electrical energy supplied to the dryer and RF heating systems during the drying process (kWh), and m_{eva} is the amount of evaporated water resulting from the difference in weight of the paddy before and after drying (kg-H$_2$O). Considering that the cost of electricity in Thailand was THB 3.2484 per kWh and the exchange rate THB/USD was 33.15:1 on 3 January 2022, the energy cost per kg of evaporated water (USD/kg-H$_2$O) was estimated using Equation (9):

$$\text{Energy cost} = \frac{(P_{dryer} + P_{rf})}{m_{eva}} \times \frac{3.2484}{33.15} \quad (9)$$

where P_{dryer} is the power supplied to the dryer system, and P_{rf} is the power supplied to the RF heating system.

2.5. Milling Quality Evaluation

2.5.1. Fissure Percentage

The dried paddy samples were cleaned and dehusked to obtain brown rice. One hundred kernels (about 2 g) were taken at random for each measurement. The measurement was performed three times for each replicate, and the mean and standard deviation were calculated. The brown rice samples were placed on a transparent plate over a fluorescent tube and a magnifying glass to visually count the fissured kernels [19]. The fissures in the kernels of brown rice were examined with the naked eye and recorded with a fissure degree of 0, 1, 2, 3, 4, and 5 or more fissures [20]. According to Shen et al. [20], the fissure percentage was introduced to quantitatively calculate the proportion of brown rice kernels with different fissure degrees. It was defined as the ratio of brown rice kernels with different fissure degrees to the total amount of selected kernels. Equation (10) illustrates the procedure for calculating the fissure percentage.

$$\text{Fissure percentage (\%)} = \frac{N_f}{N_t} \times 100 \quad (10)$$

where N_f is the total amount of brown rice kernels at a certain fissure degree (0, 1, 2, 3, 4, $\geq$5), and N_t is the total amount of the brown rice kernels.

2.5.2. Milling Yield

The milling yield was determined according to the method described in the Thai Standards for Rice [21]. A precleaning machine was used to clean the dried paddy (TPC-05, Yont Phol Dee Co., Ltd., Nakhon Sawan, Thailand). A total of 375 g of clean paddy from each replicate was collected, divided into 125 g portions, and stored in a zippered plastic bag. The measurement was repeated three times, and the mean and standard deviation were calculated. The weight of brown rice was determined after it was dehusked twice using a rubber roller husking machine (THY-05, Yont Phol Dee Co., Ltd., Nakhon Sawan, Thailand). The brown rice was then milled for 20 s in a horizontal friction-type milling machine (TFW-05, Yont Phol Dee Co., Ltd., Nakhon Sawan, Thailand). The white rice was collected, and its weight was determined. Finally, the broken white rice was sorted and separated from the head rice using a circular perforated sieve (diameter 2.4 mm) and manual picking to ensure that the broken rice was completely separated from the head rice. The head and broken rice were weighed. After the weight data were recorded, the yields of husk, bran, and head rice were calculated using Equations (11)–(13).

$$\text{Husk weight percentage (\%)} = \frac{W_{pd} - W_{br}}{W_{pd}} \times 100 \quad (11)$$

$$\text{Bran weight percentage (\%)} = \frac{W_{br} - W_{wr}}{W_{pd}} \times 100 \quad (12)$$

$$\text{Head rice weight percentage (\%)} = \frac{W_{hr}}{W_{pd}} \times 100 \quad (13)$$

where W_{pd} is the weight of rice from each measurement, W_{br} is the weight of brown rice, W_{wr} is the weight of white rice, and W_{hr} is the weight of the head rice yield.

2.5.3. White Rice Color Determination

The color value of white rice was determined with the CIE (L*, a*, b*) color scale using a chromameter (CR-400, Konica Minolta Sensing, Osaka, Japan). The whiteness index (WI) was used to calculate and compare the whiteness level of rice, as shown in Equation (14).

$$\text{Whiteness index (WI)} = 100 - \sqrt{\left[(100 - L*)^2 + a*^2 + b*^2\right]} \qquad (14)$$

where L* is lightness–darkness ($0 \leq L \leq 100$), a* is redness in a positive value, a* is greenness in a negative value, b* is yellowness in a positive value, and b* is blueness in a negative value. The total color difference ($\Delta E*$) used in repeatability testing to serve as a single number for pass/fail decision was defined in Equation (15).

$$\text{Total color difference } (\Delta E^*) = \sqrt{(\Delta L*)^2 + (\Delta a*)^2 + (\Delta b*)^2} \qquad (15)$$

where $\Delta L*$ is the L* sample minus L* reference, $\Delta a*$ is the a* sample minus a* reference, and $\Delta b*$ is the b* specimen minus b* reference. For each replicate, the measurement was repeated three times and reported as the mean and standard deviation.

2.6. Cooking Quality Evaluation

2.6.1. Elongation Ratio

Each treatment was repeated three times, and the mean and standard deviation were calculated. In each replicate, 50 white rice kernels were randomly drawn from the samples. Rice was soaked in water for 30 min before being placed on a strainer to boil. After 10 min of boiling, rice was spread on a plate, and the 30 kernels of rice grains that were elongated straightly were measured. The elongation ratio was calculated by dividing the average length of the cooked rice by the average length of the uncooked rice. The elongation ratio was calculated as follows [22]:

$$\text{Elongation ratio} = \frac{\text{Average length of cooked rice (mm)}}{\text{Average length of uncooked rice (mm)}} \qquad (16)$$

2.6.2. Apparent Amylose Content

The method specified in the Thai Standards for Rice [21] was used to determine the apparent amylose content of rice samples using a spectrophotometer (SPECORD 40, Analytik Jena, Jena, Germany) at a wavelength of 620 nm. Each treatment was repeated three times, and the results were expressed as mean and standard deviation.

2.6.3. Texture Profiles Analysis

The milled rice was washed and cooked with a rice-to-water weight ratio of 1:1.7. It was cooked to completion in 1 L household rice cookers, followed by a 10 min warming period. The cooked rice was taken directly from the center of each pot for testing. A texture analyzer (TA.XT plus, Stable Micro Systems, Surrey, UK) was used to analyze texture profiles according to the method of Champagne et al. [23]. On a glass plate, 10 cooked rice kernels were arranged in a single grain layer. A compression plate was set at the height of 5 mm above the base. To allow the plate to travel 4.9 mm, return, and repeat the test, a two-cycle program for compression, force, and distance was used. The test speed was 1 mm per second. A cylinder plunger with a diameter of 50 mm was used. Each treatment was repeated three times, and results were reported as mean and standard deviation.

2.6.4. Pasting Properties

Rice viscosity (RVA profiles) was determined using a rapid viscosity analyzer (RVA-4, Newport Scientific, New South Wales, Australia) according to AACC Method 61-02, Determination of the Pasting Properties of Rice with the Rapid Visco Analyzer, and analyzed using Thermal Cycle (TCW 2.5) software for Windows [24]. Three grams of flour from each sample was weighed into an aluminum canister at a moisture content of 14%, to which 25 mL of distilled water was added. A paddle was placed in the canister, and its blade was vigorously jogged up and down through the sample ten times. The RVA dispersed the samples by rotating the paddle at 960 rpm for the first 10 s of the test. Viscosity was then measured at a constant paddle speed of 160 rpm. The idle temperature was set at 50 °C, and the following 12.5 min test profiles were performed: (1) 50 °C was held for 1.0 min, (2) the temperature was linearly raised to 95 °C until 4.8 min, (3) the temperature was held at 95 °C until 7.5 min, (4) the temperature was linearly reduced to 50 °C until 11 min, and (5) held at 50°C until 12.5 min. Heating and cooling were performed linearly between the profile set points. Each treatment was repeated three times, and the results were expressed as mean and standard deviation.

2.7. Sensory Analysis

The triangle test was performed in accordance with the requirements of ISO 4120:2004. It was conducted to determine whether panelists, who were 24 untrained panelists who normally consumed rice on a regular basis, could distinguish the difference between cooked rice treated with RF combined hot air drying and cooked rice treated with hot air drying. Rice was washed and boiled at a rice-to-water weight ratio of 1:1.7. Rice was cooked to completion in 1 L household rice cookers, followed by a 10 min warming period. The cooked rice (1.5 g) was then served at 60 °C in a white plastic bowl with a 3-digit coding. An equal number of the six possible sequences of products A and B: ABB, AAB, ABA, BAA, and BAB were randomly distributed to panelists in groups of 6 until 24 panelists had completed the evaluation. The perceptible difference was determined using the standardized tables from ISO 4120:2004 [25].

2.8. Statistical Analysis

All experimental data were presented in terms of mean ± standard deviation (SD). Analysis was performed by analysis of variance (ANOVA) and Duncan's new multiple range test (DMRT) for comparison of means at the 5% probability level ($p < 0.05$) using SPSS software version 25 (IBM, New York, NY, USA).

3. Results and Discussion

3.1. Drying Characteristics

3.1.1. Drying Model

Figure 2 illustrates the drying rate versus the drying time for paddy dried with hot air and hot air combined with various RF heating temperatures. During the first 30 min of drying, the drying rate by hot air (HA), hot air with RF heating temperature of 38 °C (HA/RF38), hot air with RF heating temperature of 42 °C (HA/RF42), hot air with RF heating temperature of 46 °C (HA/RF46), and hot air with RF heating temperature of 50 °C (HA/RF50) were 0.0887, 0.0910, 0.1197, 0.1366, and 0.1421 kg water/kg dry matter per minute, respectively. The moisture ratio versus time for paddy dried with hot air and hot air combined with various RF heating temperatures is shown in Figure 3. After the first 30 min of drying, the moisture ratio of paddy dried by HA, HA/RF38, HA/RF42, HA/RF46, and HA/RF50 were 0.9221, 0.9228, 0.8945, 0.8807, and 0.8745, respectively. Following that, the curves demonstrated exponential decay.

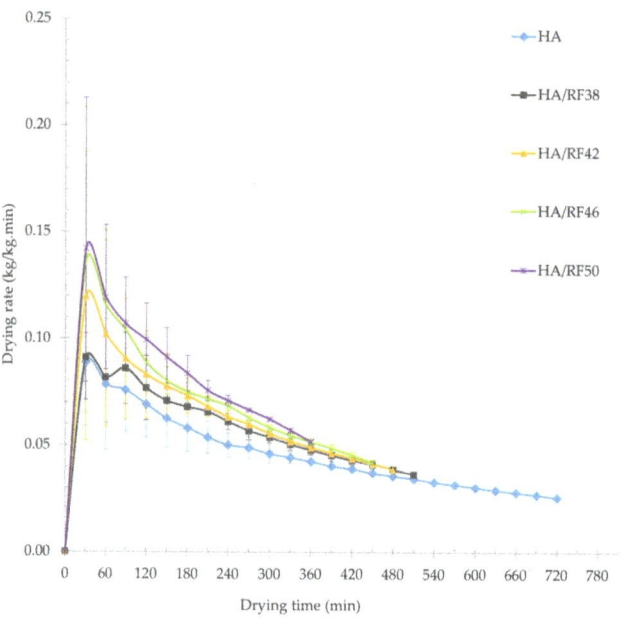

Figure 2. Drying rate versus drying time for paddy dried with hot air and hot air combined with various RF heating temperatures.

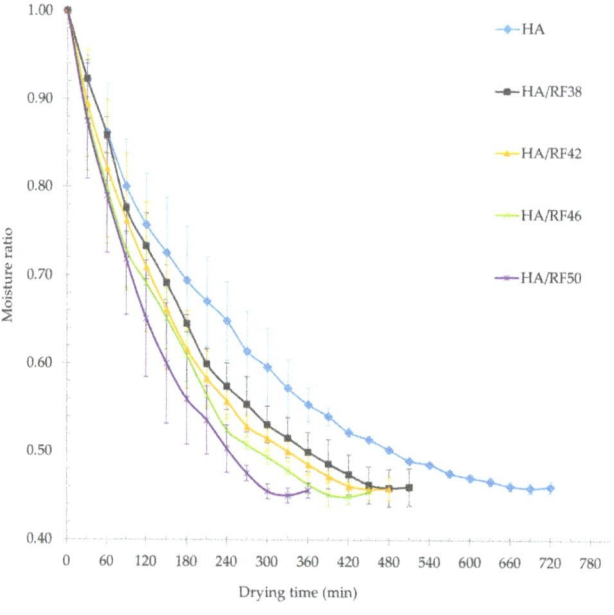

Figure 3. Drying kinetics for paddy dried with hot air and hot air combined with various RF heating temperatures.

At a final moisture content of 14%, drying paddy at HA, HA/RF38, HA/RF42, HA/RF46, and HA/RF50 took about 720, 510, 480, 450, and 360 min, respectively. Compared with HA drying time as the baseline value, the savings of drying time were 29.17%, 33.33%, 37.50%, and 50.00%, respectively. As a result, the drying time required for combined drying with HA/RF was advantageously shorter than that required for drying with HA. In all treatments, it was found that drying time was shortened when RF heating temperature increased because, in the early drying stage, the moisture content of fresh paddy was 25–26% (w.b.). The high water content in the paddy absorbed more energy at different moisture contents due to the dielectric properties of paddy seed in different moisture contents [2], and free water with high water activity accounted for most of the water content in the paddy at this early stage. The transfer and accumulation of heat accelerated the evaporation of free water in the paddy. This may indicate that the drying rate depends on the radiation intensity, which was exerted by the RF heating temperature in this study, and that the drying time was shortened with increasing radiation intensity [26,27].

The estimated values of the coefficients for selected thin-layer drying models of combined RF heating with hot air drying are shown in Table 1. In the Lewis and Henderson–Pabis models, the values of drying rate constant (k) as a function of RF heating temperature ranged between 1.3405×10^{-3} to 2.6657×10^{-3} and 1.0373×10^{-3} to 2.2164×10^{-3}, respectively. The coefficients of determination (R^2) of the Lewis model fit the curves better than the R^2 of Henderson–Pabis model, which were 0.9702–0.9776 and 0.9286–0.9420, respectively. Thus, the Lewis model was suitable for predicting the drying characteristics observed in the study. Moreover, the values of the drying rate constant (k) were found to increase with increasing RF heating temperature.

Table 1. Estimated values of the coefficients for selected thin-layer drying models and average effective moisture diffusivity (D_{eff}) of combined RF heating with hot air drying.

Treatment	Lewis		Henderson–Pabis			Average ($D_{eff} \times 10^{-9}$ m^2/s)
	k	R^2	a	k	R^2	
HA	1.3405×10^{-3}	0.9702	8.6196	1.0373×10^{-3}	0.9286	4.8156 ± 1.4550
HA/RF38	1.8522×10^{-3}	0.9775	8.9515	1.5357×10^{-3}	0.9391	6.2788 ± 2.3776
HA/RF42	1.9998×10^{-3}	0.9733	8.7751	1.6038×10^{-3}	0.9321	7.3994 ± 2.0088
HA/RF46	2.1854×10^{-3}	0.9722	8.7203	1.7436×10^{-3}	0.9297	9.0804 ± 1.6978
HA/RF50	2.6657×10^{-3}	0.9776	8.9376	2.2164×10^{-3}	0.9420	9.4426 ± 4.0670

Data of D_{eff} are expressed as mean ± SD in triplicate.

3.1.2. Effective Diffusivity

The average effective moisture diffusivity (D_{eff}) of the combined RF heating with hot air drying is shown in Table 1. The study found that the D_{eff} varied between 4.8156×10^{-3} to 9.4426×10^{-3} and raised with increasing RF heating temperature. This could be explained by the fact that the energy of RF affected the rapid temperature rise of the paddy kernels from the inside to the outside, which increased the vapor pressure and consequently accelerated the diffusion of moisture to the surface [28].

3.1.3. Energy Consumption

Table 2 shows the electricity consumption of the drying system, including the hot air dryer and the RF heating machine, the specific energy consumption (SEC), which indicates the amount of energy consumed (MJ) to evaporate the water in the paddy (kg-H$_2$O), and the specific energy cost, which indicates the electricity cost (USD) to evaporate the water per kilogram (kg-H$_2$O). The amount of SEC and the specific energy cost decreased at all RF heating temperatures, with HA/RF50 having the lowest values for water evaporation of 3.8600 ± 0.2261 MJ/kg-H$_2$O and 0.1009 ± 0.0121 USD/kg-H$_2$O, respectively. Olatunde et al. [29] reported that the specific energy consumption of 4.574–4.905 MJ/kg-H$_2$O was required for drying paddy rice in one pass from an initial moisture content of 24% to 11–13% using the MW dryer, and [30] also reported that 4.04 kWh/kg-H$_2$O (14.54 MJ/kg-H$_2$O) was required for drying parboiled paddy rice from an initial moisture content of 55.96% to 15.58% using the MW dryer. In this study, the SEC and specific energy costs showed no significant difference, while the electricity consumption of the hot air dryer decreased significantly but that of the RF heating machine increased significantly when the RF heating temperature increased. This might indicate that the power of RF, together with the power of the hot air dryer, helps to evaporate a certain amount of water from the paddy with a certain SEC. Therefore, this could be a useful observation that drying paddy with the method presented in this study requires less energy than drying with the MW dryer, as mentioned above.

Table 2. Electricity consumption, specific energy consumption, and specific energy cost after being dried under various RF heating temperatures.

Treatment	Power Consumption (kWh)		Specific Energy Consumption	Energy Cost
	Hot Air Dryer	RF Heating Machine	(MJ/kg-H_2O)	(USD/kg-H_2O)
HA	59.33 ± 3.79 [c]	0.00	4.0367 ± 0.6121	0.1098 ± 0.0167
HA/RF38	44.67 ± 4.04 [b]	10.09 ± 1.15 [a]	3.8700 ± 0.2265	0.1053 ± 0.0062
HA/RF42	40.67 ± 6.11 [b]	15.81 ± 3.21 [b]	3.9133 ± 0.4574	0.1065 ± 0.0124
HA/RF46	37.33 ± 7.51 [b]	18.53 ± 3.05 [b]	3.8600 ± 0.2261	0.1051 ± 0.0061
HA/RF50	28.67 ± 1.53 [a]	24.36 ± 2.00 [c]	3.7033 ± 0.4450	0.1009 ± 0.0121

Data are expressed as mean ± SD in triplicate. The same letter or no letter indicates no significant difference ($p > 0.05$) between the same values in a column.

3.2. Effects on Milling Qualities

3.2.1. Fissure Percentage

The effects of different RF heating temperatures on the fissure percentage of brown rice with different fissure degrees are shown in Table 3. The fissure percentage was classified into four groups, namely None (0 fissure), Few (1–2 fissures), Moderate (3–4 fissures), and Severe (≥5 fissures), as reported by [31]. The morphology of fissures in white rice with different fissure degrees is shown in Figure 4. The fissure percentage of brown rice kernels with "None" fissure degree was significantly decreased from 99.78% ± 0.44% to 85.33% ± 1.87% from HA to HA/RF50. In contrast, the fissure percentages of brown rice kernels with "Few", "Moderate", and "Severe" fissure degrees were significantly increased from 0.33% ± 0.50% to 12.44% ± 1.81%, 0.00% to 0.56% ± 0.73%, and 0.00% to 1.78% ± 0.83% from HA to HA/RF50, respectively. The "None" fissure degree considered the most important characteristic, while the "Severe", "Moderate", and "Few" fissure degrees considered the most undesirable characteristic, respectively, because fewer fissure kernels would result in a high head rice yield of milled rice. This result showed that RF heating significantly affected the fissure percentage due to the high absorption of RF energy inside the rice kernels [31] and that the temperature of the rice grain plays an important role in the development of a heat and moisture gradient inside the grain, resulting in surface hardening [32] and differential stress [33], respectively. The optimum condition for RF heating temperature was HA/RF42, as no "Moderate" and "Severe" fissure degrees were observed. Referring to the earlier study by Chitsuthipakorn and Thanapornpoonpong [12], drying paddy with hot air combined with RF heating temperatures of 45 °C, 50 °C, 55 °C, and 60 °C showed that the critical limit of RF heating temperature was HA/RF45 because the least "Moderate" and "Severe" fissure degrees were observed. Moreover, the fissure percentage result was relatively lower in this study because of the modification of the prototype machine, as mentioned in the Introduction.

Table 3. Effects of different RF heating temperatures on the fissure percentage of brown rice with different fissure degrees.

Treatment	None (0 Fissure)	Few (1–2 Fissures)	Moderate (3–4 Fissures)	Severe (≥5 Fissures)
HA	99.78 ± 0.44 [e]	0.33 ± 0.50 [a]	0.00 [a]	0.00 [a]
HA/RF38	98.67 ± 0.50 [d]	1.33 ± 0.50 [b]	0.00 [a]	0.00 [a]
HA/RF42	96.33 ± 0.71 [c]	3.89 ± 0.78 [c]	0.00 [a]	0.00 [a]
HA/RF46	95.00 ± 0.87 [b]	4.00 ± 0.87 [c]	0.22 ± 0.44 [ab]	0.78 ± 0.67 [b]
HA/RF50	85.33 ± 1.87 [a]	12.44 ± 1.81 [d]	0.56 ± 0.73 [b]	1.78 ± 0.83 [c]

Data are expressed as mean ± SD in triplicate. The same letter or no letter indicates no significant difference ($p > 0.05$) between the same values in a column.

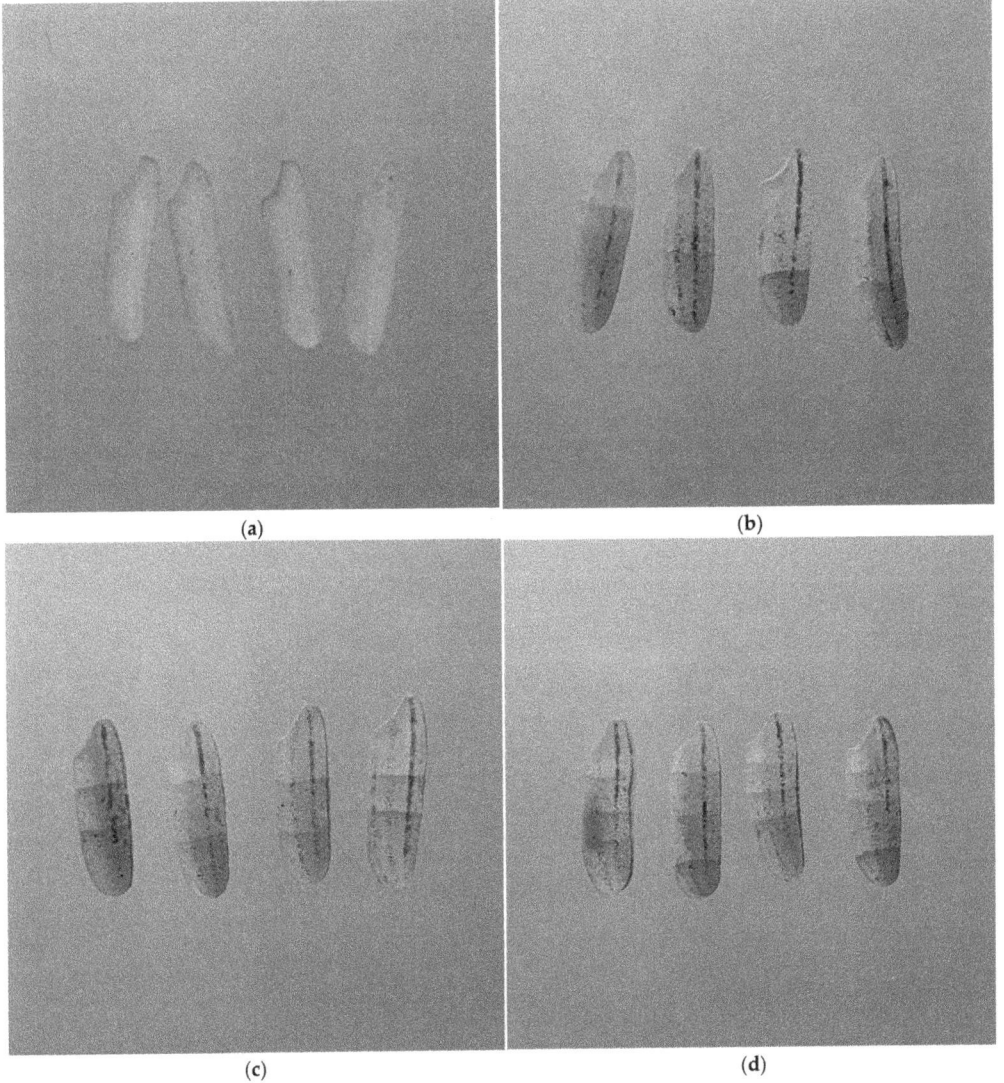

Figure 4. *Cont.*

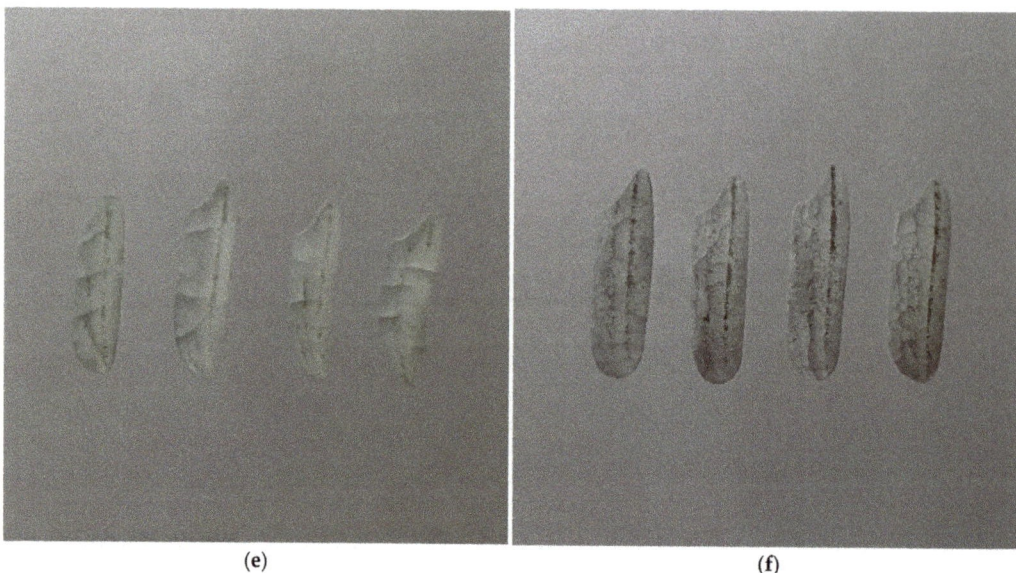

Figure 4. Morphology of different fissure degrees in white rice: (**a**) white rice with 0 fissure; (**b**) white rice with 1 fissure; (**c**) white rice with 2 fissures; (**d**) white rice with 3 fissures; (**e**) white rice with 4 fissures; (**f**) white rice with ≥5 fissures.

3.2.2. Milling Yield

The milling yield percentage determination after drying at different RF heating temperatures is shown in Table 4. It was found that the husk, brown rice, and broken rice yields significantly increased after treatment and ranged from 24.62 ± 0.78 to 25.78 ± 0.67, 74.22 ± 0.67 to 75.38 ± 0.78, and 26.22 ± 1.37 to 28.00 ± 1.79, respectively. However, head rice yield (HRY) is the most important characteristic among industrial rice millers because it has a proportional effect on the income of the enterprise, while husk and broken rice are less important because of their lower price per kg, which is about 50% of the head rice price. Thus, the higher the head rice yield, the better the drying conditions for rice millers. In this study, there was no significant difference in HRY, which could mean that all RF heating temperatures were suitable for large-scale paddy drying. With reference to the previous study by Chitsuthipakorn and Thanapornpoonpong [12], drying paddy with hot air combined with RF heating temperatures of 45 °C, 50 °C, 55 °C, and 60 °C showed that the critical limit of RF heating temperature was HA/RF50 in terms of HRY and the optimum condition was HA/RF45 in terms of HRY, color, and fissure percentage. Therefore, the aim of this study was to achieve the most suitable drying conditions by combining hot air and RF heating through a lower RF heating temperature and a narrower temperature level. Similarly, the development of a one-pass microwave heating for paddy drying was recently investigated and found that the application of microwave energy of up to 600 kJ/kg-grain to medium-grain paddy, with an initial moisture content of 23% to 24% and an additional tempering step at 60 °C for 4 h, dried the paddy to a final moisture content of 14% to 16%, depending on the rate of energy application, with HRY not significantly different from natural air at 25 °C and RH of 65% [27].

3.2.3. White Rice Color Determination

Table 5 shows the evaluation of the color of white rice after drying at different RF heating temperatures. It was found that the white rice had a slightly greenish-yellowish color after treatment. There was only a significant difference in the yellow-blue value (b*), while the lightness value (L*) and the red-green value (a*) had no significant differences. These color values changed in response to convective drying conditions, with an immediate increase in paddy temperature caused by RF energy, which accelerated the Maillard reaction and the transfer of color substances from the rice husk and rice bran to the endosperm, causing discoloration [34–37]. However, considering HA as a conventional convective method in industrial drying of paddy, the WI value was used to compare the whiteness of white rice between treatments, and there was no significant difference at WI. On the other hand, ΔE* should be a single metric to compare the distance of color difference of HA/RF from HA. A lower ΔE* indicated greater accuracy, while a higher ΔE* indicated significant deviation. The ΔE* values in this study ranged from 1.25 ± 0.53 to 1.76 ± 0.26 and showed no significant difference. Therefore, it could be concluded that all RF heating temperatures in this study did not cause any color difference and were suitable for drying paddy on a large scale.

Table 4. Effects of different RF heating temperatures on the milling yield percentage.

Treatment	Husk	Brown Rice	White Rice	Head Rice	Broken Rice	Bran
HA	25.33 ± 0.40 [b]	74.67 ± 0.40 [a]	65.07 ± 0.80 [b]	38.40 ± 2.65	26.67 ± 2.00 [a]	9.60 ± 0.89 [a]
HA/RF38	25.60 ± 0.57 [b]	74.40 ± 0.57 [a]	63.82 ± 0.96 [a]	37.51 ± 2.73	26.31 ± 1.94 [a]	10.58 ± 1.12 [ab]
HA/RF42	25.42 ± 0.78 [b]	74.58 ± 0.78 [a]	64.44 ± 1.21 [ab]	38.22 ± 1.69	26.22 ± 1.37 [a]	10.13 ± 0.69 [ab]
HA/RF46	25.78 ± 0.67 [b]	74.22 ± 0.67 [a]	63.64 ± 1.50 [a]	37.33 ± 2.62	26.31 ± 1.47 [a]	10.58 ± 1.59 [ab]
HA/RF50	24.62 ± 0.78 [a]	75.38 ± 0.78 [b]	64.62 ± 1.25 [ab]	36.62 ± 1.87	28.00 ± 1.79 [b]	10.76 ± 0.81 [b]

Data are expressed as mean ± SD in triplicate. The same letter or no letter indicates no significant difference ($p > 0.05$) between the same values in a column.

Table 5. Effects of different RF heating temperatures on the white rice color.

Treatment	L*	a*	b*	Whiteness Index	ΔE*
HA	64.51 ± 0.69	−0.54 ± 0.14 [a]	9.02 ± 0.51 [a]	63.37 ± 0.65	0.00
HA/RF38	64.15 ± 1.02	−0.41 ± 0.27 [ab]	9.95 ± 1.13 [b]	62.78 ± 1.24	1.28 ± 1.08
HA/RF42	64.07 ± 1.89	−0.41 ± 0.27 [ab]	9.75 ± 1.14 [ab]	62.76 ± 2.06	1.76 ± 1.26
HA/RF46	64.01 ± 0.81	−0.36 ± 0.32 [b]	10.09 ± 0.41 [b]	62.61 ± 0.81	1.41 ± 0.51
HA/RF50	64.61 ± 1.02	−0.38 ± 0.29 [b]	9.43 ± 0.74 [ab]	63.37 ± 0.94	1.25 ± 0.53

Data are expressed as mean ± SD in triplicate. The same letter or no letter indicates no significant difference ($p > 0.05$) between the same values in a column.

3.3. Effects on Cooking Qualities

3.3.1. Elongation Ratio

Table 6 shows the elongation ratio (ER) of cooked rice after treatments. It was found that ER increased significantly with increasing RF heating temperature, as it was also observed in accelerated aged paddy in hot air oven [38], microwave [39], and RF [4,5]. This was due to the fact that RF energy could create fissures in the rice kernel, through which water could penetrate more easily into the rice kernel, and that after treatment, the rice had uniformly closed cell walls that resisted the high pressure inside the cell during cooking [39]. In addition, such heating also resulted in various changes in protein and starch granules that affected the swelling of the cooked rice kernel, as these two components were related to the water absorption of the kernel [40,41]. Although the length expansion without increasing the width of the rice kernel was considered a desirable property of rice, the result of the longest ER in HA/RF50 showed only 2.11% longer grain than the rice dried with HA, so it should not be important that the increase of ER was beneficial for the commercial aspect.

3.3.2. Apparent Amylose Content

The apparent amylose content (AC) is shown in Table 6. It was found that there was no significant difference in AC of milled rice after treatment with RF compared to the control. This might be due to the fact that the temperatures of RF in this study were not too high to affect AC, as described by Wani et al. [42]; the H-bonds in amylose molecules can form inclusion complexes under high moisture and temperature conditions, and these H-bonds absorb more water and form clathrates with proteins and free fatty acids, which might be the reason for the decrease of AC in high moisture rice treated with RF. There were reports of AC decline in accelerated aged rice induced by RF that AC was negatively affected by moisture content, RF exposure time [3], and RF temperatures [9]. In addition, Jiao et al. [8] reported that exposure to RF of raw rice, brown rice, and milled rice at 50 °C for 5 min could cause a decrease in AC, while Hou et al. [7] argued that exposure to RF of raw rice, brown rice, and milled rice at 54 °C for 11 min showed no significant difference in AC. It could be considered that drying paddy with HA/RF at low temperature (38 °C–50 °C) could maintain the AC of rice as in HA.

Table 6. Effects of different RF heating temperatures on the elongation ratio and apparent amylose content.

Treatment	Elongation Ratio	Apparent Amylose Content (%)
HA	1.4257 ± 0.0193 [a]	26.74 ± 0.24
HA/RF38	1.4264 ± 0.0228 [a]	27.76 ± 0.94
HA/RF42	1.4377 ± 0.0169 [b]	27.07 ± 0.86
HA/RF46	1.4533 ± 0.0291 [c]	26.75 ± 0.87
HA/RF50	1.4530 ± 0.0196 [c]	27.68 ± 0.37

Data are expressed as mean ± SD in triplicate. The same letter or no letter indicates no significant difference ($p > 0.05$) between the same values in a column.

3.3.3. Texture Profiles Analysis

Table 7 shows the texture profiles of cooked rice after the treatments in terms of hardness, adhesiveness, springiness, and cohesiveness. Comparing the texture profiles of rice after RF heating and control, all the properties were not significantly different. This could mean that drying paddy with HA/RF at low temperature (38 °C–50 °C) could maintain the texture profiles of cooked rice as in HA.

Table 7. Effects of different RF heating temperatures on the textural profiles of cooked rice.

Treatment	Hardness (g)	Adhesiveness (g.sec)	Springiness (sec/sec)	Cohesiveness (g.sec/g.sec)
HA	10,400.28 ± 253.68	−145.89 ± 77.63	0.6973 ± 0.0151	0.4599 ± 0.0029
HA/RF38	10,160.72 ± 610.38	−108.92 ± 24.25	0.6560 ± 0.0354	0.4548 ± 0.0069
HA/RF42	9828.77 ± 109.48	−123.28 ± 68.41	0.6868 ± 0.0470	0.4536 ± 0.0064
HA/RF46	10,045.50 ± 337.98	−99.91 ± 69.41	0.6736 ± 0.0374	0.4572 ± 0.0223
HA/RF50	10,340.02 ± 168.02	−122.87 ± 56.59	0.6500 ± 0.0361	0.4560 ± 0.0042

Data are expressed as mean ± SD in triplicate. The same letter or no letter indicates no significant difference ($p > 0.05$) between the same values in a column.

3.3.4. Pasting Properties

The pasting properties of cooked rice after the treatments in terms of peak viscosity, trough, breakdown, final viscosity, and setback are shown in Table 8. Comparing the pasting properties of rice after RF heating and control, all properties were not significantly different. It could be proposed that drying paddy with HA/RF at low temperature (38 °C–50 °C) could maintain the pasting properties of cooked rice as in HA.

Table 8. Effects of different RF heating temperatures on the pasting properties of cooked rice.

Treatment	Peak Viscosity (cP)	Trough (cP)	Breakdown (cP)	Final Viscosity (cP)	Setback (cP)
HA	1474.78 ± 124.60	982.33 ± 108.89	492.44 ± 24.39	2195.45 ± 162.07	720.67 ± 38.11
HA/RF38	1591.33 ± 141.81	1028.89 ± 121.74	562.44 ± 24.93	2306.78 ± 200.30	715.44 ± 58.51
HA/RF42	1503.78 ± 48.10	970.56 ± 32.08	533.22 ± 76.42	2208.45 ± 92.69	704.67 ± 138.85
HA/RF46	1595.89 ± 258.83	1024.78 ± 137.36	571.11 ± 121.47	2266.33 ± 231.49	670.45 ± 32.70
HA/RF50	1768.22 ± 220.94	1068.00 ± 93.02	700.22 ± 127.95	2372.00 ± 88.76	603.78 ± 133.52

Data are expressed as mean ± SD in triplicate. The same letter or no letter indicates no significant difference ($p > 0.05$) between the same values in a column.

3.4. Sensory Analysis

A triangle test was conducted to determine whether panelists were able to distinguish cooked rice dried with hot air (control) from cooked rice dried with hot air combined with RF heating, as shown in Figure 5. It was found that cooked rice dried with HA/RF50 was significantly different from HA, and at least 6% of the population was able to perceive a difference. The proportion of the population able to perceive a difference between samples of cooked rice was calculated using a one-sided lower confidence interval, as shown in Equations (17)–(20):

$$\text{Proportion correct } (p_c) = \frac{x}{n} \quad (17)$$

$$\text{Proportion distinguished } (p_d) = 1.5 p_c - 0.5 \quad (18)$$

$$\text{Standard deviation of } p_d \ (s_d) = 1.5 \sqrt{p_c(1-p_c)/n} \quad (19)$$

$$\text{Lower confidence limit} = p_d + z_\alpha s_d \quad (20)$$

where x is the number of correct responses, n is the total number of panelists, and z_α is the critical value of the standard normal distribution ($z_\alpha = 1.64$ for a 95% confidence interval).

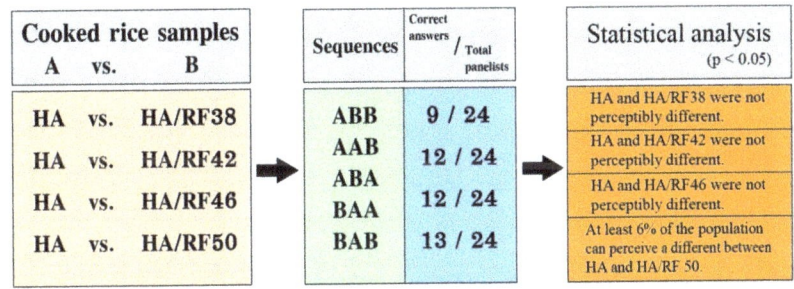

Figure 5. Triangle test to evaluate the ability of panelists to distinguish cooked rice dried with hot air (A) from cooked rice dried with hot air combined with the different RF heating temperatures (B). The statistical difference was determined using ISO 4120:2004 standardized tables.

4. Conclusions

The effects on milling and cooking qualities of milled rice at four different RF heating temperatures (38 °C, 42 °C, 46 °C, and 50 °C) combined with hot air drying (HA/RF) were compared with hot air drying (HA). The drying characteristics and moisture diffusivity showed that the higher the RF heating temperature, the shorter the observed drying time. The amount of SEC and the specific energy cost decreased when the RF heating increased. In rice drying, milling qualities are the most important factors as they are the first stage of processing. A new rice drying technology must be proven to maintain the fissure percentage, the percentage of head rice yield, and the color of milled rice compared to conventional rice drying technology. Regarding milling qualities, the optimal condition for RF heating temperature was HA/RF42 in terms of fissure percentage, while there was no significant difference between all treatments in terms of head rice yield and white rice color determination. Regarding cooking qualities, drying paddy with HA/RF at the given temperature (38 °C–50 °C) showed no significant difference with HA in terms of amylose content, texture profiles, and pasting properties. Therefore, in this study, the cooking qualities were investigated to ensure that they were also maintained, while many previous studies stated that RF heating could change these qualities depending on the intensity of RF energy used, such as accelerated ageing of rice by RF heating. In the triangle test, it was found that at least 6% of the population could perceive a difference between HA and HA/RF50. In this study, it was proposed that HA/RF42 was the optimal condition for using a continuous flow hot air dryer in combination with radio frequency heating in a large-scale process. Compared to the previous study [12], there are three advantages of this significant improvement: (1) The fissure percentage on HA/RF42 was lower than on HA/RF45. The fissure percentage was the first parameter for a rice miller to determine the efficiency of the dryer. This result shows that lower RF heating temperature can better maintain the milled rice quality. (2) The power consumption of HA/RF42 could save up to 15.19% of the power of RF heating machine. The power consumption of the hot air dryer can be minimized by using various heat sources, such as liquefied petroleum gas (LPG), rice husk and biomass. (3) With a lower RF heating temperature, the RF heating machine can be redesigned to minimize the initial investment cost of combining with a small batch hot air dryer, or the RF heating machine can be combined with a batch dryer with larger drying capacity to generate more revenue.

Author Contributions: Conceptualization, K.C. and S.-n.T.; methodology, K.C. and S.-n.T.; software, K.C.; validation, K.C. and S.-n.T.; formal analysis, K.C. and S.-n.T.; investigation, K.C.; resources, K.C.; data curation, K.C.; writing—original draft preparation, K.C.; writing—review and editing, S.-n.T.; visualization, K.C.; supervision, S.-n.T.; project administration, S.-n.T.; funding acquisition, K.C. All authors have read and agreed to the published version of the manuscript.

Funding: This research received no external funding.

Institutional Review Board Statement: Not applicable.

Informed Consent Statement: Not applicable.

Data Availability Statement: Not applicable.

Acknowledgments: This study was supported by the Postharvest Technology Innovation Center, Ministry of Higher Education, Science, Research and Innovation, Bangkok, Thailand. The prototype of the machine and paddy materials were developed and supported by Yont Phol Dee Co., Ltd., Nakhon Sawan, Thailand.

Conflicts of Interest: The authors declare no conflict of interest.

References

1. Wray, D.; Ramaswamy, H.S. Novel concepts in microwave drying of Foods. *Dry. Technol.* **2015**, *33*, 769–783. [CrossRef]
2. Pakawattana, C. Effect of Drying with Radio Frequency Heat Treatment and Hot Air Oven on Quality of Rice Seed cv. Pathum Thani 1. Master's Thesis, Chiang Mai University, Chiang Mai, Thailand, 2013.
3. Hussain, S.Z.; Iftikhar, F.; Naseer, B.; Altaf, U.; Reshi, M.; Nidoni, U.K. Effect of radiofrequency induced accelerated ageing on physico-chemical, cooking, pasting and textural properties of rice. *LWT—Food Sci. Technol.* **2021**, *139*, 110595. [CrossRef]
4. Sumreerath, P. Accelerate Aging of Paddy Rice cv. Pathum Thani 1 by Radio Frequency. Master's Thesis, Chiang Mai University, Chiang Mai, Thailand, 2010.
5. Vearasilp, S.; Chaisathidvanich, K.; Thanapornpoonpong, S.; von Hörsten, D.; Lücke, W. Aging milled rice by radio frequency heat treatment. In Proceedings of the Conference on International Research on Food Security, Natural Resource Management and Rural Development, Stuttgart, Germany, 5–7 October 2011.
6. Vearasilp, S.; Naka, J.; Thanapornpoonpong, S.; von Hörsten, D.; Lücke, W. Influence of milled rice packing methods on radio frequency heat distribution in controlling *Aspergillus flavus* and their cooking qualities. In Proceedings of the Conference on International Research on Food Security, Natural Resource Management and Rural Development, Stuttgart, Germany, 5–7 October 2011.
7. Hou, L.; Liu, Q.; Wang, S. Efficiency of industrial-scale radio frequency treatments to control *Rhyzopertha dominica* (fabricius) in rough, brown, and milled rice. *Biosyst. Eng.* **2019**, *186*, 246–258. [CrossRef]

8. Jiao, S.; Sun, W.; Yang, T.; Zou, Y.; Zhu, X.; Zhao, Y. Investigation of the feasibility of radio frequency energy for controlling insects in milled rice. *Food Bioprocess Technol.* **2017**, *10*, 781–788. [CrossRef]
9. Liu, Q.; Wang, S. Effects of various radio frequency treatment protocols on physicochemical properties and sensory quality of packaged milled rice. *LWT—Food Sci. Technol.* **2019**, *113*, 108269. [CrossRef]
10. Vearasilp, S.; Thanapornpoonpong, S.; Krittigamas, N.; Suriyong, S.; Akaranuchat, P.; von Hörsten, D. Vertical operating prototype development supported radio frequency heating system in controlling rice weevil in milled rice. *Agric. Agric. Sci. Procedia* **2015**, *5*, 184–192. [CrossRef]
11. Onwude, D.; Hashim, N.; Chen, G. Recent advances of novel thermal combined hot air drying of agricultural crops. *Trends Food Sci. Technol.* **2016**, *57*, 132–145. [CrossRef]
12. Chitsuthipakorn, K.; Thanapornpoonpong, S. Quality of milled rice from large-scale dried paddy rice by hot air combined with radio frequency heating. *Processes* **2021**, *9*, 2277. [CrossRef]
13. Giri, S.K.; Prasad, S. Drying kinetics and rehydration characteristics of microwave-vacuum and convective hot-air dried mushrooms. *J. Food Eng.* **2007**, *78*, 512–521. [CrossRef]
14. Tulek, Y. Drying kinetics of oyster mushroom (*Pleurotus ostreatus*) in a convective hot air dryer. *J. Agric. Sci. Technol.* **2011**, *13*, 655–664.
15. Fasina, O.O.; Tyler, R.T.; Pickard, M.D. Modelling the infrared radiative heating of agricultural crops. *Dry. Technol.* **1998**, *16*, 2065–2082. [CrossRef]
16. Chupawa, P.; Gaewsondee, T.; Duangkhamchan, W. Drying characteristics and quality attributes affected by a fluidized-bed drying assisted with swirling compressed-air for preparing instant red jasmine rice. *Processes* **2021**, *9*, 1738. [CrossRef]
17. Das, I.; Das, S.K.; Bal, S. Drying kinetics of high moisture paddy undergoing vibration-assisted infrared (IR) drying. *J. Food Eng.* **2009**, *95*, 166–171. [CrossRef]
18. Jafari, H.; Kalantari, D.; Azadbakht, M. Energy consumption and qualitative evaluation of a continuous band microwave dryer for rice paddy drying. *Energy* **2018**, *142*, 647–654. [CrossRef]
19. Duangkhamchan, W.; Siriamornpun, S. Quality attributes and anthocyanin content of rice coated by purple-corn cob extract as affected by coating conditions. *Food Bioprod. Process.* **2015**, *96*, 171–179. [CrossRef]
20. Shen, L.; Zhu, Y.; Wang, L.; Liu, C.; Liu, C.; Zheng, X. Improvement of cooking quality of germinated brown rice attributed to the fissures caused by microwave drying. *J. Food Sci. Technol.* **2019**, *56*, 2737–2749. [CrossRef] [PubMed]
21. Ministry of Agriculture and Cooperatives, Thailand (MOAC). *Thai Agricultural Standard: Thai Rice*; National Bureau of Agricultural Commodity and Food Standards: Bangkok, Thailand, 2017. Available online: https://www.acfs.go.th/standard/download/Thai-Rice_60.pdf (accessed on 10 November 2021).
22. Juliano, B.O.; Perez, C.M. Results of a collaborative test on the measurement of grain elongation of milled rice during cooking. *J. Cereal Sci.* **1984**, *2*, 281–292. [CrossRef]
23. Champagne, E.; Lyon, B.; Min, B.; Vinyard, B.; Bett, K.; II, F.; Webb, B.; McClung, A.; Moldenhauer, K.; Linscombe, S.; et al. Effects of postharvest processing on texture profile analysis of cooked rice. *Cereal Chem.* **1998**, *75*, 181–186. [CrossRef]
24. Yan, C.-J.; Li, X.; Zhang, R.; Sui, J.M.; Liang, G.H.; Shen, X.P.; Gu, S.L.; Gu, M.H. Performance and inheritance of rice starch rva profile characteristics. *Rice Sci.* **2005**, *12*, 39–47.
25. Valdez-Meza, E.E.; Raymundo, A.; Figueroa-Salcido, O.G.; Ramírez-Torres, G.I.; Fradinho, P.; Oliveira, S.; de Sousa, I.; Suárez-Jiménez, M.; Cárdenas-Torres, F.I.; Islas-Rubio, A.R.; et al. Pasta enrichment with an amaranth hydrolysate affects the overall acceptability while maintaining antihypertensive properties. *Foods* **2019**, *8*, 282. [CrossRef] [PubMed]
26. Das, I.; Das, S.; Bal, S. Drying performance of a batch type vibration aided infrared dryer. *J. Food Eng.* **2004**, *64*, 129–133. [CrossRef]
27. Smith, D.L. Development of a One Pass Microwave Heating Technology for Rice Drying and Decontamination. Master's Thesis, University of Arkansas, Fayetteville, AR, USA, 2017.
28. Hashimoto, A.; Kameoka, T. Effect of infrared irradiation on drying characteristics of wet porous materials. *Dry. Technol.* **1999**, *17*, 1613–1626. [CrossRef]
29. Olatunde, G.A.; Atungulu, G.G.; Smith, D.L. One-pass drying of rough rice with an industrial 915 MHz microwave dryer: Quality and energy use consideration. *Biosyst. Eng.* **2017**, *155*, 33–43. [CrossRef]
30. Smith, D.L. Heat and Mass Transfer in Parboiled Rice during Heating with 915 MHz Microwave Energy and Impacts on milled Rice Properties. Doctoral Dissertation, University of Arkansas, Fayetteville, AR, USA, 2020.
31. Shen, L.; Wang, L.; Zheng, C.; Liu, C.; Zhu, Y.; Liu, H.; Liu, C.; Shi, Y.; Zheng, X.; Xu, H. Continuous microwave drying of germinated brown rice: Effects of drying conditions on fissure and color, and modeling of moisture content and stress inside kernel. *Dry. Technol.* **2021**, *39*, 669–697. [CrossRef]
32. Olatunde, G.A.; Atungulu, G.G. Milling behavior and microstructure of rice dried using microwave set at 915 MHz frequency. *J. Cereal Sci.* **2018**, *80*, 167–173. [CrossRef]
33. Cnossen, A.G.; Jiménez, M.J.; Siebenmorgen, T.J. Rice fissuring response to high drying and tempering temperatures. *J. Food Eng.* **2003**, *59*, 61–69. [CrossRef]
34. Horrungsiwat, S.; Therdthai, N.; Ratphitagsanti, W. Effect of combined microwave-hot air drying and superheated steam drying on physical and chemical properties of rice. *Int. J. Food Sci. Technol.* **2016**, *51*, 1851–1859. [CrossRef]
35. Inprasit, C.; Noomhorm, A. Effect of drying air temperature and grain temperature of different types of dryer and operation on rice quality. *Dry. Technol.* **2001**, *19*, 389–404. [CrossRef]

36. Palamanit, A.; Musengimana Sugira, A.; Soponronnarit, S.; Prachayawarakorn, S.; Tungtrakul, P.; Kalkan, F.; Raghavan, V. Study on quality attributes and drying kinetics of instant parboiled rice fortified with turmeric using hot air and microwave-assisted hot air drying. *Dry. Technol.* **2020**, *38*, 420–433. [CrossRef]
37. Rordprapat, W.; Nathakaranakule, A.; Tia, W.; Soponronnarit, S. Comparative study of fluidized bed paddy drying using hot air and superheated steam. *J. Food Eng.* **2005**, *71*, 28–36. [CrossRef]
38. Rosniyana, A.; Hashifah, M.A.; Shariffah Norin, S.A. Effect of heat treatment (accelerated ageing) on the physicochemical and cooking properties of rice at different moisture contents. *J. Trop. Agric. Food Sci.* **2004**, *32*, 155–162.
39. Le, T.; Songsermpong, S.; Rumpagaporn, P.; Suwanagul, A.; Wallapa, S. Microwave heating for accelerated aging of paddy and white rice. *Aust. J. Crop Sci.* **2014**, *8*, 1348–1358.
40. Zhao, S.; Xiong, S.; Qiu, C.; Xu, Y. Effect of microwaves on rice quality. *J. Stored Prod. Res.* **2007**, *43*, 496–502. [CrossRef]
41. Zhou, Z.; Robards, K.; Helliwell, S.; Blanchard, C. Ageing of stored rice: Changes in chemical and physical attributes. *J. Cereal Sci.* **2002**, *35*, 65–78. [CrossRef]
42. Wani, A.A.; Singh, P.; Shah, M.A.; Schweiggert-Weisz, U.; Gul, K.; Wani, I.A. Rice starch diversity: Effects on structural, morphological, thermal, and physicochemical properties—A review. *Compr. Rev. Food Sci. Food Saf.* **2012**, *11*, 417–436. [CrossRef]

Article

Hot Air-Assisted Radio Frequency (HARF) Drying on Wild Bitter Gourd Extract

Chang-Yi Huang [1], Yu-Huang Cheng [2] and Su-Der Chen [2,*]

[1] Department of Biotechnology and Animal Science, National Ilan University, Number 1, Section 1, Shen-Lung Road, Yilan City 26041, Taiwan; huaallen@gmail.com
[2] Department of Food Science, National Ilan University, Number 1, Section 1, Shen-Lung Road, Yilan City 26041, Taiwan; rin520520ever@gmail.com
* Correspondence: sdchen@niu.edu.tw; Tel.: +886-920518028; Fax: +886-39351892

Abstract: Wild bitter gourd (*Momordica charantia* L. var. *abbreviata* S.) is a kind of Chinese herbal medicine and is also a vegetable and fruit that people eat daily. Wild bitter gourd has many bioactive components, such as saponin, polysaccharide, and protein, and the extract is used to adjust blood sugar in patients with diabetes. The objective of this study was to investigate simultaneous hot air-assisted radio frequency (HARF) drying and pasteurization for bitter gourd extract, and then to evaluate its effects on blood sugar of type II diabetic mice. The results showed that the solid–liquid ratio of the wild bitter gourd powder to water was 1:10 and it was extracted using focused ultrasonic extraction (FUE) for only 10 min with 70 °C water. Then, 1 kg of concentrated bitter gourd extract was mixed with soybean fiber powder at a ratio of 2:1.1. It was dried by HARF, and the temperature of the sample could reach above 80 °C in only 12 min to simultaneously reduce moisture content (wet basis) from 58% to 15% and achieve a pasteurization effect to significantly reduce the total bacterial and mold counts. Type II diabetic mice induced by nicotinamide and streptozocin (STZ) for two weeks and then were fed four-week feeds containing 5% RF-dried wild gourd extract did not raise fasting blood glucose. Therefore, the dried powder of wild bitter gourd extracts by HARF drying had a hypoglycemic effect.

Keywords: wild bitter gourd; ultrasonic extraction; radio frequency (RF); drying; diabetic mice

1. Introduction

Wild bitter gourd (*Momordica charantia* L. var. *abbreviata* S.), also called wild bitter melon, is the Taiwanese endemic species of bitter gourds and features effective hypoglycemic components, such as saponin, polysaccharide, and peptide. Diabetic patients can consume fresh bitter gourd juice to reduce their blood glucose concentration and prevent postprandial hyperglycemia [1]. Bitter gourd saponins also promote glycogen storage and insulin secretion [2] and are involved in activating AMP-activated protein kinase phosphorylation to regulate energy metabolism [3]. Bitter gourd proteins can bond with insulin receptors to regulate blood sugar metabolism [4], and inhibit the activities of α-glucosidase and α-amylase to impede the degradation of starch and lower glucose content [5].

Traditional extraction methods used for plant extraction include mechanical stirring [6], boiling, and Soxhlet extraction [7]. Although traditional extraction methods are relatively inexpensive, they are subjected to the following limitations: lower extraction efficiency, longer extraction time, and larger solvent consumption [8]. In order to overcome these problems and to improve the extraction efficiency, ultrasonic extraction (UE) can be used. UE has three combined advantageous effects: cavitation, pressure, and thermodynamics. When ultrasonic waves propagate in the extraction liquid, ultrasonic vibration instantly causes the pressurization and decompression of the liquid, pushing the medium and generating cavitation. When countless small vacuum bubbles in the extraction liquid burst, high temperature and strong impact are instantly generated; therefore, the components of the extract can be separated to achieve an effective extraction efficiency. UE is a simple and inexpensive technology that can be easily applied to reduce solvent consumption, the extraction temperature, and time required to improve the extraction efficiency. Focused ultrasonic extraction (FUE) can promote and enhance biologically active ingredients from the plant in the extraction solvent. Therefore, the FUE method yields the highest total phenol content, total flavonoid content, total antioxidant capacity, and DPPH free radical scavenging capacity [9].

The plant extracts are mixed with encapsulates, such as maltodextrin, gum, or soy protein; they are directly sprayed under pressure to raise the surface area of the droplets using high-temperature hot air drying (over 100 °C), which is commonly used in the pharmaceutical industry. Most of the heat of hot air is used to enhance moisture evaporation, and the temperature of the dried powder can be controlled at 60~70 °C. The bitter gourd aqueous extract is mixed with encapsulate (maltodextrin and gum) as an encapsulating agent in 1:1.5, 1:2, and 1:3 ratios, and studied by spray drying [10–12]. The optimal inlet and outlet temperatures of spray drying have been determined to be 140 °C and 80 °C, respectively [11].

Moreover, most plant extracts are sterilized at a high temperature to avoid excessively high bacterial counts or mold counts prior to the final freeze-drying procedure. Traditional freeze drying uses a hot plate to conduct heat into the frozen food for ice sublimation. Freeze-dried encapsulated extracts feature a lower degradation and longer shelf life compared with spray-dried samples [13]. Although the quality of the freeze-dried product is good, the process is time consuming and energy consuming, which results in the manufacturing process being expensive.

Radio frequency (RF) waves are part of the electromagnetic spectrum. Three RF frequencies (13.56, 27.12, and 40.68 MHz) are allowed to be used in science, medicine, and industry by the US Federal Communications Commission (FCC) to avoid interference with other communication systems. The mechanisms of RF heating are due to the migration of ions and the rotation of polarized molecules, resulting in friction that generates heat. RF heating is an effective thermal treatment for various food and agricultural products used for different purposes, such as disinfestation, pasteurization, thawing, enzyme inactivation, and drying [14,15].

When RF waves pass through food, the water molecules in the food produce dipole friction to generate heat, causing the food to heat up quickly. RF energy leads to simultaneous water evaporation, pasteurization, and hot air-assisted water vapor transfer. Therefore, hot air-assisted radio frequency (HARF) drying can overcome the drawbacks of traditional hot air drying, which is both time and energy consuming [15]. Moreover, the developed RF vacuum drying method has been successfully applied for the drying of kiwi fruit for lower-temperature RF heating [16].

RF treatment has been increasingly studied in food drying [14] and pasteurization [17]. However, there remains a lack of studies on the use of HARF heating to dry the extracted solution directly. This may be due to the high dielectric loss factor of the extract solution, which induces an exceeding RF power and leads to an unstable process. The main barrier of HARF drying is the non-uniform heating, especially causing areas with large cold spots and overheating in parts of the high moisture materials, which results in undesirable qualities in the final products [18]. Therefore, a two-stage drying process was developed to overcome the difficulties of RF drying. For example, one study showed the moisture of mango slices was reduced from 88% to 40% by the first stage of hot air drying at 60 °C for 5 h, then the HARF drying was applied in the second stage of drying to further reduce the moisture content from 40% to 18% within 45 min. Therefore, the length of time needed for combination hot air and HARF drying is apparently lower than that of hot air drying (8 h) or vacuum drying (7 h). The overall quality of mango slices after the two-stage drying process was better than that of mango dried by hot air drying and close to that of mango dried by vacuum drying [17]. Furthermore, HARF drying has also been used in the final stage of drying in pre-dried carrot cubes (250 g) to reduce the moisture content from 40% to 10% using 4 h of drying [19].

Various encapsulating agents, such as polysaccharides, gum, lipids, and proteins, are added into the aqueous extract during spray and freeze drying [13]. Currently there is no research regarding drying aqueous extract using HARF drying. Therefore, soybean fiber powder was added into the extract solution as an encapsulating agent for reducing moisture content to solve the HARF drying problem. The objective of this study was to use HARF drying for the simultaneous drying and pasteurizing of the concentrated bitter gourd extract, and then to evaluate the hypoglycemic effect by adding 5% in the feed for type II diabetic mice.

2. Materials and Methods

2.1. Materials

Hot air dried wild bitter gourd was obtained from Asakusa Agriculture Processing Co. (Hualien, Taiwan). Bovine serum albumin (BSA), Coomassie brilliant blue G-250, 1,1-Diphenyl-2-picryl hydrazyl (DPPH), ascorbic acid (Vitamin C), glacial acetic acid synthetic and antioxidant butylated hydroxyl anisole (BHA) were bought from Sigma Chemical Co. (St. Louis, MO, USA). Ginsenoside Rg1 was bought from ChromaDex, Corp. (Los Angeles, CA, USA). Aerobic count plate (3M Petrifilm 6400), yeast, and mold count plates (3M Petrifilm 6477, 500 EA/CS), and papain (2000 FCCU/mg, Decken Biotech, Taichung, Taiwan) were acquired. Soybean fiber powder was bought from Prime Creative International CO., LTD (Hsinchu, Taiwan). Experimental male mice (BALB/c strain), MFG feed, and BETA chip were purchased from BioLASCO Taiwan Co. Ltd (Taipei, Taiwan). Streptozocin (STZ) and Nicotinamide were purchased from Sigma-Aldrich® (St. Louis, MO, USA)

2.2. Equipment

A focused ultrasonic extractor (20k Hz, 1400 W, Ever Great Ultrasonic Co., New Taipei City, Taiwan), spectrophotometer (Model U-2001, Hitachi Co., Tokyo, Japan), benchtop centrifuge (HERMLE Z300, Gosheim, Germany), mini-protein system (Bio-Rad, CA, USA), RF with hot air equipment (40.68 MHz, 10 kW, Yh-Da Biotech Co., LTD., Yilan, Taiwan), oven (Channel DCM-45, Yilan, Taiwan), digital pocket refractometer (Pocket, 3810, PAL-1, ATAGO Corp., Tokyo, Japan), and multifunctional infrared thermometer (Testo104-IR, Hot Instruments Co., LTD., New Taipei, Taiwan) were employed.

2.3. Focused Ultrasonic Extraction (FAE) and Hot Water Extraction (HWE) of Wild Bitter Gourd

The wild bitter gourd was ground using a 60 mesh, and 50 g of the ground wild bitter gourd was mixed with 500 mL of reverse osmosis (RO) water. Samples were extracted from the solutions at 20 kHz, 1400 W, and 70 °C for 10 min by FAE or using 100 °C hot water extraction for 1 h.

The solid content of the extract was converted from the measurement of °Brix by an analogue refractometer. The extract was centrifuged at 6000 rpm for 5 min and stored at 4 °C for analysis.

2.4. Hot Air-Assisted Radio Frequency (HARF) Drying Extract from Wild Bitter Gourd

The extract obtained by FUE from wild bitter gourd was concentrated six times under vacuum conditions to raise the solid content from 4°Brix to 24°Brix before HARF drying for time-saving, energy-saving, and high-efficiency purposes.

In this study, soybean fiber powder was used as encapsulate and mixed with extracts from wild bitter gourd to overcome the issue of the sample being too wet. The samples were prepared with 1 kg of wild gourd concentrated extract were homogeneously mixed with soybean fiber powder in three ratios of (A) 2:1, (B) 2:1.1, and (C) 2:1.2, respectively. Then, they were put into polypropylene (PP) plastic containers (31.3 × 6.9 × 20.5 cm³) and treated at a HARF equipment (Figure 1). HARF drying was carried out at 10 cm gap between the parallel electrode plates and 100 °C hot air. The surface temperatures of the samples were measured at three points (at the center and 5 cm on either side of the center), and the weights by balance were measured at the time interval of 30 s to obtain the temperature profile and drying curve.

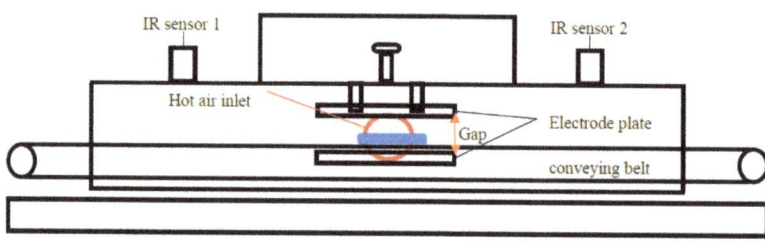

Figure 1. Schematic view of 10 kW, 40.68 MHz continuous radio frequency (RF) (10 kW, 40.68 MHz) with hot air heating system.

2.5. Analytical Methods

2.5.1. Protein Content Measurement

Testing was conducted using the Bradford protein-binding assay by mixing 800 µL of wild bitter gourd extract with 200 µL of protein assay reagent. The solution was left idle for 10 min of reaction time. The absorbance at the 595 nm wavelength was measured three times, and the mean value was calculated. The BSA standard curve was employed to determine the protein concentration in the sample.

2.5.2. Total Saponins Measurement

The saponins of the sample was measured by vanillin-perchloric acid colorimetry, and this experiment was modified from the method of [20]. The 50 mL extract was mixed well with 70% ethanol for 10 min of reaction. Then, it was centrifuged (6000 rpm, 5 min) to obtain 0.2 mL of the supernatant, into which 0.20 mL of 5% vanillin-glacial acetic acid solution and 0.80 mL of perchloric acid were then added, and the mixture was heated in a 60 °C water bath for 15 min. When it was cooled, 5 mL of glacial acetic acid was added for 20 min reaction and measured the absorbance at a wavelength of 548 nm three times. The mean value was calculated. The ginsenoside (Rg1) standard curve was employed to determine the total saponins concentration in the sample.

2.5.3. DPPH Free Radical Scavenging Ability Test

For this test, 2 mL of the supernatant was evenly mixed with 2 mL of 0.2 mM DPPH MeOH solution and left idle away from light at room temperature for 30 min. The absorbance at the 517 nm wavelength was measured and applied to the following equation to calculate the DPPH scavenging ability. [21] The result was compared with that for the control group comprising 5 mg/mL of ascorbic acid and BHA.

2.5.4. Microbiological Test

The extracts (1 mL) were each diluted 10, 100, and 1000 times with sterilized water. Subsequently, 0.5 mL of each diluted extract was cultured at 35 °C for 72 h and counted on an aerobic count plate for a total bacteria test, yeast count, and mold count. The results were observed and the bacterial colonies were counted.

2.6. Hypoglycemic Effect of Wild Bitter Gourd Extract on STZ-Induced Mice

2.6.1. Experimental Animal and Raising Environment

Twenty-four 6-week-old BALB/c male mice (BioLASCO Taiwan Co., Ltd., Yilan, Taiwan) were quarantined for 1 week and acclimated in polyethylene cages at the experimental animal room of National Ilan University, where the temperature was 22 ± 2 °C, the relative humidity was $60 \pm 5\%$, and the light cycle was 12 h light/12 h dark. They were fed with free water and MFG feed. The wood shavings (Beta Chips) and shredded aspen shavings were purchased from BioLASCO Taiwan Co. Ltd. and changed every 3 d. The body weights of all mice were measured before the experiment, and weekly thereafter. The fasting blood glucose of all mice was measured by ACCU-CHEK active blood glucose meter after one week of adaptation to being randomly assigned to control and treatment groups (8 mice in each group) based on body weight.

2.6.2. Preparation of Feed 5% Bitter Gourd Extract Powder

The concentrated concentrate of bitter gourd extract was mixed with soybean fiber powder in the ratio of 2:1.1, and the bitter gourd extract was dried by RF energy, added to the original mouse feed and crushed, and a little sterile water was added to make a thick shape and then mixed homogeneously; and after cold air drying, the mouse feed containing 5% of the bitter gourd extract powder was made.

2.6.3. Regulated Blood Glucose Function Evaluation Method

According to the evaluation of blood sugar regulation method by the Ministry of Health and Welfare, Executive Yuan. Except for the normal group, all mice were intraperitoneally injected with Nicotinamide (230 mg/kg) and then STZ (75 mg/kg) for two weeks. Then, the experimental mice were divided into three groups of eight mice each (normal, STZ, and STZ-fed extract). The STZ mice were fed a diet containing 5% of dried extract powder from wild bitter gourd bitter, and the normal diet mice and STZ mice were used as control groups. Because each group of eight mice was not fed one mouse in one cage, it was difficult to obtain the accurate data on water and feed intake of each mouse, and the data are not shown here. During the test period, the changes in body weight of each mouse were recorded, and fasting blood glucose values were measured at the initial week and final fourth week to provide a basis for evaluation of the hypoglycemic effect.

2.7. Statistical Analysis

The test results were expressed as mean $\pm$ SD and analyzed using the Statistical Package for Social Science 14.0 (SPSS Inc., Data Statistical Analysis Corporation, Chicago, IL, USA). Differences between the data were examined through one-way analysis of variance, and the significance of the differences was investigated through Duncan's multiple range test ($\alpha = 0.05$).

3. Results and Discussion

3.1. Focused Ultrasonic Extraction (FUE) of Wild Bitter Gourd

Table 1 shows the extraction qualities of the wild bitter gourd powder mixed with water at a solid to liquid ratio of 1:10. The extraction ratios of the extract from ultrasonic extraction at 70 °C for 10 min and conventional hot water extraction at 100 °C for 60 min were 22.4% and 21%, respectively. The contents of total saponins and total proteins extracted by hot water were higher than those of the extract by ultrasonic extraction. However, the scavenge DPPH free radicals extract by ultrasonic was 79.2% and was significantly higher than 59.6% of hot water extract.

Table 1. Effect of focused ultrasonic extraction (FUE) and hot water extraction (HWE) on yield, active components, and antioxidant activity of wild bitter gourd.

Extraction Method	FUE	HWE
Extraction temperature (°C)	70	100
Extraction time (min)	10	60
Extraction yield (%)	22.4 ± 0.1 *	21.0 ± 0.0
Total saponins (mg/g)	0.288 ± 0.002	0.296 ± 0.022 *
Total proteins (mg/g)	1.198 ± 0.025	1.302 ± 0.032 *
Scavenging DPPH ability (%)	79.2 ± 0.1 *	59.6 ± 0.1

Data are expressed as mean ± S.D. (n = 3). Means with * in the same row were significantly different ($p < 0.05$). Scavenging DPPH ability (%) of 10 mg/mL Vitamin C and BHA were 95.5 ± 0.1 and 95.3 ± 0.1, respectively.

Garude et al. [9] contended that the phenol and flavonoid contents in bitter gourd peel extracts by ultrasonic extraction were higher than those in extracts by hot water extraction. The ultrasonic water extract of fresh bitter gourd exhibited a strong antioxidant ability of scavenging DPPH free radicals and a higher inhibition of α-amylase and α-glycosidase activity [22].

Therefore, traditional high-temperature long-duration extraction degraded the antioxidant properties of the bioactive components; however, short-duration and low-temperature ultrasonic extraction could avoid damaging the bioactive components and antioxidant properties of the extract. Compared with traditional extraction, FUE took only 10 min; therefore, it could also achieve the time-saving and energy-saving effects. Considering the cost of extraction, previous studies have recommended that the optimal conditions for the focused ultrasonic extraction of wild bitter gourd powder are mixed with water in a solid–liquid ratio of 1:10 at 70 °C for 10 min to obtain 6°Brix. The extraction yield has been shown to be higher than that of bitter gourd aqueous mixture made using a shaking water bath with a solid–liquid ratio of 1:20 at 40 °C water for 15 min, where only a 2% total solid content was obtained [10].

3.2. Hot Air-Assisted Radio Frequency (HARF) Drying of Extract from Wild Bitter Gourds

Therefore, the moisture content decreased with the increasing level of additional soybean fiber powder. Figure 2 shows the temperature profiles and drying curves of these three samples during 10 kW radio frequency (RF electrode gap of 10 cm) combined with 100 °C hot air drying. With the same loading of 1 kg of sample, the drying time was reduced from 14 to 10 min as the ratio of mixed soybean fiber powder increased. In addition, the surface temperature of the sample quickly reached above 70 °C during the first 4 min of RF heating, then the temperature of sample gradually increased to 80 °C, as most RF-induced heat was absorbed for latent heat of water evaporation.

Table 2 shows that the period of constant drying rate appeared during the RF drying. Increasing the soybean fiber powder addition from (A) 2:1 to (B) 2:1.1 and (C) 2:1.2 caused higher drying rates of 36.881, 43.836, and 53.968 g/min, respectively. The moisture contents of the final products were 14.2%, 16.7%, and 12%, respectively. The HARF drying time may be extended by 1 min more to further reduce the moisture content for storage. The level of additional soybean fiber powder in the original extract solution was about 33~37.5% in HARF drying. The encapsulate (maltodextrin and gum powder 1:1) addition in the bitter gourd aqueous extract was about 20~25% [11,12] by spray drying with inlet and outlet temperatures of 140 °C and 80 °C, respectively [10]. In this study, the final temperatures of samples were also near 80 °C, and the drying time was much less than that of spray drying and freeze drying, because the feed rate of spray drying was only 10 mL/min [10] and freeze drying required at least 24 h.

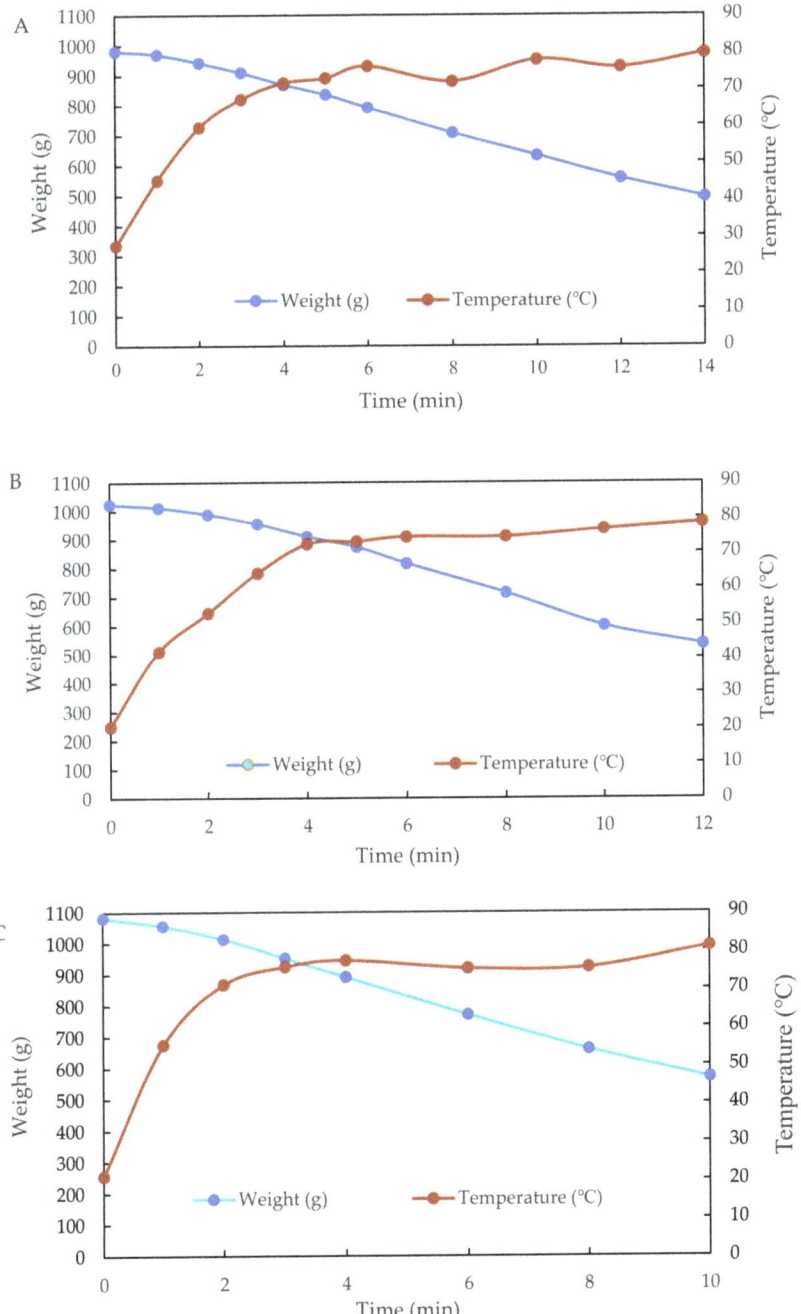

Figure 2. Temperature profiles and drying curves of wild bitter gourd extract mixed with soybean fiber: (**A**) 2:1, (**B**) 2:1.1, and (**C**) 2:1.2 during HARF heating.

Table 2. Drying rate of wild bitter gourd extract (E) mixed with soybean fiber powder (F) by 10 kW radio frequency heating.

E:F	Loading (kg)	Linear Regression Equation	R^2	Rate (g/min)
2:1	1.0	W = −36.881t + 1007.8	0.996	36.881
2:1.1	1.0	W = −43.836t + 1066.4	0.984	43.836
2:1.2	1.0	W = −53.968t + 1103.2	0.995	53.968

W is weight of sample (g), and t is drying time (min) in linear regression equation.

3.3. HARF Pasteurizing of Wild Bitter Gourd Extract

The analyses of total bacterial counts are shown in Table 3. The original total bacterial counts for ultrasonic extract from the wild bitter gourd and soybean fiber powder were 523,000 CFU/mL and 9400 CFU/g, respectively. However, after HARF drying, the counts became only 30 CFU/mL in the 2:1.1 ratio of mixture of wild bitter gourd extract to soybean fiber. Both the mold and yeast counts were significantly reduced after the HARF drying. It was found that RF induced heating for the samples to achieve a pasteurization effect (above 70 °C) within the first 4 min. Then, the rest time of HARF heating could keep the temperature in a range of 70~80 °C due to the induced heat absorbed as latent heat for water evaporation. The moisture contents of the samples were also reduced from 58% to 15%, and the mold and yeast counts were significantly reduced. Therefore, HARF heating could proceed simultaneously for the drying and pasteurizing of the extracts.

Table 3. Total counts and mold and yeast counts for wild bitter gourd extract, soybean fiber powder, and RF product.

Wild Bitter Gourd Sample	Extract from Wild Bitter Gourd	Soybean Fiber Powder	RF-Dried Extract Fiber Product
		Total bacterial colonies	
CFU/mL	523,000	9400	30
Picture (1:10)			
		Mold and yeast counts	
CFU/mL	>1000	30	<10
Picture (1:10)			

3.4. Hypoglycemic Effect of Radio Frequency Dried Wild Bitter Gourd Extract

The average water intake of STZ mice was 2.8 g at week 0, which was lower than that of normal mice (3.8 g); there was no significant difference between the three groups after one week of feeding. Furthermore, the average feed intake of STZ mice was 3 g at week 0, which was slightly lower than the 3.2 g of normal mice. Similarly, there was no significant difference in average feed intake of about 3.2 g between the three groups after one week of feeding. The STZ mice had litter less body weight, especially in the STZ-fed extract group (Table 4) and the higher blood glucose 105~107 mg/dL than normal mice (74.88 mg/dL)

(Table 5). The mice were injected with nicotinamide and STZ to induce them to become type II diabetic mice, which had higher blood glucose levels at beginning. However, it did not completely destroy the pancreas; the consequences would thus be minimal.

Table 4. The changes of body weight (g) in mice for feeding RF-dried bitter gourd extract.

Group	0 Week	1 Week	2 Weeks	3 Weeks	4 Weeks
Normal	23.05 ± 1.64 [b]	23.69 ± 1.65 [a]	24.03 ± 1.63 [a]	24.61 ± 1.21 [a]	24.95 ± 1.57 [a]
STZ	22.58 ± 0.88 [ab]	22.94 ± 1.27 [a]	23.67 ± 1.38 [a]	24.01 ± 1.45 [a]	24.78 ± 1.32 [a]
STZ + Extract	21.69 ± 0.93 [a]	23.66 ± 2.09 [a]	24.16 ± 2.49 [a]	24.90 ± 2.30 [a]	25.35 ± 2.06 [a]

Data are expressed as mean ± S.D. ($n = 8$). 0 week: after 2 weeks injecting nicotinamide and STZ; 1~4 weeks: free feeding for 1~4 weeks. [a,b] Means at different times with different superscript letter in the same column were significantly different ($p < 0.05$).

Table 5. The changes of fasting blood glucose (mg/dL) in mice for feeding RF-dried bitter gourd extract.

Group	0 Week	4 Weeks
Normal	74.88 ± 6.20 [b]	90.00 ± 6.50 [c]
STZ	107.00 ± 6.21 [a]	115.13 ± 5.69 [a]
STZ + Extract	105.63 ± 4.34 [a]	105.38 ± 4.07 [b]

Data are expressed as mean ± S.D. ($n = 8$). 0 week: after 2 weeks of injection of nicotinamide and STZ; 4 weeks: free feeding for 4 weeks. [a–c] Means at different times with different superscript letter in the same column were significantly different ($p < 0.05$).

After four weeks of treatment, the average body weight of these three groups of mice increased to 24.78~25.35 g with no significant difference ($p > 0.05$). However, after one week of feeding with RF-dried bitter gourd extract, the average body weight of the STZ-fed extract group mice was the same as that of the normal group. The average body weight of the STZ-fed extract group mice for the next three weeks was even slightly higher than that of the normal group ($p > 0.05$) (Table 4).

After four weeks of feeding, the blood glucose of mice in the normal group was 90 mg/dL as the feeding time increased, while the blood glucose of mice in the STZ group increased from 107 mg/dL to 115 mg/dL, but the blood glucose of mice in the STZ-fed extract group did not increase and remained at 105 mg/dL. Although the 5% RF-dried extracts from the STZ mice did not reach the blood glucose level of normal mice, they could control the increase in blood glucose. This suggests that the dried powder of ultrasonic extracts by HARF had a hypoglycemic effect.

The MCP was isolated from the water-soluble polysaccharide form of the bitter gourd and was orally administered once a day though water after 3 d of alloxan-induction type II diabetic mice at 300 mg/kg body weight for 28 d, exhibiting a hypoglycemic effect. Hence, MCP can be incorporated as a supplement in health care food, drugs, and/or combined with other hypoglycemic medicines [23].

4. Conclusions

In this study, wild bitter gourd powder at a solid–liquid ratio of 1:10 was extracted with ultrasonication for 10 min with hot water at 70 °C. The total protein and saponin contents were 1.198 mg/g and 0.288 mg/g, respectively. The ability of these samples to scavenge DPPH free radicals was better than that of the 100 °C hot water extracts. HARF for drying the mixture of 1 kg of concentrated wild bitter gourd extract and soybean fiber powder (at a ratio of 2:1.1) took only 4 min to achieve the pasteurization effect due to the temperature being above 70 °C, and the moisture content of the sample was reduced from 58% to 15% by a total of 12 min of HARF heating. This indicates that using ultrasonic extraction and HARF drying to prepare extracts for the herb and pharmaceutical industries can save time and energy. Type II diabetic mice were fed with a free diet containing 5% RF-dried wild bitter gourd extract for four weeks and their blood sugar levels did not rise.

As a result, the dried powder of ultrasonic wild bitter gourd extracts by HARF heating also had a hypoglycemic effect.

Author Contributions: S.-D.C.: supervision and project administration; C.-Y.H.: preparation of extracted sample, data analysis, and writing of original draft; Y.-H.C.: assisting with the animal experiment. All authors have read and agreed to the published version of the manuscript.

Funding: This research received no external funding.

Institutional Review Board Statement: Not applicable.

Informed Consent Statement: This experiment was reviewed and approved by the Laboratory Animal Care and Use Team of National Ilan University, project number 109-5.

Data Availability Statement: The data and samples presented in this study are available on request from the corresponding author. Data is contained within the article.

Acknowledgments: We thank the Department of Biotechnology and Animal Science, National Ilan University, for providing animal housing for mice experiments. We thank Ying-Chen Yang for her assistance in the animal experiment. We thank Ching-Tung Kuo for checking the English manuscript draft.

Conflicts of Interest: The authors declare no conflict of interest.

References

1. Oishi, Y.; Sakamoto, T.; Udagawa, H. Inhibition of increases in blood glucose and serum neutral fat by *Momordica charantia* saponin fraction. *Biosci. Biotechnol. Biochem.* **2007**, *71*, 735–740. [CrossRef] [PubMed]
2. Zhu, Y.; Dong, Y.; Qian, X. Effect of superfine grinding on antidiabetic activity of bitter melon powder. *Int. J. Mol. Sci.* **2012**, *13*, 14203–14218. [CrossRef] [PubMed]
3. Wang, Q.; Wu, X.; Shi, F.; Liu, Y. Comparison of antidiabetic effects of saponins and polysaccharides from *Momordica charantia* L. in STZ-induced type 2 diabetic mice. *Biomed. Pharmacother.* **2019**, *109*, 744–750. [CrossRef] [PubMed]
4. Lo, H.; Ho, T.; Lin, C. *Momordica charantia* and its novel polypeptide regulate glucose homeostasis in mice via binding to insulin receptor. *J. Agric. Food Chem.* **2013**, *63*, 2461–2468. [CrossRef] [PubMed]
5. Ahmad, Z.; Zamhuri, K.F.; Yaacob, A. In vitro anti-diabetic activities and chemical analysis of polyepetide-k and oil isolated from seeds of *Momordica charantia* (bitter gourd). *Molecules* **2012**, *17*, 9631–9640. [CrossRef]
6. Cai, Z.; Qu, Z.; Lan, Y.; Zhao, S.; Ma, X.; Wan, Q.; Jing, P.; Li, P. Conventional, ultrasound-assisted, and accelerated-solvent extractions of anthocyanins from purple sweet potatoes. *Food Chem.* **2016**, *197*, 266–272. [CrossRef]
7. Gil-Chávez, G.J.; Villa, J.A.; Fernando Ayala-Zavala, J.; Basilio Heredia, J.; Sepulveda, D.; Yahia, E.M.; González-Aguilar, G.A. Technologies for extraction and production of bioactive compounds to be used as nutraceuticals and food ingredients: An overview. *Compr. Rev. Food Sci. Food Saf.* **2013**, *12*, 5–23. [CrossRef]
8. Caldas, T.W.; Mazza, K.E.; Teles, A.S.; Mattos, G.N.; Brígida, A.I.S.; Conte-Junior, C.A.; Borguini, R.G.; Godoy, R.L.; Cabral, L.M.; Tonon, R.V. Phenolic compounds recovery from grape skin using conventional and non-conventional extraction methods. *Ind. Crops Prod.* **2018**, *111*, 86–91. [CrossRef]
9. Garude, H.S.; Ade, V.R.; Gadhave, R.K. Comparative study of various methods for extraction of antioxidants compound from bitter gourd peel. *Adv. Life Sci.* **2016**, *5*, 11189–11192.
10. Tan, S.P.; Kha, T.C.; Parks, S.; Stathopoulos, C.; Roach, P.D. Optimising the encapsulation of an aqueous bitter melon extract by spray-drying. *Foods* **2015**, *4*, 400–419. [CrossRef]
11. Tan, S.P.; Tuyen, C.K.; Parks, S.E.; Stathopoulos, C.E.; Roach, P.D. Effects of the spray-drying temperatures on the physiochemical properties of an encapsulated bitter melon aqueous extract powder. *Powder Technol.* **2015**, *281*, 65–75. [CrossRef]
12. Raj, N.; Priya, B. Encapsulation of bitter gourd (*Momordica charantia* L.) extract by spray drying technique. *Biosci. Biotechnol. Res. Asia* **2016**, *13*, 1189–1193. [CrossRef]
13. Antigo, J.L.D.; Bergamasco, R.D.C.; Madrona, G.S. Effect of pH on the stability of red beet extract (*Beta vulgaris* L.) microcapsules produced by spray drying or freeze drying. *Food Sci. Technol.* **2018**, *38*, 72–77. [CrossRef]
14. Zhou, X.; Wang, S. Recent developments in radio frequency drying of food and agricultural products: A review. *Dry. Technol.* **2019**, *37*, 271–286. [CrossRef]
15. Marra, F.; Zhang, L.; Lyng, J.G. Radio frequency treatment of food: Review of recent advances. *J. Food Eng.* **2009**, *91*, 497–508. [CrossRef]
16. Zhou, X.; Ramaswamy, H.; Qu, Y.; Xu, R.; Wang, S. Combined radio frequency-vacuum and hot air drying of kiwifruits: Effect on drying uniformity, energy efficiency and product quality. *Innov. Food Sci. Emerg. Technol.* **2019**, *56*, 102182. [CrossRef]
17. Zhang, L.; Lan, R.; Zhang, B.; Erdogdu, F.; Wang, S. A comprehensive review on recent developments of radio frequency treatment for pasteurizing agricultural products. *Crit. Rev. Food Sci. Nutr.* **2021**, *61*, 380–394. [CrossRef]

18. Zhu, H.; Li, D.; Ma, J.; Du, Z.; Li, P.; Li, S.; Wang, S. Radio frequency heating uniformity evaluation for mid-high moisture food treated with cylindrical electromagnetic wave conductors. *Innov. Food Sci. Emerg. Technol.* **2018**, *47*, 56–70. [CrossRef]
19. Gong, C.; Liao, M.; Zhang, H.; Xu, Y.; Miao, Y.; Jiao, S. Investigation of hot air–assisted radio frequency as a final-stage drying of pre-dried carrot cubes. *Food Bioproc. Technol.* **2020**, *13*, 419–429. [CrossRef]
20. Li, J.; Huang, Y.; Liu, N.; Chen, S.; Liu, X. Study on stability of momordicosides. *Food Sci.* **2008**, *10*, 109–111. (In Chinese)
21. Xu, B.J.; Chang, S.K.C. A comparative study on phenolic profiles and antioxidant activities of legumes as affected by extraction solvents. *J. Food Sci.* **2007**, *72*, 159–166. [CrossRef] [PubMed]
22. Yan, J.K.; Yu, Y.B.; Wang, C.; Cai, W.D.; Wu, L.X.; Yang, Y.; Zhang, H.N. Production, physicochemical characteristics, and in vitro biological activities of polysaccharides obtained from fresh bitter gourd (*Momordica charantia* L.) via room temperature extraction techniques. *Food Chem.* **2021**, *337*, 127798. [CrossRef] [PubMed]
23. Xu, X.; Shan, B.; Liao, C.H.; Xie, J.H.; Wen, P.W.; Shi, J.Y. Anti-diabetic properties of *Momordica charantia* L. polysaccharide in alloxan-induced diabetic mice. *Int. J. Biol. Macromol.* **2015**, *81*, 538–543. [CrossRef] [PubMed]

Article

Radio Frequency Drying Behavior in Porous Media: A Case Study of Potato Cube with Computer Modeling

Xiangqing Chen [1,2], Yu Liu [1,2], Ruyi Zhang [1,2], Huacheng Zhu [3], Feng Li [1,2], Deyong Yang [4] and Yang Jiao [1,2,*]

1. College of Food Science and Technology, Shanghai Ocean University, Shanghai 201306, China
2. Engineering Research Center of Food Thermal-Processing Technology, Shanghai Ocean University, Shanghai 201306, China
3. College of Electronics and Information Engineering, Sichuan University, Chengdu 610065, China
4. College of Engineering, China Agricultural University, Beijing 100083, China
* Correspondence: yjiao@shou.edu.cn

Abstract: To study the mechanism of heat and mass transfer in porous food material and explore its coupling effect in radio frequency (RF) drying processes, experiments were conducted with potato cubes subjected to RF drying. COMSOL Multiphysics® package was used to establish a numerical model to simulate the heat and mass transfer process in the potato cube and solved with finite element method. Temperature history at the sample center and the heating pattern after drying was validated with experiment in a 27.12 MHz RF heating system. Results showed the simulation results were in agreement with experiments. Furthermore, the temperature distribution and water vapor concentration distribution were correspondent with water distribution in the sample after RF drying. The water concentration within the food volume was non-uniform with a higher water concentration than the corner, the maximum difference of which was 0.03 g·cm^{-3}. The distribution of water vapor concentration in the sample was similar to that of water content distribution since a pressure gradient from center to corner allowed the mass transfer from the sample to the surrounding in the drying process. In general, the moisture distribution in the sample affected the temperature and water vapor concentration distribution since the dielectric properties of the sample were mainly dependent on its moisture content during a drying process. This study reveals the mechanism of RF drying of porous media and provides an effective approach for analyzing and optimizing the RF drying process.

Keywords: radio frequency drying; finite element method; heat transfer; mass transfer; water vapor concentration

1. Introduction

Drying is one of the oldest food-preservation processes, which reduces water content and water activity to prolong food shelf-life [1]. Hot air drying is the most popular drying method because of its wide applicability and low cost. However, one of its disadvantages is that heat transfer is from the outer surface to the center of sample, which is in the opposite direction of mass transfer. This usually results in surface hardening, lowering the mass transfer rate and product quality [2]. Novel drying technologies are still needed due to the consumers' demand for a high-quality food product, which retains most of its original color, flavor, and nutrients, with a higher rehydration rate, better appearance, and moisture uniformity. Moreover, food manufacturers are continuously searching for high-efficiency and energy-saving drying technologies and equipment [3].

Recently, radio frequency (RF)-assisted vacuum drying, heat pump drying, and hot air drying were explored and showed a significant effect [4–6]. Previous studies have demonstrated that the application of RF heating technology in drying could shorten drying time and maintain a good quality of products. Wang et al. [7] studied the effects of carrot slices in hot-air-assisted RF drying. Results showed hot-air-assisted RF drying reduced 30% of the drying time compared to solely hot air drying and maintained the good quality of carrot slices. Zhang et al. [8] adopted a two-stage drying method to treat mango slices, which utilized hot air to dry mango slices to 40% (w.b.) in the first stage and then used hot-air-assisted RF heating (HA-RF) to continue to dry the product to 18% (w.b.). Results demonstrated that the developed two-stage drying process took less time (5 h) compared to vacuum drying (7 h) and hot air drying (8 h) and maintained good product quality.

Various factors influence the electric field distribution and the temperature distribution in the product during RF drying, including the shape and size of the sample, electrode gaps, and the properties of materials, etc. [9,10]. The uneven distribution of temperature significantly affects the moisture distribution and ultimately affects the product quality [11]. In RF drying experiments, it is difficult to visualize the effects of different factors on the interior of the product, such as electric field, temperature, water concentration, and water vapor concentration [12]. In addition, experimental approaches have some disadvantages, such as long cycles, high economic cost, low efficiency, and operating difficulties. Compared to the experimental method, computer simulation is an economical, flexible, and intuitive way to reveal the invisible parameters of products in a complex drying process. Computer simulation has already been proven to be effective in predicting and optimizing parameters in RF heating processes [13]. Therefore, the establishment of a mathematical model of RF drying performance of porous material based on multiphase simultaneous transportation is of great significance to simulate the RF drying performance on food products.

The RF drying process couples heat, mass, and electromagnetic energy conversion to heat. Additionally, many foods are plant-based, with intracellular and intercellular water, which adds complexity to the system description. Analyzing the combined effects would assist in fully understanding the RF drying mechanism and facilitating the drying process design. Jia et al. [14] developed a one-dimensional mathematical model to describe the transport phenomena of wood during continuous RF combined with vacuum drying. Their study suggested that the dry zone traveled from the end to the center of the wood and the temperature within the wood decreased with water loss. Hou et al. [15] developed a 3D multiphase porous model to simulate heat and mass transfer of kiwifruit slices under RF–vacuum drying with COMSOL Multiphysics®. The results of the simulation agreed well with those of the experiment. The sample temperature at the center was below that at the corners and edges, and the lowest temperature was at the center for a single kiwifruit slice. However, the distribution of moisture content was opposite to that of temperature. The drying rate increased rapidly and then decreased slowly until the end. However, there has not been a 3D numerical model established to comprehensively visualize the distribution of electromagnetic field, moisture, vapor concentration, and temperature within food samples during RF drying at atmospheric pressure. To promote the application of RF drying technology in the food-processing industry, there is a significant need to establish a 3D multi-phase porous model to assist food-processing protocol development. In this study, a mathematical model was established with COMSOL Multiphysics® software package to analyze the temperature, moisture, and vapor concentration distribution within a potato cube as a model food throughout the RF drying process.

The specific objectives of this study were: (1) to develop and solve a fundamental-based mathematical model for an RF drying process with an example sample of potato cube; (2) to validate the model with an experiment by comparing the temperature–time history, surface temperature distribution, and water content with modeling results; and (3) to predict the electromagnetic field intensity, moisture distribution, vapor concentration, and temperature distribution within samples during RF drying.

2. Materials and Methods

2.1. Model Development

Basic assumptions:
Within all the domains in the model:
(1). All three phases (gas, liquid, solid) in the food sample were continuous media;
(2). All gases were ideal gases;

(3). All fluid, including gas and liquid, were with the same pressure;
(4). Evaporation and condensation occurred within the whole food matrix rather than only on the surface;
(5). Volume shrinkage was not considered; i.e., the food porosity was set as a constant;
(6). Gravity was neglected.

Definition of components in a porous medium:

In a potato cube sample, the internal structure was described as a porous medium as shown in Figure 1, with solid (s), water (w), and gas (g) phases. The gas phase include both water vapor (v) and air (a), which were both fluid (f). The volumetric fraction of solid and fluid in the sample was expressed as [16]:

$$\Delta V = \Delta V_s + \Delta V_f \tag{1}$$

where V is the volume of the potato sample (m³), V_s is the volume of solid (m³), and V_f is the volume of fluid (m³). The porosity of the porous medium can be defined as the volume fraction occupied by fluid, which was expressed as [17]:

$$\varnothing = \frac{\Delta V_f}{\Delta V} = \frac{\Delta V_w + \Delta V_g}{\Delta V} \tag{2}$$

where $\varnothing$ is the porosity (–). The volumetric concentration of different phases is related to its saturation, which was expressed as [16]:

$$S_i = \frac{\Delta V_i}{\Delta V_p} = \frac{\Delta V_i}{\varnothing \Delta V} \quad i = w, g \tag{3}$$

where S_i is the saturation ratio (–), and V_p is the volume of the pore (m³).

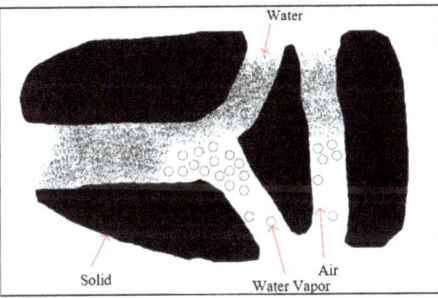

Figure 1. Structural scheme of porous media.

2.1.1. Governing Equations

Maxwell equations are the fundamental base of electromagnetic wave propagation and electromagnetic energy conversion. Since the magnetic properties in foods are usually negligible, and the energy conversion from electromagnetic energy to heat is much faster than the speed of heat transfer, Laplace's equation can be used to describe the energy conversion in RF heating [18]:

$$-\nabla \cdot ((\sigma + j2\pi f \varepsilon_0 \varepsilon') \nabla V) = 0 \tag{4}$$

where σ is the electrical conductivity of the food material (S·m⁻¹); j is the imaginary part of the square root of -1; $j = \sqrt{-1}$; ε_0 is the permittivity of electromagnetic waves in free space, which is 8.854×10^{-12} F·m⁻¹; ε' is the relative dielectric constant of the food material (–); and V is the electric potential across the electrode gap (V).

The power conversion from electromagnetic energy to heat is described by the following equation [19]:

$$Q = 2\pi f \varepsilon_0 \varepsilon'' \left|\vec{E}\right|^2 \tag{5}$$

where Q is the electromagnetic power conversion in the sample to be heated (food) per unit volume (W·m^{-3}), f is the working frequency of the RF equipment (Hz), ε'' is the relative loss factor of the food material (–), and $\vec{E}$ is the electric field intensity in the food material (V·m^{-1}).

2.1.2. Mass Transfer

The concentration of liquid water, water vapor, and gas can be estimated with the microscopic mass balance of different liquid phases as [16]:

$$\frac{\partial c_i}{\partial t} + \nabla \overline{n}_i = -\dot{I} \tag{6}$$

$$i = w, g, v$$

where c is the concentration of component (mol·m^{-3}), n is the mass flux per unit volume (mol·m^{-3}·s^{-1}), and $\dot{I}$ is the rate of evaporation or condensation due to phase change (mol·m^{-3}·s^{-1}). The liquid water in the pores of food has capillary force, which is in an opposite direction to the internal pressure and hinders the water migration. Thus, the net pressure P_w on the liquid water can be expressed as [16]:

$$P_w = P - P_c \tag{7}$$

where P_w is the capillary pressure (Pa), P is vapor pressure (Pa), and P_c is net pressure (Pa). Water flux ($\overline{n}_w$) is expressed as follows based on Darcy's law [16]:

$$\overline{n}_w = -\rho_w \frac{k_{in,w} k_{r,w}}{\mu_w} \nabla P_w = \rho_w \frac{k_{in,w} k_{r,w}}{\mu_w} \nabla P + D_{w,cap} \nabla c_w \tag{8}$$

where $D_{w,cap}$ is the capillary diffusion rate (m^3·s^{-1}), $D_{w,cap} = \rho_w \frac{k_{in,w} k_{r,w}}{\mu_w} \cdot \frac{\partial P_c}{\partial c_w}$, k_{in} is the intrinsic permeability (m^2), and k_r is the relative permeability. The gas flux ($\overline{n}_g$), including air and water vapor, can be obtained by Darcy's law [16]:

$$\overline{n}_g = -\rho_g \frac{k_{in,g} k_{r,g}}{\mu_g} \nabla P \tag{9}$$

The mass flux of water vapor ($\overline{n}_v$) is dominated by both the pressure-driven mass flow and the diffusivity of the air at the boundary [20]:

$$\overline{n}_v = -\rho_v \overline{v}_g - \left(\frac{C_g^2}{\rho_g}\right) M_v M_a D_{bin} \nabla x_v \tag{10}$$

where M is the molecular mass, and C_g is the mass fraction.

The mass fraction of water vapor can be obtained from water vapor mass conservation and converted to water vapor concentration (c_v). Thus, the gas concentration (c_g) can be determined with the following equations [16]:

$$c_v = w_v c_g$$

$$w_a = 1 - w_v$$

$$c_a = w_a c_g \tag{11}$$

The phase changes in a drying process, including the water evaporation and condensation, due to the high pressure generation within the food matrix during drying can be represented as [21]:

$$\dot{I} = K_{evap} \frac{M_v}{RT}(P_{v,eq} - P_v) S_g \varnothing \tag{12}$$

where $P_{v,eq}$ is the equilibrium pressure of water vapor in food (Pa), P_v is the calculated instant water vapor pressure from Darcy's law (Pa), M_v is molecular mass of water vapor, R is universal gas constant (kJ·kmol^{-1}·K), and K_{evap} is an evaporation rate constant (1000 s^{-1}) [22].

2.1.3. Momentum Balance

During drying, the permeability of water and gas in porous media is relatively low, and thus, the momentum conservation can be expressed by Darcy's law. The pressure gradient is the main driven force of water and gas transport in a drying process. The diffusion rate of water and gas calculated following Darcy's law and pressure gradient is expressed as [16]:

$$\bar{n}_i = -\frac{k_{in,i} k_{r,i}}{\mu_i} \nabla P \quad i = w, g, v \tag{13}$$

2.1.4. Energy Balance

Assuming all the components, including solid, water, and gas, are all in energy equilibrium, the microscopic energy balance can be expressed as follows [15]:

$$\rho_{eff} c_{p,eff} \frac{\partial T}{\partial t} + \sum_{i=w,v,a} \left(n_i \nabla (C_{p,i} T) \right) = \nabla \left(k_{eff} \nabla T \right) - \lambda \dot{I} + Q_{mic} \tag{14}$$

where the evaporation rate ($\dot{I}$, in a unit of mol·m^{-3}·s^{-1}) and RF energy absorption rate (Q_{mic}, in a unit of J·s^{-1}) are all functions of time and space; λ is the latent heat of vaporization (J·kg^{-1}).

Effective parameters, including density (ρ_{eff}), specific heat ($c_{p,eff}$), and thermal conductivity (k_{eff}), are used for describing the volumetric properties of food material [15]:

$$\rho_{eff} = (1 - \varnothing)\rho_s + \varnothing(S_w \rho_w + S_g \rho_g) \tag{15}$$

$$c_{p,eff} = m_s c_{p,s} + m_w c_{p,w} + m_g (m_a c_{p,a} + m_v c_{p,v}) \tag{16}$$

$$k_{eff} = (1 - \varnothing)k_s + \varnothing(S_w k_w + S_g (w_v k_v + w_a k_a)) \tag{17}$$

where ρ_{eff} is the effective density (m^3·kg^{-1}), $c_{p,eff}$ is the effective heat capacity (kJ·kg^{-1}·K^{-1}), and k_{eff} is the effective thermal conductivity (W·m^{-1}·K^{-1}).

2.1.5. Dielectric Properties

The dielectric properties of food are significantly influenced by water content in a drying process. A developed equation, namely the "Landau and Lifshitz, Looyenga equation (LLLE)", was used to estimate dielectric properties of potato samples with various water content and at different temperatures [23]:

$$\varepsilon^{\frac{1}{3}} = \sum_{i=s,w,g} v_i \varepsilon_i^{\frac{1}{3}} \tag{18}$$

where v_i is the volume fraction of phase i, and the ε_i is the dielectric properties of phase i.

2.1.6. Permeability

The permeability of porous materials is influenced by the properties of the pores inside the material and the transport properties of the fluid phase. The total permeability of a material is the product of its intrinsic permeability (which depends on structural parameters such as porosity) and relative permeability (which depends on the relative saturation of the fluid phase) as follows [16]:

$$k_{tot,i} = k_{in,i} k_{r,i} \quad i = w, g \tag{19}$$

where $k_{tot,i}$ is total diffusivity of phase i (m^2); $k_{in,i}$ is intrinsic permeability of phase i (m^2); $k_{r,i}$ is relative permeability of phase i (–).

The intrinsic permeability of dry solid is difficult to obtain theoretically and experimentally [16]. Thus, in this study, the intrinsic permeability of the potato solid was estimated as 1×10^{-16} m^2 based on a trial-and-error method. An assumed diffusion coefficient was firstly assumed and brought into the model to calculate the drying curve, and the result was compared with the experimental drying curve. If the mean square error was higher than 0.05, the diffusion coefficient value was modified accordingly, and the above process was repeated until the error met the requirement. As reported, the range of intrinsic permeability values of raw potatoes was $10^{-15} - 10^{-17}$ m^2 [16].

The intrinsic permeability of gas is [24]:

$$k_{in,g} = k_{in,w}\left(1 + \frac{0.15 k_{in,w}^{-0.37}}{P}\right) \tag{20}$$

The relative permeability of water is [25]:

$$k_{r,w} = \begin{cases} \left(\dfrac{S_w - 0.09}{1 - 0.09}\right)^3, & S_w > 0.09 \\ 0 & S_w < 0.09 \end{cases} \tag{21}$$

where S_w is the saturation ratio of water (–).

The relative permeability of gas is [25]:

$$k_{r,g} = \begin{cases} 1 - 1.1 S_w, & S_w > 0.09 \\ 0 & S_w < 0.09 \end{cases} \tag{22}$$

2.1.7. Initial and Boundary Conditions

The voltage of the top electrode was initially set according to the calculated results. The value was tuned with a trial and error method and finalized when predicted results were matched with the experimental temperature pattern and evaporation rate of samples [26]. The bottom electrode was set as grounded ($V = 0$). The metal enclosure of the RF cavity was regarded as the ideal conductor and was defined as electrical insulation ($\nabla E = 0$). Other initial and boundary conditions are listed in Table 1.

Table 1. Input parameters used in simulation.

Symbol	Parameter	Value	Unit	Reference
f	Frequency	27.12	MHz	
V	Electrode voltage	2750	V	
		Solid		
ε'_s	Dielectric constant	2.4	-	
ε''_s	Dielectric loss factor	0.4	-	
k_s	Thermal conductivity	0.21	W·m^{-1}·K^{-1}	[27]
ρ_s	Density	1528	kg·m^{-3}	[28]
C_{ps}	Specific heat	1650	J·kg^{-1}·K^{-1}	[27]

Table 1. Cont.

Symbol	Parameter	Value	Unit	Reference
		Water		
ε'_w	Dielectric constant	$-0.2833T + 80.67$ [a]	-	[29]
ε''_w	Dielectric loss factor	$0.05T + 20$ [a]	-	[29]
k_w	Thermal conductivity	0.59	W·m^{-1}·K^{-1}	[27]
ρ_w	Density	998	kg·m^{-3}	
C_{pw}	Specific heat	4182	J·kg^{-1}·K^{-1}	[16]
μ_w	Viscosity	0.998×10^3	Pa s	[16]
$k_{in,w}$	Intrinsic permeability	1×10^{-16}	m^2	
$k_{r,w}$	Relative permeability	Equation (21)		
		Water vapor		
k_v	Thermal conductivity	0.026	W·m^{-1}·K^{-1}	[27]
ρ_v	Density	Ideal gas [b]	kg·m^{-3}	
C_{pv}	Specific heat	2062	J·kg^{-1}·K^{-1}	[16]
		Air		
k_a	Thermal conductivity	0.026	W·m^{-1}·K^{-1}	[27]
ρ_a	Density	Ideal gas [b]	kg·m^{-3}	
C_{pa}	Specific heat	1006	J·kg^{-1}·K^{-1}	[27]
		Gas (air + water vapor)		
μ_g	Viscosity	1.8×10^{-5}	Pa s	[16]
$k_{in,g}$	Intrinsic diffusivity of gas	Equation (20)		[24]
$k_{r,g}$	Relative diffusivity of gas	Equation (22)		
		Other		
$D_{w,cap}$	Capillary diffusion rate	$10^{-16} \exp(-2.8 + 2M)$ [c]	m^2·s^{-1}	[21]
$P_{v,eq}$	Equilibrium vapor pressure	$P_{sat}\exp(-0.0267M^{-1.656} + 0.0107e^{-1.287M}M^{-1.513}\ln(P_{sat}))$ [c, d]		
h_t	Convective heat transfer coefficient	20	W·m^{-2}·K^{-1}	[30]
h_m	Mass transfer coefficient	2×10^{-7}	m·s^{-1}	[31]
ρ_a	Density of air	Ideal gas [b]	kg·m^{-3}	
D_{bin}	Diffusion rate of water vapor in air	$\frac{2.13}{P}\left(\frac{T}{273}\right)^{1.8}(S_g\varnothing)^{3-\varnothing}/\varnothing$ [a, d]	m^2·s^{-1}	
P_{amb}	Atmospheric pressure	1.01×10^5	Pa	
λ	Latent heat of evaporation	2.26×10^6	J·kg^{-1}	
K_{evap}	Evaporation rate constant	1000	s^{-1}	[21]
$\varnothing$	Porosity	0.88		
$c_{w,0}$	Initial water concentration	797.4	kg·m^{-3}	
$w_{w,0}$	Mass fraction of water vapor	0.026		
T_0	Initial temperature of sample	285.65	K	
T_{amb}	Ambient temperature	283.15	K	

[a] all the temperatures (T) in the table are in a unit of °C; [b] all the ideal gas densities in the table are 1.205 kg·m^{-3}; [c] all the moisture contents (M) in the table are in a unit of (kg water/kg dry solid); [d] all the pressures in the table are in a unit of Pa.

When the water vapor on the surface of the potato reaches saturation, the water inside the potato will transfer directly to the surface in liquid form. The boundary flux of water can be expressed as [16]:

$$\dot{J}_{\bar{n},w|surf} = h_m \varnothing S_w(\rho_v - \rho_{v,amb}) + c_w \bar{v}_{\bar{n},w} \quad \text{(when } S_w = 1\text{)} \tag{23}$$

where $\dot{J}_{\bar{n},w|surf}$ is the boundary flux of water a phase (kg·m^{-2}·s^{-1}); h_m is the mass transfer coefficient (m·s^{-1}); ρ_v is the vapor density on the sample surface (kg·m^{-3}); $\rho_{v,amb}$ is the vapor density in ambient air (kg·m^{-3}); $\bar{v}_{\bar{n},w}$ is the water velocity on the sample boundaries (m·s^{-1}).

Water vapor can be blown out of the surface if the internal pressure within the food is too high. Thus, the boundary flux of water vapor was set as [16]:

$$\dot{J}_{\bar{n},v|surf} = h_m \varnothing S_g(\rho_v - \rho_{v,amb}) + c_g \bar{v}_{\bar{n},g} \tag{24}$$

where $\dot{J}_{\bar{n},v|surf}$ is the boundary flux of the vapor phase (kg·m^{-2}·s^{-1}).

The total amount of heat transfer on the sample surface is composed of the following four items: heat convection between food and the surrounded air, the heat loss due to water evaporation on the food surface, the heat loss due caused by water vapor leaving the food surface, and the heat loss due to water loss as mass transfer. The accumulated heat was calculated as [16]:

$$q|_{surf} = h_t\left(T_{surf} - T_{amb}\right) - \lambda \dot{J}_{\bar{n},w|surf} - \sum_{i=w,v}\left(\dot{J}_{\bar{n},i|surf}\right)C_{p,v}T_{surf} - c_w \bar{v}_{\bar{n},w}C_{p,w} \tag{25}$$

where $q|_{surf}$ is the heat transfer on the sample boundary (W·m^{-2}); h_t is the convective heat transfer coefficient (W·m^{-2}·K^{-1}); T_{surf} is the temperature of the sample surface (K); T_{amb} is the temperature of ambient air (K).

2.1.8. Computer Simulation

There modules including "electric current (AC/DC)", "heat transfer in the solid", and "dilute phase flow" were selected and coupled with the parameter of water content and temperature in COMSOL Multiphysics® (COMSOL Multiphysics 5.2, COMSOL Inc., Burlington, MA, USA). The scheme of the RF heater is described in Figure 2. A potato cube (40 × 40 × 40 mm) was placed at the center of the bottom electrode, with an electrode gap between the top and bottom electrode of 100 mm. After geometry creation, the material properties and appropriate initial/boundary conditions were set up for all domains according to experimental conditions, which was discussed in Section 2.1.7. Tetrahedral mesh was applied to all domains. The mesh size was selected as "extremely fine" for the potato sample, and the maximum size of the mesh was set as 0.00242 m. The determination of mesh size was based on the comprehensive consideration of mesh convergence criteria and saving computing resources. That is, the maximum temperature difference between the temperature–time curves of the sample before and after continuous refinement of the mesh was less than 1%. The mesh size before refinement was selected considering the computational resources and the accuracy of calculation. The total meshes include 152,638 domain elements, 7699 boundary elements, and 495 edge elements. The Multifrontal Massively Parallel Solver (MUMPS) was selected with the time step set as 0.001 s, and the backward difference iteration method was adopted for solving the model. The temperature field, the moisture content distribution, the water vapor concentration, and the temperature–time history of the sample center were obtained, and the changes of each value in the drying process were compared, respectively. The simulation process was performed on a Dell workstation with two dual-core Intel Xeon CPU 2.60 GHz processors and 128 GB RAM with the Microsoft Windows Server (Microsoft Corporation, Redmond, WA, USA) 2012 R2 standard operating system installed. The total computing time was 572 min.

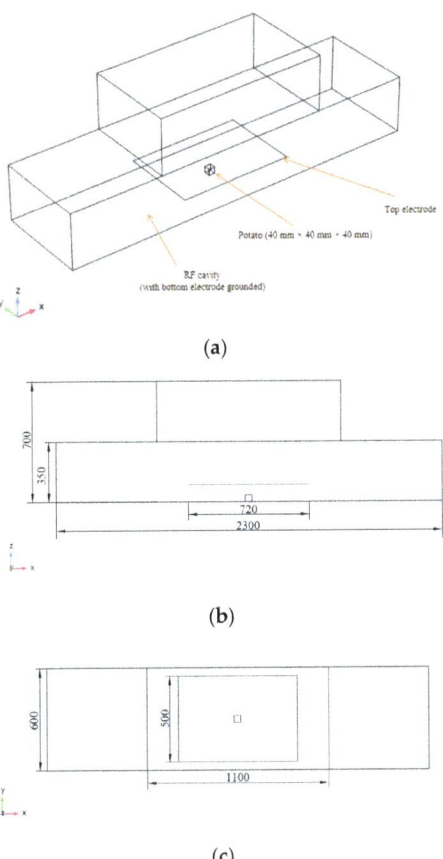

Figure 2. Geometric model of RF drying heater and placement of potato: (**a**) layout, (**b**) front view, and (**c**) top view (all dimensions are in mm).

2.2. Experimental Validation

2.2.1. Material Preparation

Fresh potatoes (*Solanum tuberosum* L., locally planted in Shanghai) without dents and mold areas were purchased from a local supermarket (Lingang, Shanghai, China). Potatoes were equilibrated in a 10 °C, clean, and dry environment for storage for 24 h. Before the experiment, three potatoes were rinsed, peeled, and dried with tissue papers. From each potato, a cube sample was cut from the center with an identical dimension of 40 × 40 × 40 mm and weight of 70 g for drying experiments.

2.2.2. Dielectric Properties Determination

Dielectric properties of the food sample were necessary input parameters in the modeling process, which were influenced significantly by the temperature variation and moisture content of the samples. As the most popular dielectric properties measuring technique, the transmission line method and open-coaxial probe method were used to measure the dielectric properties of food. However, few measurement techniques were found to be successful for measuring porous materials [32,33]. It was due to the non-uniform geometry of the porous material, the uneven pore size distribution, and the incomplete contact between the material surface and the dielectric probe [23]. Thus, the dielectric properties of potato powder and water were measured, respectively, and an estimation equation, namely "Landau and Lifshitz, Looyenga equation (LLLE)" was employed to predict the dielectric properties of potatoes at different water contents and different temperatures following Equation (18) in Section 2.1.5. Since the dielectric properties of potato powder change little with the temperature at 27.12 MHz in the drying temperature range (12~50 °C), temperature effect was ignored in the dielectric properties measurement of potato powders.

Potato cubes (100 g) were placed in a hot-air drying chamber (GZX-9076 MBE, Boxun, Shanghai, China) at 110 °C and sampled every 2 h until reaching a constant weight (weight change between two measurements was <0.1 g). The dried potato samples were grounded to potato powder with an automatic grinder (JYS-M01, Jiuyang, Chengdu, China) for dielectric properties determination. Dried potato powders were placed into a self-designed sample holder until full, and an open-ended probe (Agilent N1501A, Agilent Technologies Inc., San Jose, CA, USA) connected to a network analyzer (Agilent E5071C, Agilent Technologies Inc., San Jose, CA, USA) was utilized for measurement. The detailed dielectric properties measurement procedure can be found in the literature [34]. The experiment was replicated three times, and an average value was used in the modeling process.

2.2.3. RF Drying Experiment

A 12 kW, 27.12 MHz, free-running type RF heater (GJD-12A-27-JY, Huashijiyuan High-Frequency Equipment, Cangzhou, Hebei, China) was used for model validation. The electrode gap was set as 100 mm. Three potato cubes were placed at the center of the bottom electrode with a space of 100 mm. Before the drying experiment, the initial temperatures of all potato sample cubes were verified with an infrared camera (FLIR A655sc, Wilsonville, OR, USA) as 10.0 ± 1.0 °C. The initial weights of potato samples were also measured. The center temperatures of potato samples during drying were recorded by inserting fiber optic sensors (HQ-FTS-D1F00, Heqi guangdian, Xi'an, China) into the geometrical center. Sample weights were measured, and temperature profiles of sample surfaces were captured every 5 min throughout the RF drying process, which required a temporary pause of the process. When treatment was suspended, the sample was quickly taken to an infrared camera for obtaining thermal images, then transferred to a electronic balance for weighing, and eventually put back into the original position of the RF cavity to continue drying. This process was completed within 10 s to minimize the heat loss. The drying process lasted for 720 min until the sample moisture content reached 0.35 g·cm^{-3}. The experiment was replicated three times.

The dry basis water content in potato samples was calculated as follows [35]:

$$X_i = \frac{w_i - w_{end}}{w_{end}} \tag{26}$$

where X_i is the water content in potato samples at the time i (g·g^{-1}), w_i is the potato weight at the time i (g), and the w_{end} is the absolute dry weight of potato samples (g).

3. Results and Discussion

3.1. Electromagnetic Field Distribution

The visualized electromagnetic field distribution within the potato sample is plotted at the time of 0, 6, and 12 h for comparison in Figure 3. A certain angle of deflection of the electric field was observed at the edges, especially the corners of the potato cube sample, throughout the drying process, which is usually the main cause of edge heating. This was possibly attributed to the sample size being smaller than the top electrode, and the electric field was deflected at the edges at the surface of potato sample [36]. The deflection angle was reducing as the drying process continued, which indicated that the RF drying process led to faster water evaporation at the corners and enhanced the electric field intensity at the edges.

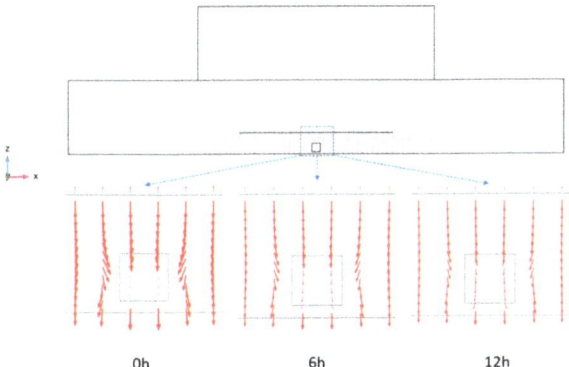

Figure 3. Electric field distribution in the central section of the radio frequency heater cavity (y = 300 mm).

3.2. Temperature Distribution

Figure 4 shows the temperature distribution at the potato cube surfaces from both computer simulation and experiment during RF drying. Results showed a similar distribution pattern between simulation and experiment results. The center temperature increased faster than that at the edges, which was possibly caused by a higher moisture content at the centers, with the electromagnetic field focused on this zone, and resulted in a higher temperature. This is the moisture auto-balancing advantage for RF drying since it prevents the sample from overheating [19]. Similar results were also reported by Wang et al. [37] and Zhou et al. [38]. Within the 1st h of drying, a significant temperature increase at the sample surface was found, and the temperature difference between simulation and experiment was less than 2.6 °C. At this stage, results demonstrated that the simulated temperature distribution pattern matched the experimental one well. Within the 1~7 h of drying, the surface temperature increased by only ~2 °C. The energy absorption was not reflected in the temperature increase because of water evaporation [38]. In general, the experimental results were also in good agreement with the simulation results. However, the hot region started moving to the left boundary from the center, while the simulation result was still in the center of the sample surface as shown in 7 h temperature diagrams. This might be due to uneven water transfer in the actual drying process caused by uneven shape or accidental thermal-runaway phenomena. After 7 h treatment, sample temperature started to increase significantly again, and the drying process entered the falling rate period, which indicated insufficient water in the sample for continuous evaporation [39]. The hot spot area continued to move to the left boundary of the sample in this stage, while the simulation results were still in the central position. The discrepancy between simulation and experimental results was because the measurement errors of material properties were exaggerated during drying experiments.

After the drying experiment, the potato sample was quickly cut into halves with a thin blade, and the temperature profile of the cross-sectional surface was captured with an infrared thermal imager. Figure 5 compares the temperature distribution on the sample cross-sectional surface between the experimental and the simulated results. The internal temperature of the sample was reduced to a certain extent due to cutting. The temperature drop was ~3.2 °C from the comparison of fiber optic sensor reading (51.2 °C) and the infrared camera capture at the center point (48.0 °C). An interesting phenomenon was observed: the temperature at the center of the sample where the fiber sensor was inserted was higher than that of the other areas of the sample, as shown in Figure 5. The possible reason is that there was a little gap between the fiber optic sensor and the potato sample during insertion, resulting in a localized heating phenomena.

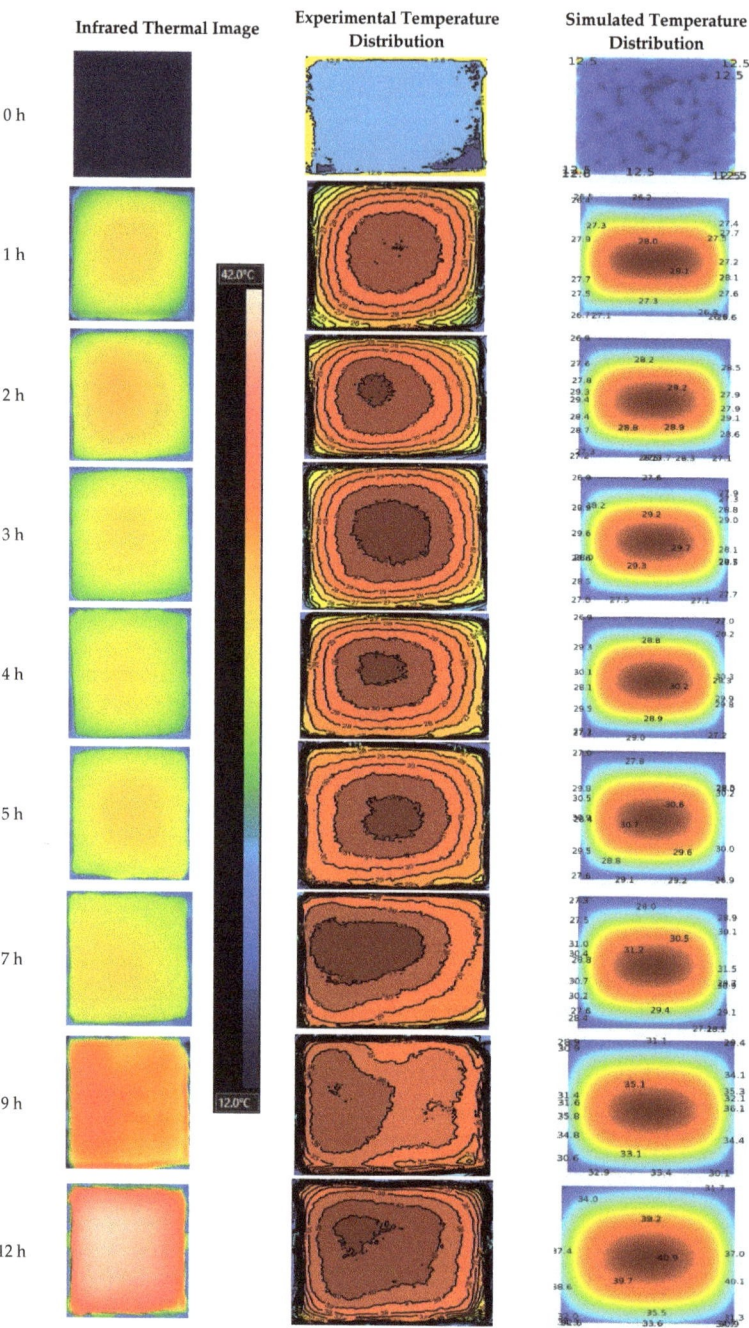

Figure 4. Experiment and simulated temperature profiles of potatoes after RF drying.

The time–temperature history curves of the sample center for both experiment and simulation results are shown in Figure 6. The curve could be divided into three stages: In stage I (0–1 h), the temperature increased rapidly from 12.0 to 34.0 °C, this could be attributed to the electromagnetic energy was mainly used for temperature elevation of samples; in stage II (1–7 h), the temperature of sample increased slowly by 4.0 °C, which was due to the usage of energy for water evaporation; in stage III (7–12 h), the sample temperature increased rapidly from 38.0 to 51.2 °C due to the water removal, and the specific heat capacity of the sample decreased with the water loss [40]. The experimental results were consistent with the simulated results in the trend. The temperature of the experimental value was higher than that of the simulated value, and the temperature difference between the two was <5.5 °C. This may be due to the pore shrinkage that occurred within the volume of sample during the drying process, which was not considered in the simulation. Drying would result in surface hardening and volume shrinkage due to the loss of water. As the water transfer from the internal part of the sample to the surrounding air was limited, a relatively high water content and higher dielectric loss factor in the sample would be retained and ultimately result in a high temperature inside the sample volume [41].

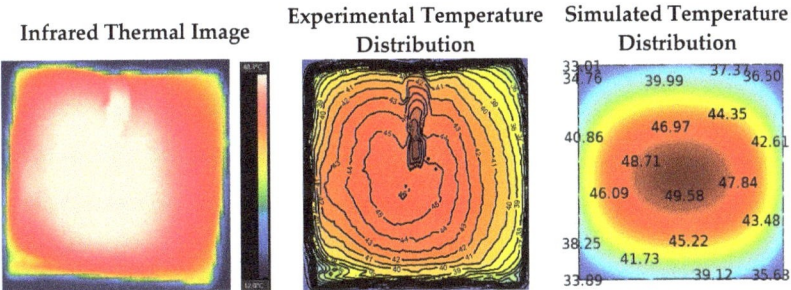

Figure 5. Temperature distribution on the cross-sectional (the zx-section at the center) surface of potato sample from simulated and experimental results.

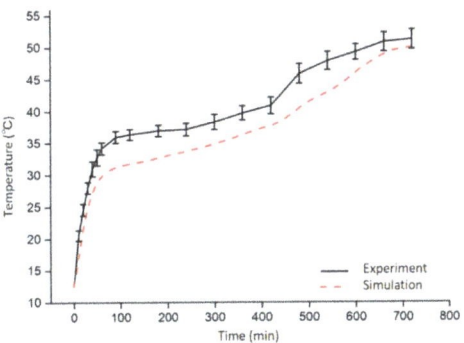

Figure 6. Time–temperature curve of the center point of potato during RF drying from both experiment and simulation.

3.3. Moisture Migration

The removal rate and distribution of moisture content distribution are important parameters in evaluating a drying process. Figure 7 shows the reduction amount of water content (d.b.) with time from both experimental and simulation results. During the drying process, the water content showed a linear decline from both experiment and simulation, and the drying rate had little discrepancy. During the 12 h drying period, the experimental and simulated values were highly consistent, and the simulated error of moisture content (d.b.) was less than 0.18. During the RF drying process, the drying rate was constant in the previous part, then gradually decreased to the end. This is in line with an earlier report for carrot slices dried with combined microwave and vacuum treatment [39]. The initial dry basis moisture content of the potato was 4.56 (d.b.) and decreased to 1.75 (d.b.) after 12 h drying. This indicates that RF drying is effective for potatoes.

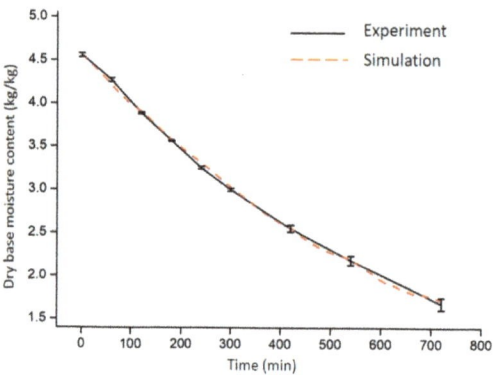

Figure 7. Moisture content of potato samples during RF drying from both simulation and experiment.

Figure 8 shows the water distribution inside the sample from modeling after 12 h of drying. The water concentration in the sample is non-uniform, and the maximum difference of water concentration within the volume is 0.03 g·cm^{-3}. The moisture distributions within the sample corresponded to the temperature–time history inside the sample during drying: the central area had a higher water concentration than the corner. A similar water concentration distribution pattern was also reported for kiwifruit slices in RF–vacuum drying [15,38]. Figure 9 shows that the moisture content of samples decreases with increased drying time, and the drying rate decreases at the final drying period (600–720 min). The main reason is that the moisture within the sample decreased at the falling rate period, which leads to a slower heating rate with the decrease of dielectric loss factor [42]. The initial water concentration at the sample center was 0.90 g·cm^{-3} and decreased to 0.35 g·cm^{-3} after 12 h drying.

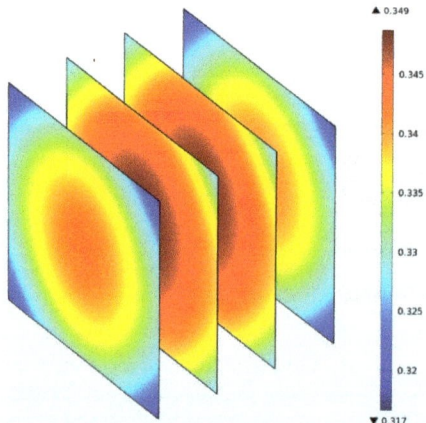

Figure 8. Distribution of water concentration (mg·cm^{-3}) in potato samples from simulation (t = 12 h).

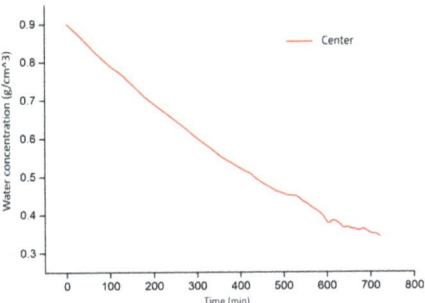

Figure 9. Water concentration at the center of potato samples during RF drying.

3.4. Vapor Migration

Figures 10 and 11 show the water vapor concentration distribution in potato samples after drying and the water vapor concentration at the center point of the potato sample during drying over time from modeling. The distribution of water vapor concentration is related to the distribution of both the temperature and water content within the sample since higher temperature led to a higher evaporation rate of water vapor [43]. Results indicate that the water vapor concentration at the sample center is greater than that at the corners because water vapor increased as temperature increased, and a pressure gradient from center to corner would guarantee the mass transfer from the sample to the surrounding area in the drying process. It can be seen from Figure 11 that the trend of water vapor concentration variation at the sample center corresponds to the change of temperature at the same spot, as shown in Figure 5. This indicates the temperature and water vapor concentration in the center of the sample increase as the electromagnetic energy continually converts to heat energy. This is due to the selective heating mechanism of RF heating: the larger moisture content results in a higher energy localization, which facilitates the drying process [14]. Moreover, since the speed of water migration from the inside to the surface of the sample was relatively stable, the variation of the evaporation speed was reflected on the concentration of water vapor.

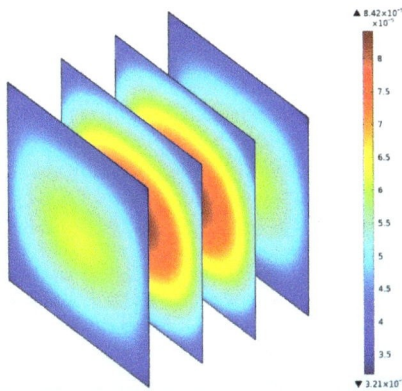

Figure 10. Distribution of vapor concentration (mg·cm^{-3}) in potato samples during RF drying from simulation (t = 12 h).

From comparison, the developed model was effective and reliable in predicting temperature and moisture distribution in a model food system.

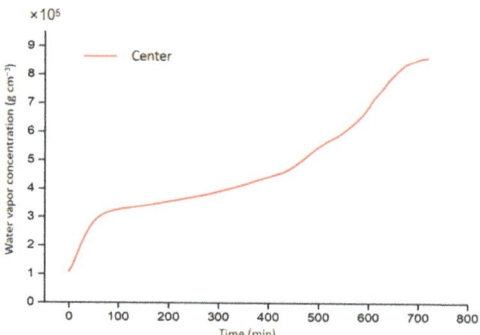

Figure 11. The water vapor content in potato samples with RF drying duration from simulation.

4. Conclusions

In this study, a numerical model of RF drying was established and solved by coupling electromagnetic heating, heat transfer, and mass transfer in a potato cube sample, and the temperature, moisture content, and vapor variation and distribution in the sample in RF drying were described. The prediction model was proven to be reliable and effective through experiments. The distribution of water concentration and water vapor concentration corresponds to the temperature distribution. The results of RF drying potato showed a relatively constant drying rate throughout the RF drying process; thus, it could be used as a supplemental method for traditional drying processes to improve the drying rates and quality of products. The numerical model developed in this research could be used as a tool to explore the mechanism of RF drying processes and also to provide a convenient and economic method for food drying.

To promote industrial applications of RF drying, it is suggested that future research could focus on the following directions:

(1) One of the major challenges of RF heating is non-uniform heating, which results in non-uniform quality for dried products. Products with irregular geometry need particular attention in RF drying since the drying uniformity was significantly influenced by sample shapes and sizes.

(2) The properties of fluids and coefficients in mass and heat transfer, such as porosity, viscosity, permeability, diffusion coefficient, etc., had significant influences on heat and mass transfer processes in the RF drying process. To improve the accuracy of the simulation, sensitivity analyses of these parameters were needed, and the significance of influence of each parameter should be evaluated.

(3) It would be necessary to broadly investigate the combination of RF drying with some other technologies (freeze drying, vacuum, cold air, etc.) to further improve the quality of the final dried product. In a combined drying process, the hurdle effect needs to be explored and emphasized for optimizing the processes.

Author Contributions: Conceptualization, X.C., R.Z. and Y.J.; methodology, X.C., R.Z. and Y.J.; experiment, X.C., R.Z., F.L. and Y.J.; software, X.C., Y.L. and R.Z.; formal analysis, X.C., R.Z. and Y.J.; writing—original draft preparation, X.C., R.Z. and Y.J.; validation, X.C., R.Z. and F.L.; resources, Y.J.; data curation, X.C., R.Z. and Y.J.; reviewing and editing, X.C., D.Y., H.Z. and Y.J.; visualization, Y.J.; supervision, D.Y., H.Z. and Y.J.; project administration, Y.J.; funding acquisition, Y.J. All authors have read and agreed to the published version of the manuscript.

Funding: This work was financially supported by the China National Science Foundation (31801613).

Data Availability Statement: The data presented in this study are available in this article.

Acknowledgments: The authors acknowledge China National Science Foundation (31801613) for its financial support to this research.

Conflicts of Interest: No potential conflict of interest was reported by the authors.

Nomenclature

V	Voltage (V)
V_i	Volume of component i (m^3)
f	Frequency (Hz)
Q	RF power conversed to thermal energy (W·m^{-3})
E	Electric field intensity (V·m^{-1})
c_i	Concentration of a fluid phase, i (mol·m^{-3})
n_i	Mass flux per unit volume of a fluid phase, i (mol·m^{-3})
$\hat{I}$	Rate of evaporation (mol·m^{-3}·s^{-1})
P	Vapor pressure (Pa)
P_w	Capillary pressure (Pa)
P_c	Net pressure (Pa)
$D_{w,cap}$	Capillary diffusivity of a fluid phase, i (m^2·s^{-1})
$k_{in,i}$	Intrinsic permeability of a fluid phase, i (m^2)
$k_{r,i}$	Relative permeability
$k_{tot,i}$	Total permeability of a fluid phase, i
D_{bin}	Diffusion rate of water vapor in air (m^2·s^{-1})
M	Moisture content (kg water/kg dry solid)
M_a	Molecular weight of air
M_v	Molecular weight of vapor
X_i	Water content in potato samples at the time i (g·g^{-1})
w_i	Potato weight at the time i (g)
w_{end}	Absolute dry weight of potato samples (g)
$\bar{v}$	Velocity (m·s^{-1})
S_i	Saturation of fluid phase, i (−)
$C_{p,i}$	Specific heat of component i (J·kg^{-1}·K^{-1})
$P_{v,ep}$	Equilibrium vapor pressure (J·kg^{-1})
K_{evap}	Evaporation rate constant (s^{-1})
h_t	Heat transfer coefficient (W·m^{-2}·K^{-1})
h_m	Mass transfer coefficient (m·s^{-1})
P_{amb}	Atmospheric pressure (Pa)
t	Time (min)
Δt	Time interval (min)
T	Temperature (°C)
R	Universal gas constant (kJ·kmol^{-1}·K)
k_i	Thermal conductivity of component i (W·m^{-1}·K^{-1})
Greek symbols	
μ_i	Viscosity of component i (Pa·s)
ε_0	Free space permittivity (8.86 × 10^{-12} F·m^{-1})
ε'	Relative dielectric constant (−)
ε''	Relative dielectric loss factor (−)
σ	Electrical conductivity (S·m^{-1})
ϕ	Porosity (−)
∇	Gradient operator
λ	Latent heat of vaporization (J·kg^{-1})
ρ_i	Density of component i (kg·m^{-3})
Subscript	
a	Air
s	Solid
w	Water
f	Fluid
g	Gas
v	Vapor
eff	Effective
surf	Surface

amb	Ambient air
eq	Equilibrate
r	Relative
in	Intrinsic
0	Time $t = 0$
tot	Total
sat	Saturate

References

1. Shewale, S.R.; Rajoriya, D.; Bhavya, M.L.; Hebbar, H.U. Application of radiofrequency heating and low humidity air for sequential drying of apple slices: Process intensification and quality improvement. *LWT Food Sci. Technol.* **2020**, *135*, 109904. [CrossRef]
2. Pan, Z.; Shih, C.; McHugh, T.H.; Hirschberg, E. Study of banana dehydration using sequential infrared radiation heating and freeze-drying. *LWT Food Sci. Technol.* **2008**, *41*, 1944–1951. [CrossRef]
3. Wang, H.; Zhang, M.; Mujumdar, A.S. Comparison of Three New Drying Methods for Drying Characteristics and Quality of Shiitake Mushroom (*Lentinus edodes*). *Dry. Technol.* **2014**, *32*, 1791–1802. [CrossRef]
4. Gong, C.; Liao, M.; Zhang, H.; Xu, Y.; Miao, Y.; Jiao, S. Investigation of Hot Air–Assisted Radio Frequency as a Final-Stage Drying of Pre-dried Carrot Cubes. *Food Bioprocess Technol.* **2020**, *13*, 419–429. [CrossRef]
5. Liu, S.; Wang, H.; Ma, S.; Dai, J.; Zhang, Q.; Qin, W. Radiofrequency-assisted hot-air drying of Sichuan pepper (Huajiao). *LWT Food Sci. Technol.* **2020**, *135*, 110158. [CrossRef]
6. Wang, W.; Wang, W.; Wang, Y.; Yang, R.; Tang, J.; Zhao, Y. Hot-air assisted continuous radio frequency heating for improving drying efficiency and retaining quality of inshell hazelnuts (*Corylus avellana* L. cv. Barcelona). *J. Food Eng.* **2020**, *279*, 109956. [CrossRef]
7. Wang, C.; Kou, X.; Zhou, X.; Li, R.; Wang, S. Effects of layer arrangement on heating uniformity and product quality after hot air assisted radio frequency drying of carrot. *Innov. Food Sci. Emerg. Technol.* **2021**, *69*, 102667. [CrossRef]
8. Zhang, H.; Gong, C.; Wang, X.; Liao, M.; Yue, J.; Jiao, S. Application of hot air-assisted radio frequency as second stage drying method for mango slices. *J. Food Process Eng.* **2018**, *42*, e12974. [CrossRef]
9. Tiwari, G.; Wang, S.; Tang, J.; Birla, S.L. Analysis of radio frequency (RF) power distribution in dry food materials. *J. Food Eng.* **2011**, *104*, 548–556. [CrossRef]
10. Zhang, S.; Lan, R.; Zhang, L.; Wang, S. Computational modelling of survival of Aspergillus flavus in peanut kernels during hot air-assisted radio frequency pasteurization. *Food Microbiol.* **2020**, *95*, 103682. [CrossRef]
11. Huang, Z.; Zhu, H.; Yan, R.; Wang, S. Simulation and prediction of radio frequency heating in dry soybeans. *Biosyst. Eng.* **2015**, *129*, 34–47. [CrossRef]
12. Topcam, H.; Gogus, F.; Ozbek, H.N.; Elik, A.; Yanik, D.K.; Dalgic, A.C.; Erdogdu, F. Hot air-assisted radio frequency drying of apricots: Mathematical modeling study for process design. *J. Food Sci.* **2022**, *87*, 764–779. [CrossRef] [PubMed]
13. Mao, Y.; Wang, S. Recent developments in radio frequency drying for food and agricultural products using a multi-stage strategy: A review. *Crit. Rev. Food Sci. Nutr.* **2021**. [CrossRef] [PubMed]
14. Jia, X.; Zhao, J.; Cai, Y. Mass and heat transfer mechanism in wood during radio frequency/vacuum drying and numerical analysis. *J. For. Res.* **2016**, *28*, 205–213. [CrossRef]
15. Hou, L.; Zhou, X.; Wang, S. Numerical analysis of heat and mass transfer in kiwifruit slices during combined radio frequency and vacuum drying. *Int. J. Heat Mass Transf.* **2020**, *154*, 119704. [CrossRef]
16. Zhu, H.; Gulati, T.; Datta, A.K.; Huang, K. Microwave drying of spheres: Coupled electromagnetics-multiphase transport modeling with experimentation. Part I: Model development and experimental methodology. *Food Bioprod. Process.* **2015**, *96*, 314–325. [CrossRef]
17. Karageorgiou, V.; Kaplan, D. Porosity of 3D biomaterial scaffolds and osteogenesis. *Biomaterials* **2005**, *26*, 5474–5491. [CrossRef]
18. Birla, S.L.; Wang, S.; Tang, J. Computer simulation of radio frequency heating of model fruit immersed in water. *J. Food Eng.* **2008**, *84*, 270–280. [CrossRef]
19. Alfaifi, B.; Tang, J.; Rasco, B.; Wang, S.; Sablani, S. Computer simulation analyses to improve radio frequency (RF) heating uniformity in dried fruits for insect control. *Innov. Food Sci. Emerg. Technol.* **2016**, *37*, 125–137. [CrossRef]
20. Armstrong, R.C.R. Byron Bird: The integration of transport phenomena into chemical engineering. *AIChE J.* **2014**, *60*, 1219–1224. [CrossRef]
21. Halder, A.; Dhall, A.; Datta, A.K. An Improved, Easily Implementable, Porous Media Based Model for Deep-Fat Frying: Part I: Model Development and Input Parameters. *Food Bioprod. Process.* **2007**, *85*, 209–219. [CrossRef]
22. Halder, A.; Dhall, A.; Datta, A.K. An Improved, Easily Implementable, Porous Media Based Model for Deep-Fat Frying: Part II: Results, Validation and Sensitivity Analysis. *Food Bioprod. Process.* **2007**, *85*, 220–230. [CrossRef]
23. Liu, Y.; Tang, J.; Mao, Z. Analysis of bread dielectric properties using mixture equations. *J. Food Eng.* **2009**, *93*, 72–79. [CrossRef]
24. Tanikawa, W.; Shimamoto, T. Comparison of Klinkenberg-corrected gas permeability and water permeability in sedimentary rocks. *Int. J. Rock Mech. Min. Sci.* **2009**, *46*, 229–238. [CrossRef]
25. Bear, J. *Dynamics of Fluids in Porous Media*; American Elsevier Publishing Company: New York, NY, USA, 1972.

26. Alfaifi, B.; Tang, J.; Jiao, Y.; Wang, S.; Rasco, B.; Jiao, S.; Sablani, S. Radio frequency disinfestation treatments for dried fruit: Model development and validation. *J. Food Eng.* **2014**, *120*, 268–276. [CrossRef]
27. Choi, Y.H. Effects of Temperature and Composition on the Thermal Properties of Foods. Ph.D. Thesis, Purdue University, West Lafayette, IN, USA, 1986.
28. Farkas, B.; Singh, R.; Rumsey, T.R. Modeling heat and mass transfer in immersion frying. I, model development. *J. Food Eng.* **1996**, *29*, 211–226. [CrossRef]
29. Datta, A.K. *Handbook of Microwave Technology for Food Application*; CRC Press: New York, NY, USA, 2001.
30. Rakesh, V.; Datta, A.K. Microwave puffing: Determination of optimal conditions using a coupled multiphase porous media–Large deformation model. *J. Food Eng.* **2011**, *107*, 152–163. [CrossRef]
31. Guine, R.P.F.; Brito, M.F.S.; Ribeiro, J.R.P. Evaluation of Mass Transfer Properties in Convective Drying of Kiwi and Eggplant. *Int. J. Food Eng.* **2017**, *13*, 20160257. [CrossRef]
32. Keskin, S.O.; Sumnu, G.; Sahin, S. A study on the effects of different gums on dielectric properties and quality of breads baked in infrared-microwave combination oven. *Eur. Food Res. Technol.* **2006**, *224*, 329–334. [CrossRef]
33. Sumnu, G.; Datta, A.K.; Sahin, S.; Keskin, S.O.; Rakesh, V. Transport and related properties of breads baked using various heating modes. *J. Food Eng.* **2007**, *78*, 1382–1387. [CrossRef]
34. Jiao, Y.; Tang, J.; Wang, S. A new strategy to improve heating uniformity of low moisture foods in radio frequency treatment for pathogen control. *J. Food Eng.* **2014**, *141*, 128–138. [CrossRef]
35. Atungulu, G.G.; Olatunde, G.; Sadaka, S. Impact of rewetting and drying of rough rice on predicted moisture content profiles during in-bin drying and storage. *Dry. Technol.* **2018**, *36*, 468–476. [CrossRef]
36. Huang, Z.; Datta, A.K.; Wang, S. Modeling radio frequency heating of granular foods: Individual particle vs. effective property approach. *J. Food Eng.* **2018**, *234*, 24–40. [CrossRef]
37. Wang, Y.; Zhang, L.; Gao, M.; Tang, J.; Wang, S. Pilot-Scale Radio Frequency Drying of Macadamia Nuts: Heating and Drying Uniformity. *Dry. Technol.* **2014**, *32*, 1052–1059. [CrossRef]
38. Zhou, X.; Xu, R.; Zhang, B.; Pei, S.; Liu, Q.; Ramaswamy, H.S.; Wang, S. Radio Frequency-Vacuum Drying of Kiwifruits: Kinetics, Uniformity, and Product Quality. *Food Bioprocess Technol.* **2018**, *11*, 2094–2109. [CrossRef]
39. Cui, Z.-W.; Xu, S.-Y.; Sun, D.-W. Microwave–vacuum drying kinetics of carrot slices. *J. Food Eng.* **2004**, *65*, 157–164. [CrossRef]
40. Rakesh, V.; Datta, A.K.; Walton, J.H.; McCarthy, K.L.; McCarthy, M.J. Microwave combination heating: Coupled electromagnetics-multiphase porous media modeling and MRI experimentation. *AIChE J.* **2012**, *58*, 1262–1278. [CrossRef]
41. Abbasi Souraki, B.; Mowla, D. Simulation of drying behaviour of a small spherical foodstuff in a microwave assisted fluidized bed of inert particles. *Food Res. Int.* **2008**, *41*, 255–265.
42. Zhou, X.; Wang, S. Recent developments in radio frequency drying of food and agricultural products: A review. *Dry. Technol.* **2018**, *37*, 271–286. [CrossRef]
43. Darvishi, H.; Azadbakht, M.; Rezaeiasl, A.; Farhang, A. Drying characteristics of sardine fish dried with microwave heating. *J. Saudi Soc. Agric. Sci.* **2013**, *12*, 121–127.

Article

Thermal Inactivation Kinetics and Radio Frequency Control of *Aspergillus* in Almond Kernels

Yu Gao [1], Xiangyu Guan [1], Ailin Wan [2], Yuan Cui [2], Xiaoxi Kou [1], Rui Li [1,*] and Shaojin Wang [1,3,*]

1. College of Mechanical and Electronic Engineering, Northwest A&F University, Xianyang 712100, China; gaoyu2001@nwafu.edu.cn (Y.G.); xiangyuguan@nwafu.edu.cn (X.G.); kouxiaoxi@nwafu.edu.cn (X.K.)
2. College of Food Science and Engineering, Northwest A&F University, Xianyang 712100, China; wanailin221@nwafu.edu.cn (A.W.); cuiyuan0507@nwafu.edu.cn (Y.C.)
3. Department of Biological Systems Engineering, Washington State University, Pullman, WA 99164-6120, USA
* Correspondence: ruili1216@nwafu.edu.cn (R.L.); shaojinwang@nwafu.edu.cn (S.W.); Tel./Fax: +86-29-8709-2391 (R.L. & S.W.)

Abstract: Mold infections in almonds are a safety issue during post-harvest, storage and consumption, leading to health problems for consumers and causing economic losses. The aim of this study was to isolate mold from infected almond kernels and identify it by whole genome sequence (WGS). Then, the more heat resistant mold was selected and the thermal inactivation kinetics of this mold influenced by temperature and water activity (a_w) was developed. Hot air-assisted radio frequency (RF) heating was used to validate pasteurization efficacy based on the thermal inactivation kinetics of this target mold. The results showed that the two types of molds were *Penicillium* and *Aspergillus* identified by WGS. The selected *Aspergillus* had higher heat resistance than the *Penicillium* in the almond kernels. Inactivation data for the target *Aspergillus* fitted the Weibull model better than the first-order kinetic model. The population changes of the target *Aspergillus* under the given conditions could be predicted from Mafart's modified Bigelow model. The RF treatment was effectively used for inactivating *Aspergillus* in almond kernels based on Mafart's modified Bigelow model and the cumulative lethal time model.

Keywords: almond kernels; *Aspergillus*; radio frequency; thermal inactivation kinetics; verification

Citation: Gao, Y.; Guan, X.; Wan, A.; Cui, Y.; Kou, X.; Li, R.; Wang, S. Thermal Inactivation Kinetics and Radio Frequency Control of *Aspergillus* in Almond Kernels. *Foods* 2022, 11, 1603. https://doi.org/10.3390/foods11111603

Academic Editor: Francisco J. Morales

Received: 7 May 2022
Accepted: 27 May 2022
Published: 29 May 2022

Publisher's Note: MDPI stays neutral with regard to jurisdictional claims in published maps and institutional affiliations.

Copyright: © 2022 by the authors. Licensee MDPI, Basel, Switzerland. This article is an open access article distributed under the terms and conditions of the Creative Commons Attribution (CC BY) license (https://creativecommons.org/licenses/by/4.0/).

1. Introduction

Almonds are rich in unsaturated fatty acids, a variety of vitamins and trace elements, and are accepted and loved by consumers around the world. Global almond production in 2020 was approximately 4.14 million metric tons reported by the Food and Agriculture Organization (FAO, Rome, Italy), and the United States, Spain, Australia, Iran and Turkey are the top five product-consuming countries [1]. However, potential mold contamination in almonds is considered a very serious food safety problem all around the world. Molds in low moisture foods can survive for quite a long period and may grow quickly once the storage environment becomes appropriate, thereby causing great quality degradations and economic losses. Therefore, it is of great significance and urgency to eliminate molds in almonds and almond products during storage, production and processing.

RF heating has already been applied to control the population of insect pests and pathogens in a wide variety of agricultural products owing to its characteristics of volumetric heating, deep penetration, short treatment time, no chemical residues, and no noteworthy quality loss [2–7]. Proper RF treatment parameters (heating temperature and time) can effectively avoid safety problems caused by insufficient heating and food quality deterioration made by excessive heating [8]. Since different molds have different heat resistance under different environmental factors and food compositions [9–11], the detailed information on almond molds and their heat resistances influenced by temperature and a_w is limited. Therefore, it is essential to identify the almond mold species and evaluate the thermal inactivation kinetics of molds influenced by temperature and a_w before RF validation [12].

The thermal inactivation kinetics of mold is usually determined under isothermal conditions. The test cells developed by our laboratory [13] may provide nearly isothermal conditions with fast heating or cooling rates and good heating uniformity and could be potentially used for acquiring thermal inactivation kinetics of target mold inoculated in almond kernel flour. The first-order kinetic, Weibull [14–16], and Mafart's modified Bigelow models [17,18] were applied to describe the thermal inactivation kinetics of molds after thermal treatments under isothermal conditions. For real practical thermal treatments with non-isothermal performances, the inactivation rate of the target microorganism was evaluated by the cumulative thermal lethal time model [8,19,20].

In the actual process of RF pasteurization, the effects of the non-isothermal treatment stage during heating up and the isothermal treatment stage during holding should be comprehensively considered regarding mold inactivation. The cumulative thermal lethal model is useful to guide the development of the RF treatment protocol and further determine the total RF process time for achieving the required inactivation level of mold in almonds.

The objectives of this study were: (1) to isolate mold from infected almond kernels in cold storage conditions and identify mold species using the whole genome sequence (WGS); (2) to compare the heat resistance of the isolated molds and develop the thermal inactivation kinetic models of the selected more thermal-resistant mold (*Aspergillus*) as influenced by three temperatures and three a_w levels; and (3) to verify the inactivation rate of the target mold in almonds when subjected to hot air-assisted RF treatments using the developed thermal inactivation kinetic model.

2. Materials and Methods

2.1. Sample Preparation

About 30 kg raw and dried almond kernels (Nonpareil) were bought from Paramount Farming Company (Modesto, CA, USA). The incomplete and damaged almond kernels were eliminated, then the polyethylene bags were used for sealing intact almond kernels, and the refrigerator (BD/BC-297KMQ, Media Refrigeration Division, Hefei, China) at 4 ± 1 °C was used for storing these almond kernels. The almond kernels' initial moisture content (MC) was $3.91 \pm 0.12\%$ on wet basis (w.b.), which was determined by a moisture analyzer (HE53, Mettler-Toledo, Shanghai, China). The MC of almond kernels was adjusted to three different levels of MC or a_w by directly adding pre-calculated distilled water for studying the effect of MC or a_w levels on molds' thermal inactivation efficacy. The adjusted almond kernels were sealed into polyethylene bags for at least 5 d at 4 °C and shaken at least 3 times each day to obtain the almond kernels with a sufficiently even MC distribution. After the almond kernels were adjusted to the predetermined MC, the kernels were grounded with a grinder until their flour could pass through No.18 sieve (Aperture size was 1 mm, corresponding to 16 Taylor sieve). The water activity meter (Aqua Lab 4 TE, Decagon Devices, Inc., Pullman, WA, USA) was used for measuring the a_w of almond kernels.

2.1.1. Isolation of Spoilage Molds

About 200 g almond kernels were randomly selected from samples stored in the refrigerator and then their MC was adjusted to 10.11% (w.b.) and stored in a 25 °C incubator (LRH-250, Zhujiang, Guangdong, China) for 15 d, moldy almond kernels appeared. About 25 g moldy almond kernels were immersed into 95% ethyl alcohol for sterilization, and then put into 225 mL normal saline and shaken fully for about 30 min [13]. Next, the suspension was transferred into Potato Dextrose Agar (PDA; Beijing Land Bridge Technology Co., Ltd., Beijing, China) and Czapek Yeast Extract Agar (CYA; Beijing Land Bridge Technology Co., Ltd., Beijing, China) media, respectively. All the media were monitored for about 5 d in a biochemical incubator (LTH-100, Shanghai Longyue Instrument Equipment Co., Ltd., Shanghai, China) at 29 ± 0.5 °C. Two single pure isolated colony types were obtained by conducting gradient dilution and streaking plate method on mixed colonies and then identified by WGS.

2.1.2. Preparation of Mold Suspension

The molds of *Penicillium* and *Aspergillus* were identified by WGS. The strains of *Penicillium* and *Aspergillus* were cultivated on CYA and PDA media, respectively. The two strains on different media were incubated at 29 ± 0.5 °C for 5 d in the biochemical incubator. Conidia were gently scraped off from the surfaces of the 5-day-old cultures using a spreader after pouring sterile 0.85% isotonic NaCl solution on the cultivated agar. The conidia's population was adjusted to 1×10^{10} CFU/mL in both suspensions for further use.

2.2. Thermal Treatment

Custom-designed test cells were used for conducting the isothermal treatment (Figure 1), which were successfully used for studying the thermal inactivation kinetics of *Penicillium* in chestnuts [13]. These test cells' detailed information can be found in Hou et al. [21]. Before inoculation, the test cells and almond kernel flour were sterilized at 121 °C and 105 °C for 20 min and 10 min by a vertical autoclave (LMQ.C, Shinva Medical Instrument Co., Ltd., Shandong, China), respectively. Then, 0.88 ± 0.03 g almond kernel flour was put into test cells and 20 µL mold suspension was inoculated into almond kernel flour. Then the test cells were left inside a biosafety hood at 25 °C for 1 h to achieve moisture equilibrium before hot water treatments. After that, the test cells were immersed and heated in a preheated water bath (YT-10A, Beijing Yatai Cologne Experimental Technology Development Center, Beijing, China). The treatment time started from the moment when the central temperature of the suspension reached the set temperature value, and the temperature fluctuation did not exceed ±0.5 °C, which could be considered near-ideal isothermal conditions. The sample temperature was monitored from one cell filled with uninoculated almond kernel flour by type-T thermocouples (HH-25TC, Omega Engineering Ltd., Stamford, CT, USA).

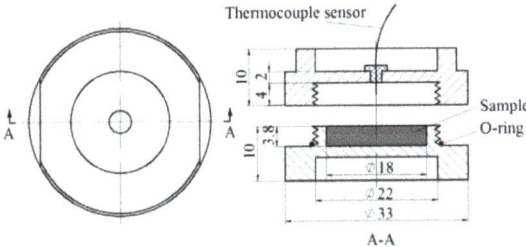

Figure 1. Schematic view of a test cell with 18 mm diameter and 3.8 mm height (All dimensions are in mm) (Adapted from Hou et al. [13]).

Based on the preliminary results, 62 °C + 5 min, 65 °C + 3 min, and 68 °C + 1 min were selected for comparing the heat resistance of two molds isolated from moldy almond kernels. Then, the higher heat resistance mold in almond kernels was chosen to obtain the thermal inactivation kinetics for further RF pasteurization validation. Three a_w levels of 0.657, 0.854, and 0.923 corresponded to sample MC of 5.82%, 10.11%, and 13.85% w.b. were used to determine the a_w effect on inactivation of the target molds at three target temperatures. To achieve at least 4 log reductions of the target mold for thermal inactivation kinetic determination and further for RF pasteurization validation, 59, 62 and 65 °C for a_w of 0.923, 62, 65 and 68 °C for a_w of 0.854, or 65, 68 and 71 °C for a_w of 0.657 were selected. For comparing the a_w influence on the target mold inactivation, 65 °C was included at each a_w.

After holding different time intervals, the test cells with inoculated almond kernel flour were placed into cold water ($\leq 4\ °C$) over 3 min before further analysis. One test cell with inoculated almond kernel flour without thermal treatment served as control. The total population of colonies in the control and heat-treated samples was counted and compared for evaluating thermal treatment effects.

Almond kernel flour was scraped into sterile 0.85% NaCl solution and shaken for at least 3 min. A total of 100 μL of the solution was then added to 0.9 mL sterile NaCl solution for 10-fold serial dilutions until suitable countable numbers were reached. Finally, 100 μL of each dilution was evenly spread on the cultivated agar and 29 °C incubation for about 2 d. Colony counts were obtained by plate counting.

2.3. Thermal Inactivation Kinetics Model

The thermal inactivation kinetics was described by the first-order kinetic and the Weibull distribution. The equation of the first-order kinetic model was presented below [17]:

$$\log \frac{N}{N_0} = -\frac{t}{D} \quad (1)$$

where N and N_0 are the mold populations (CFU/g) at time t and initial time, t means isothermal treatment holding time (min), and D is a decimal time (min) for 1 log reduction of the microbial population at a required temperature (°C).

The equation of the Weibull distribution model was described as follows [22–24]:

$$\log \frac{N}{N_0} = -(\frac{t}{\delta})^p \quad (2)$$

where δ-value is a scale parameter that primarily represents the survival curve steepness. The p-value is a shaped parameter, and may be linear ($p = 1$) or nonlinear ($p < 1$ or $p > 1$). The suitability of the models can be evaluated by the coefficient of determination (R^2) and root mean square error (RMSE).

2.4. Effects of Temperature and a_w on Thermal Inactivation Kinetic Model

The secondary model was usually applied to characterize the influence of temperature (T) or a_w of the samples on the parameters of kinetic model [25,26]. The model of simplified Mafart's modified Bigelow in references [8,17] was depicted as follows:

$$\log \frac{D}{D_{ref}} = -\frac{(T - T_{ref})}{z_T} - \frac{(a_w - a_{wref})}{z_{a_w}} \quad (3)$$

where D_{ref} is the decimal time (min) reducing 1 log population at T_{ref} (65 °C) and a_{wref} (1.00), z_T and z_{aw} are temperature (°C) and a_w increments, respectively, required to reach 90% D-value reduction of target microorganisms.

2.5. Determining Cumulative Time–Temperature Effects

The lethal effect of heating up and isothermal time can be explored from the cumulative lethal time model during the whole thermal treatment. At a reference temperature T_{ref} (°C), the equivalent total lethal time M_{ref} (min) for a specific temperature–time history of $T(t)$ can be calculated by the cumulative thermal inactivation rate of this thermal treatment using the following integral equation [19,27]:

$$M_{ref} = \int_0^t 10^{\frac{T(t)-T_{ref}}{z}} dt \quad (4)$$

where z is in the thermal inactivation time curve, the temperature difference (°C) required for reducing 1 log population. The z-value can be calculated based on the following equation [28]:

$$z = \frac{T_2 - T_1}{\log D_1 - \log D_2} \quad (5)$$

where D_1 and D_2 are the decimal reduction times (min) of target molds under temperatures (°C) of T_1 and T_2. The z-value could be defined as the ratio of the difference in the log D-values to the difference in the exposure temperatures.

2.6. RF Pasteurization Validation

2.6.1. Inoculated Almond Samples

The a_w of almond kernel samples was adjusted to 0.657, 0.854 and 0.923, respectively. Each almond kernel sample with different a_w was first exposed to ultraviolet lights for at least 1 h and turned over every 30 min [13]. Then, about 5 g (5 ± 0.2 g) sterilized almond kernels with different a_w were put in sterile polyethylene bag (5 × 7 cm^2), and 20 µL target mold suspension was inoculated into the sterile bag. All the bags were rubbed at least 3 min by hand to make the suspension evenly attached to the almond kernels' surface [29,30]. Inoculated almond kernels were left for 12 h at 23 ± 2 °C inside a biosafety hood to achieve a sufficient moisture equilibration and then wrapped in sterile filter paper, and tied with a rubber band [31]. The final populations of target mold on different a_w almond kernel samples were achieved at 10^7 CFU/g.

2.6.2. Selection of Electrode Gap

Each 1.5 kg of almond kernel with a_w of 0.657, 0.854 and 0.923 were placed homogeneously into the uncovered five-layer container, respectively (300 g almond kernel for each layer). Detailed information on a five-layer container can be found in Li et al. [32]. Then, the five-layer container containing 1.5 kg pretreated almond kernels was placed vertically above the bottom electrode of the RF system (Figure 2) to obtain the general relationship between the electrode gap and current (I, A). The RF system's detailed information can be found in Wang et al. [33]. Based on the anode current (I, A) shown on the RF system screen, the output RF power (P, kW) was calculated according to the equation of $P = 5 \times I - 1.5$ recommended by the manufacturer, and the heating rates of the almond kernels were estimated [34,35]. The heating rate of each location and the location of the coldest spot were determined by inserting probes into the almond kernels through pre-drilled holes in five representative locations (A–E) (Figure 3) using a fiber optic temperature sensor system (HQ-FTS-D120, Heqi Technologies Inc., Xian, China). According to the similar heating rate around 6.7 °C/min during RF heating, the electrode gaps of 10.5 cm, 12.5 cm and 13.0 cm were finally selected for the almond kernels with a_w of 0.657, 0.854 and 0.923, respectively.

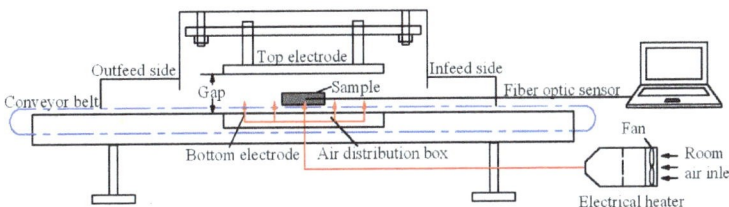

Figure 2. Schematic view of the pilot-scale 6 kW, 27.12 MHz RF system (Adapted from Wang et al. [33]).

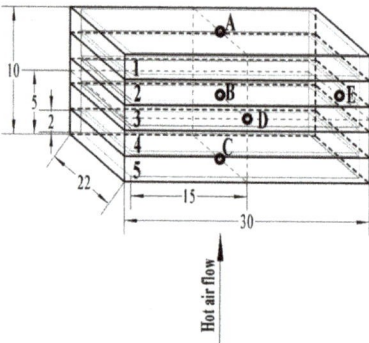

Figure 3. Five-layer (1–5) container for sample temperature measurements with five positions (A–E) and pre-drilled holes (all dimensions are in cm) (Adapted from Li et al. [32]).

2.6.3. RF Pasteurization Verification

Based on the target mold's thermal inactivation kinetics in almond kernels, the hot air-assisted RF system was used for pasteurization verification. The temperatures of 71, 68, and 65 °C were selected, respectively, as the target holding temperatures of almond kernels with a_w of 0.657, 0.854 and 0.923 for pasteurization validation. Every four packs of filter-paper-wrapped inoculated almond kernels with three different a_w were placed at cold point (point B of Layer 3, Figure 3) in the five-layer container, respectively [32]. Then, the five-layer container was placed above the bottom electrode of the hot air-assisted RF system and heated with the appropriate electrode gap until the temperature of the cold spot reached the target value. The RF system was then switched off and the almond kernels were kept at the target temperature only by hot air. To ensure the heating uniformity during RF pasteurization, the position order of the five-layer container was rearranged from L1, L2, L3, L4, and L5 to the order of L5, L4, L3, L2, and L1 according to Li et al. [32]. The hot air holding temperatures of almond kernels with a_w of 0.657, 0.854 and 0.923 were set to 74, 71, and 68 °C, respectively, slightly above the target temperature based on the thermal loss during heating [13].

According to the different D-values of target mold under different a_w and temperature levels, the packs were taken out at different time intervals, sealed with polyethylene bags, and then immersed in cold water below 4 °C for at least 3 min for fully cooling. The pack wrapped in inoculated but no-treated almond kernels was conducted for plate counting to detect the total numbers of molds before thermal treatment. Specifically, the almond kernels were put into normal saline (10 mL) and shaken for 3 min sufficiently. The target mold suspensions were then diluted by gradient dilution and appropriate dilutions were selected to count the population of the sample colonies. The validation test was repeated three times for each a_w.

2.7. Statistical Analysis

Each trial was performed for three biologically separate replicates. Analysis of variance (ANOVA) and Tukey's test ($p \leq 0.05$) were used for evaluating the statistical significance of differences. SPSS statistics 21.0 software (IBM, Armonk, NY, USA) was used for performing model fitting and parameter estimations.

3. Results and Discussion

3.1. Spoilage Molds Isolated from Almond Kernels

Colonies appeared after three to four days of inoculation on both CYA and PDA media. A cyan mold and a black mold were separated and purified in CYA media and PDA media, respectively. The cyan mold was identified as *Penicillium* and the black mold was identified as *Aspergillus* after WGS by the identification mechanism (Sangon Biotech Co., Ltd., Shanghai, China).

3.2. Selection of the More Thermal-Resistant Mold

Table 1 showed the population reductions of *Penicillium* and *Aspergillus* inoculated into almond kernel flour with an a_w of 0.854 under three combinations of heating temperature and time. The population reductions of *Penicillium* were higher than those of *Aspergillus* ($p \leq 0.05$), suggesting that the selected *Aspergillus* had higher heat resistance than the selected *Penicillium* in almond kernels. Therefore, *Aspergillus* was selected as the target mold to explore the influence of different a_w levels and temperatures on the thermal inactivation kinetics.

Table 1. Population reductions (mean ± SD, log CFU g^{-1}) of *Penicillium* and *Aspergillus* inoculated in almond kernel flour with an a_w of 0.854 under the three treatment conditions.

Types of Molds	Temperature (°C) + Holding Time (min)		
	62 °C + 7 min	65 °C + 3 min	68 °C + 1 min
Penicillium	2.07 ± 0.07 [a,*]	3.52 ± 0.21 [a]	1.55 ± 0.13 [a]
Aspergillus	1.42 ± 0.12 [b]	2.21 ± 0.11 [b]	1.10 ± 0.17 [b]

* Different letters in the same column indicate that there were significant differences in the values of the population reductions with $p < 0.05$ between the two molds.

3.3. Primary Model

Table 2 presented the D-, δ- and p-values of the two models for the target *Aspergillus* in almond kernel flour at three a_w and temperature levels. The Weibull model's coefficients of determination (R^2 = 0.988–0.998) were higher than those (0.935–0.992) of the first-order kinetic model, and the Weibull model's root mean square errors (RMSE = 0.056–0.181) were lower than those (0.153–0.503) of the first-order kinetic model. The Weibull model was more appropriate for describing the survival curves of the target *Aspergillus* in almond kernels when compared with the first-order kinetic. All the Weibull model's p-values were less than 1, indicating a tailing behavior of the curves. This might be due to the fact that with the temperature increasing, the surviving mold had stronger heat resistance, or was more adaptable with treatment time [36]. Dong [37] and Zhang et al. [8] also reported similar results in *Clostridium sporogenes* and *Aspergillus flavus*.

Table 2. D-, δ- and p-values of the two models for the target *Aspergillus* in almond kernel flour at three a_w and temperature levels using test cells.

Moisture Content (% w.b.)	a_w	Temperature (°C)	First-Order Model			Weibull Model			
			D (min)	R^2	RMSE	δ (CI 95%) [a]	p (CI 95%)	R^2	RMSE
5.82	0.657	65	21.82	0.992	0.153	19.28 (12.21–26.34)	0.92 (0.65–1.20)	0.992	0.150
		68	7.28	0.966	0.356	3.73 (2.21–5.25)	0.70 (0.54–0.85)	0.995	0.135
		71	2.10	0.968	0.315	1.15 (0.54–1.75)	0.70 (0.48–0.92)	0.993	0.151
10.11	0.854	62	7.09	0.980	0.222	4.64 (3.59–5.69)	0.76 (0.64–0.88)	0.997	0.082
		65	2.29	0.946	0.285	1.10 (0.82–1.37)	0.59 (0.49–0.69)	0.998	0.056
		68	1.05	0.991	0.167	0.85 (0.53–1.18)	0.88 (0.63–1.13)	0.993	0.149
13.85	0.923	59	5.43	0.979	0.249	3.27 (2.00–4.55)	0.73 (0.55–0.92)	0.995	0.118
		62	2.45	0.953	0.331	1.14 (0.34–1.94)	0.63 (0.36–0.89)	0.988	0.168
		65	0.48	0.935	0.503	0.17 (0.03–0.30)	0.59 (0.38–0.80)	0.992	0.181

[a] CI 95%: Confidence Interval.

At a specific a_w value, the D-values were dependent on the sample temperature, that was, when the temperature was higher, the shorter time needed for achieving the target *Aspergillus'* inactivation rate. As an example, at a_w of 0.854, when the temperature was 62 °C, the D-value was 7.09 min, but the D-values dropped to 2.29 min and 1.05 min at 65 °C and 68 °C, respectively. The Weibull model's δ-values also decreased with the temperature increase, which indicated that as the temperature increased, the target *Aspergillus'* thermal inactivation rate increased. For example, at 62 °C, the δ-value was 4.64 min when a_w was 0.854 but sharply declined to 1.10 min at 65 °C and 0.85 min at 68 °C. The tendency was in agreement with *Acidovorax citrulli* on watermelon seeds [26], *E. coli* ATCC 25922 in mashed potato [38], and *Salmonella enterica* in goat's milk caramel [39]. The target *Aspergillus* inactivation from the first-order kinetic and the Weibull models affected by temperature under a_w of 0.854 were shown in Figure 4. The slope of the curves increased with the increase in temperatures, and also showed that the lower the temperature, the more obvious the tailing effect, which is corresponding to Table 2.

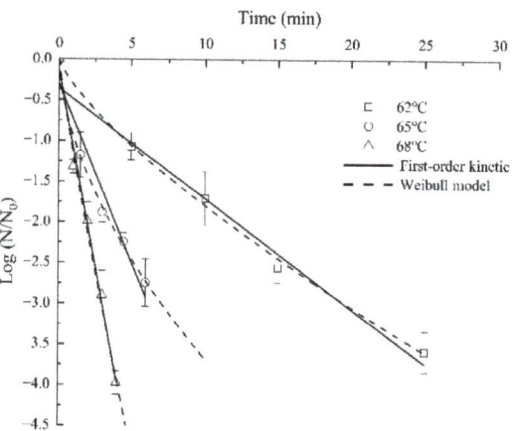

Figure 4. The target *Aspergillus* inactivation from the first-order kinetic and the Weibull models affected by temperature under a_w of 0.854.

The D-values and the δ-values both decreased with the increase in a_w at the same temperature. For example, when the temperature was 65 °C and a_w was 0.657, the D-values were 21.82 min and the δ-values were 19.28 min. However, when a_w increased to 0.854 and 0.923, the D-values were reduced to 2.29 min and 0.48 min, and δ-values also decreased to 1.10 min and 0.17 min, respectively. Zhang et al. [40] also displayed that the thermal treatment time could be effectively shortened and the ideal microbial inactivation level could be achieved in a short time with the increase in a_w levels. For example, the time required to reduce the populations of the target *Aspergillus* in almond kernel flour by 4 log at 65 °C calculated from the Weibull model, 77.12 min, 4.40 min and 0.68 min were needed when the a_w of almond kernels was 0.657, 0.854 and 0.923, respectively. Figure 5 showed the survival curves of *Aspergillus* at 65 °C with a_w of 0.657, 0.854 and 0.923, by fitting with first-order kinetic and Weibull models. The survival curve of *Aspergillus* with a_w of 0.923 was relatively straight, and the survival curves of *Aspergillus* with a_w of 0.854 and 0.657 were slightly upward.

According to the data in Table 2, the *p*-value of the shape parameter appeared to be independent of a_w and temperature, which is consistent with the previous results [18,41]. The re-estimated δ'-values at the mean of survival curves with the *p*-value fixed to 0.70 are shown in Table 3. The re-estimated δ'-values ranged from 0.23 min to 13.40 min, which were influenced by the test temperature and sample a_w as explained by Possas et al. [42].

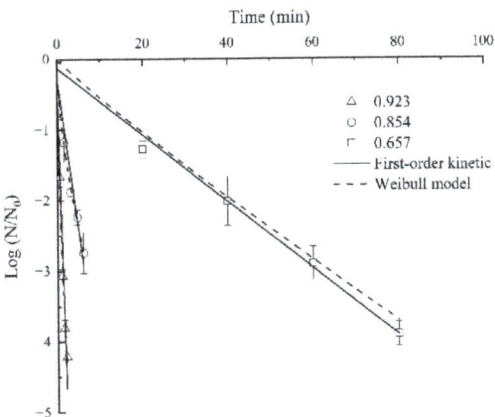

Figure 5. Survival curves of *Aspergillus* at 65 °C with a_w of 0.657, 0.854 and 0.923, by fitting with first-order kinetic and Weibull models.

Table 3. The re-estimated δ'-values at the mean of survival curves with the *p*-value fixed to 0.70.

Moisture Content (% w.b.)	a_w	Temperature (°C)	δ' (CI 95%) [a]	R^2	RMSE
5.82	0.657	65	13.40 (10.95–15.85)	0.973	0.238
		68	3.76 (3.47–4.06)	0.995	0.117
		71	1.14 (1.04–1.25)	0.993	0.131
10.11	0.854	62	4.12 (3.79–4.45)	0.995	0.098
		65	1.37 (1.23–1.50)	0.991	0.103
		68	0.62 (0.53–0.72)	0.980	0.213
13.85	0.923	59	3.04 (2.81–3.28)	0.995	0.108
		62	1.35 (1.18–1.52)	0.985	0.162
		65	0.23 (0.20–0.26)	0.984	0.216

[a] CI 95%: Confidence Interval.

3.4. Secondary Model

Table 4 presented the D_{ref}, z_{aw}, and z_T values of Mafart's modified Bigelow model calculated at 65 °C using the data from first-order kinetic and the Weibull model for the thermal inactivation of *Aspergillus* inoculated into the almond kernels. The Mafart's modified Bigelow model conforms to the first-order kinetic model ($R^2 \geq 0.932$ with RMSE ≤ 0.150), or the Weibull model for related *p*-value ($R^2 \geq 0.853$ with RMSE ≤ 0.256) and for single *p*-value ($R^2 \geq 0.907$ with RMSE ≤ 0.182). Combined with the estimated parameters from Table 4 and Equation (3), the thermal inactivation results of the target *Aspergillus* under any given treatment temperature and a_w conditions within the experimental limits can be predicted.

Table 4. Calculated D_{ref}, z_{aw}, and z_T values of Mafart's modified Bigelow model at 65 °C for the thermal inactivation of *Aspergillus* inoculated into the almond kernels.

Parameter	First-Order Kinetic Model	Weibull Model	
		δ	δ'
D_{ref} or δ_{ref} (min)	0.326	0.140	0.173
Z_T (°C)	6.660	6.130	6.493
z_{aw}	0.189	0.169	0.185
R^2	0.932	0.853	0.907
RMSE	0.150	0.256	0.182

3.5. Electric Current under Different Electrode Gaps

The relationship between electric current and electrode gap without conveyor belt movement and hot air-assisted heating was shown in Figure 6. In the five-layer container, the electric current gradually decreased as the electrode gap increased from 10.5 cm to 19.0 cm, which is similar to the previous research results [43,44]. Because of the same output power calculated by the same electric currents, the 10.5 cm, 12.5 cm and 13.0 cm electrode gaps of almond kernels with a_w of 0.657, 0.854, and 0.923 were selected, respectively, to achieve similar heating rates in RF heating process. The heating rates measured by the fiber optic temperature sensor system under the corresponding electrode gap were 6.54 ± 0.12, 6.84 ± 0.16 and 6.65 ± 0.17 °C/min, respectively.

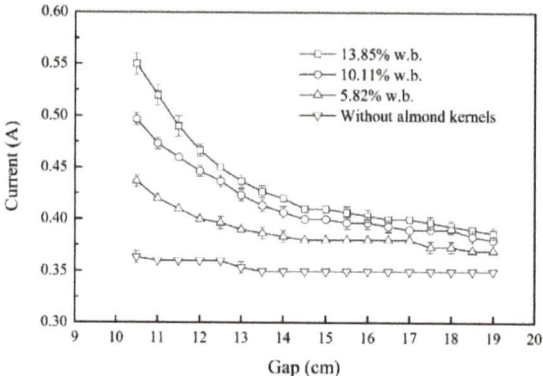

Figure 6. The relationship between electric current and electrode gap for almond kernels with three different a_w levels without conveyor belt movement and hot air-assisted heating.

3.6. Cumulative Lethal Effect of Aspergillus

The target molds' thermal inactivation kinetics is built under isothermal conditions. However, in practical production and application, most thermal treatment processes were of non-isothermal characteristics. The average temperature–time history of 1.5 kg almond kernels with 0.854 a_w (10.11% w.b. MC) in the five positions (A–E) of the five-layer container with a 12.5 cm electrode gap was shown in Figure 7. To design the RF treatment processes for almond kernels' pasteurization according to the thermal inactivation kinetics of the target *Aspergillus*, the heating up processes should be transformed into isothermal processes based on the cumulative lethal effect model depicted in Equation (4). The target *Aspergillus'* z-value was estimated to be 7.41 °C when a_w was 0.854 according to the data in Table 2. At the reference temperature of 68 °C, the equivalent lethal time M_{ref} of the RF heating up process curve (from 25 °C to 68 °C) was the area of the shaded part (0.471 min) in Figure 7. When a_w values were 0.657 and 0.923, the cumulative thermal lethal time during heating up were 0.392 and 0.367 min at the reference temperature of 71 °C and 65 °C, respectively. In a certain thermal process, when the temperature increases, the cumulative thermal curve becomes steeper and steeper, which was the same as the result obtained by Zhang et al. [8]. Theoretically, the $D_{68°C}$-value of *Aspergillus* in almond kernels with an a_w value of 0.854 was 1.05 min (shown in Table 1). To obtain 4 log reductions of *Aspergillus*, almond kernels need to be heated continuously at 68 °C for 4.20 min. As shown in Figure 7, there would be an additional 3.73 min holding time required in this thermal process to obtain the 4 log reductions of the target *Aspergillus*.

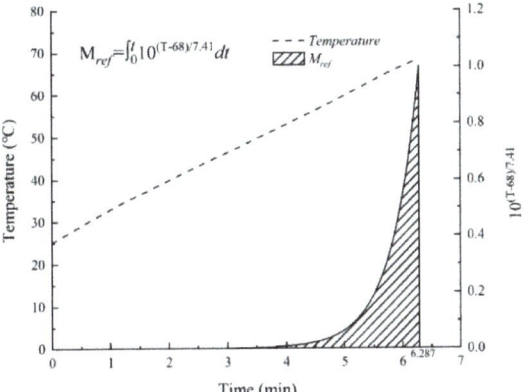

Figure 7. Average temperature–time history of five locations (A–E) in Figure 3 of RF heating from 25 to 68 °C, and the equivalent lethal time M_{ref} for this heating up curve of *Aspergillus* inoculated in almond kernels with a_w of 0.854 at 68 °C.

3.7. RF Treatment Verification

Figure 8 showed the experimental data for verifying almond kernels' RF pasteurization levels and the predicted survival curve for the target *Aspergillus* inoculated into almond kernel flour with a_w of 0.854 at 68 °C by combining the Weibull model with the Mafart's modified Bigelow equation. The time shown on the abscissa in Figure 8 was the sum of the cumulative thermal lethal time calculated by the heating up process and the time of the isothermal thermal process. The results showed that the RF pasteurization verification time was slightly longer than the time predicted by the combined cumulative thermal lethal time and the isothermal heating time.

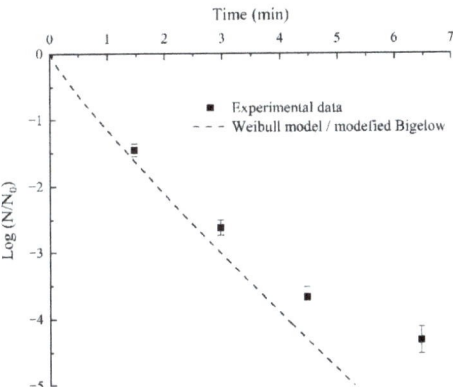

Figure 8. Experimental data and predicted survival curves for the target *Aspergillus* inoculated into almond kernel flour with 0.854 a_w at 68 °C by combining the Weibull model with Mafart's modified Bigelow equation.

The longer time required in validated RF pasteurization may be due to the difference in the particle size. When obtaining thermal inactivation kinetics of *Aspergillus*, the *Aspergillus* suspension was inoculated in the almond kernel flour, while the *Aspergillus* suspension was inoculated on the whole almond kernels when validated in the RF system. These results were the same as those in previous research. For example, Fine et al. [45] found that the *Saccharomyces cerevisiae* in larger size wheat flour exhibited higher heat resistance. Zhang et al. [18] also observed that the *E. coli* ATCC 25922 inoculated in pepper powder behaved more thermal resistant with the increase in the particle size of pepper powder. In addition, the *Aspergillus* in the almond kernels may be more thermal resistant than in the almond kernel flour because it takes time for central heat to diffuse to the surface of almond kernels.

For validating RF pasteurization, as the heating time was prolonged, the MC of almond kernels gradually declined, which enhanced the heat resistance of *Aspergillus*. This phenomenon was consistent with that in a previous study. For example, Li et al. [32] found that the heat resistance of *E. coli* ATCC 25922 inoculated in almond kernels increased with the increase in RF heating time. Chen et al. [46] also found that the rapid evaporation of water on the hard-shell surface of hazelnuts with a shell led to the unsatisfactory inactivation effect of *Salmonella*.

4. Conclusions

Penicillium and *Aspergillus* were identified from moldy almond kernels by WGS. The selected *Aspergillus* had higher heat resistance than the *Penicillium* in almond kernels. The thermal inactivation kinetics of *Aspergillus* in almond kernel flour affected by temperature and a_w was studied and then fitted by using the first-order kinetics and Weibull models. The Weibull model was more appropriate when characterizing the survival curves of the target *Aspergillus* in almond kernels due to the higher coefficients of determination and lower root mean square errors. The D_{ref}, z_{aw}, and z_T values from Mafart's modified Bigelow model were calculated and used for predicting the thermal inactivation of *Aspergillus* under any given treatment temperature and a_w conditions. The predicted thermal inactivation kinetic models were verified by RF heating in combination with the cumulative thermal lethal model. The results showed that RF pasteurization verification time was slightly longer than the time predicted by the combined cumulative thermal lethal time and the isothermal heating time due to the different particle sizes and other possible factors. Future studies may focus on the effect of real-time moisture content change on microbial heat resistance in almond kernels under RF treatment.

Author Contributions: Y.G. conducted the experiment, analyzed data, and wrote the first version of the manuscript; X.G. helped to analyze the data; A.W. and Y.C. assisted in conducting the experiments; X.K. also helped to analyze the data; R.L. and S.W. are the PI of the project, guided the experimental design and revised the manuscript. All authors have read and agreed to the published version of the manuscript.

Funding: This research was supported by research grants from the China Postdoctoral Science Foundation (2021M692656) and the Experimental Technology Research and Laboratory Management Innovation Project in 2021 (SY20210215) supported by Northwest A&F University.

Data Availability Statement: The data presented in this study are available in this article.

Conflicts of Interest: The authors declare no conflict of interest.

References

1. FAOSTAT. Food and Agriculture Organization of the United States. 2022. Available online: https://www.fao.org/faostat/en/#data/QCL (accessed on 16 March 2022).
2. Cheng, T.; Tang, J.M.; Yang, R.; Xie, Y.C.; Chen, L.; Wang, S.J. Methods to obtain thermal inactivation data for pathogen control in low-moisture foods. *Trends Food Sci. Technol.* **2021**, *112*, 174–187. [CrossRef]
3. Hou, L.X.; Liu, Q.Q.; Wang, S.J. Efficiency of industrial-scale radio frequency treatments to control *Rhyzopertha dominica* (Fabricius) in rough, brown, and milled rice. *Biosyst. Eng.* **2019**, *186*, 246–258. [CrossRef]
4. Yu, D.; Shrestha, B.L.; Baik, O.D. Temperature distribution in a packed-bed of canola seeds with various moisture contents and bulk volumes during radio frequency (RF) heating. *Biosyst. Eng.* **2016**, *148*, 55–67. [CrossRef]
5. Verma, T.; Chaves, B.D.; Irmak, S.; Subbiah, J. Pasteurization of dried basil leaves using radio frequency heating: A microbial challenge study and quality analysis. *Food Control.* **2021**, *124*, 107932. [CrossRef]
6. Ballom, K.; Dhowlaghar, N.; Tsai, H.C.; Yang, R.; Tang, J.M.; Zhu, M.J. Radiofrequency pasteurization against *Salmonella* and *Listeria monocytogenes* in cocoa powder. *LWT-Food Sci. Technol.* **2021**, *145*, 111490. [CrossRef]
7. Ling, B.; Ouyang, S.H.; Wang, S.J. Radio-frequency treatment for stabilization of wheat germ: Storage stability and physicochemical properties. *Innov. Food Sci. Emerg.* **2019**, *52*, 158–165. [CrossRef]

8. Zhang, S.; Zhang, L.H.; Lan, R.G.; Zhou, X.; Kou, X.X.; Wang, S.J. Thermal inactivation of *Aspergillus flavus* in peanut kernels as influenced by temperature, water activity and heating rate. *Food Microbiol.* **2018**, *76*, 237–244.
9. Nevarez, L.; Vasseur, V.; Le Dréan, G.; Tanguy, A.; Guisle-Marsollier, I.; Houlgatte, R.; Barbier, G. Isolation and analysis of differentially expressed genes in *Penicillium glabrum* subjected to thermal stress. *Microbiology* **2008**, *154*, 3752–3765. [CrossRef]
10. Ozturk, S.; Liu, S.X.; Xu, J.; Tang, J.M.; Chen, J.R.; Singh, R.K.; Kong, F.B. Inactivation of *Salmonella* Enteritidis and *Enterococcus faecium* NRRL B-2354 in corn flour by radio frequency heating with subsequent freezing. *LWT-Food Sci. Technol.* **2019**, *111*, 782–789. [CrossRef]
11. Syamaladevi, R.M.; Tang, J.M.; Villa-Rojas, R.; Sablani, S.; Carter, B.; Campbell, G. Influence of water activity on thermal resistance of microorganisms in low-moisture foods: A review. *Compr. Rev. Food Sci. Food Saf.* **2016**, *15*, 353–370. [CrossRef]
12. Ozturk, S.; Kong, F.B.; Singh, R.K. Evaluation of *Enterococcus faecium* NRRL B-2354 as a potential surrogate of *Salmonella* in packaged paprika, white pepper and cumin powder during radio frequency heating. *Food Control.* **2020**, *108*, 106833.
13. Hou, L.X.; Kou, X.X.; Li, R.; Wang, S.J. Thermal inactivation of fungi in chestnuts by hot air assisted radio frequency treatments. *Food Control.* **2018**, *93*, 297–304. [CrossRef]
14. Lin, B.Y.; Zhu, Y.F.; Zhang, L.H.; Xu, R.Z.; Guan, X.Y.; Kou, X.X.; Wang, S.J. Effect of physical structures of food matrices on heat resistance of *Enterococcus faecium* NRRL-2356 in wheat kernels, flour and dough. *Foods.* **2020**, *9*, 1890. [CrossRef] [PubMed]
15. Liu, S.X.; Ozturk, S.; Xu, J.; Kong, F.B.; Gray, P.; Zhu, M.J.; Sablani, S.S.; Tang, J.M. Microbial validation of radio frequency pasteurization of wheat flour by inoculated pack studies. *J. Food Eng.* **2018**, *217*, 68–74. [CrossRef]
16. Lopez-Galvez, F.; Posada-Izquierdo, G.D.; Selma, M.V.; Perez-Rodriguez, F.; Gobet, J.; Gil, M.I.; Allende, A. Electrochemical disinfection: An efficient treatment to inactivate *Escherichia coli* O157:H7 in process wash water containing organic matter. *Food Microbiol.* **2012**, *30*, 146–156. [CrossRef]
17. Villa-Rojas, R.; Tang, J.M.; Wang, S.J.; Gao, M.X.; Kang, D.H.; Mah, J.H.; Gray, P.; Sosa-Morales, M.E.; Lopez-Malo, A. Thermal inactivation of Salmonella enteritidis PT 30 in almond kernels as influenced by water activity. *J. Food Prot.* **2013**, *76*, 26–32. [CrossRef]
18. Zhang, B.H.; Zhang, L.H.; Cheng, T.; Guan, X.Y.; Wang, S.J. Effects of water activity, temperature and particle size on thermal inactivation of *Escherichia coli* ATCC 25922 in red pepper powder. *Food Control.* **2020**, *107*, 106817. [CrossRef]
19. Tang, J.; Ikediala, J.N.; Wang, S.; Hansen, J.D.; Cavalieri, R.P. High-temperature-short-time thermal quarantine methods. *Postharvest Biol. Technol.* **2000**, *21*, 129–145. [CrossRef]
20. Hou, L.X.; Wu, Y.; Wang, S.J. Thermal death kinetics of *Cryptolestes pusillus* (Schonherr), *Rhyzopertha dominica* (Fabricius), and *Tribolium confusum* (Jacquelin du Val) using a heating block system. *Insects* **2019**, *10*, 119. [CrossRef]
21. Hou, L.X.; Ling, B.; Wang, S.J. Kinetics of color degradation of chestnut kernel during thermal treatment and storage. *Int. J. Agric. Biol. Eng.* **2015**, *8*, 106–115.
22. Mafart, P.; Couvert, O.; Gaillard, S.; Leguerinel, I. On calculating sterility in thermal preservation methods: Application of the Weibull frequency distribution model. *Int. J. Food Microbiol.* **2002**, *72*, 107–113. [CrossRef]
23. Ruiz-Hernández, K.; Ramírez-Rojas, N.Z.; Meza-Plaza, E.F.; García-Mosqueda, C.; Jauregui-Vázquez, D.; Rojas-Laguna, R.; Sosa-Morales, M.E. UV-C treatments against *Salmonella* Typhimurium ATCC 14028 in inoculated peanuts and almonds. *Food Eng. Rev.* **2021**, *13*, 706–712. [CrossRef]
24. van Boekel, M. On the use of the Weibull model to describe thermal inactivation of microbial vegetative cells. *Int. J. Food Microbiol.* **2002**, *74*, 139–159. [CrossRef]
25. Gil, M.M.; Miller, F.A.; Brandão, T.R.S.; Silva, C.L.M. Mathematical models for prediction of temperature effects on kinetic parameters of microorganisms' inactivation: Tools for model comparison and adequacy in data fitting. *Food Bioprocess. Technol.* **2017**, *10*, 2208–2225. [CrossRef]
26. Guan, X.; Lin, B.; Xu, Y.; Bai, S.; Li, R.; Wang, S. Thermal inactivation kinetics for *Acidovorax citrulli* on watermelon seeds as influenced by seed component, temperature, and water activity. *Biosyst. Eng.* **2021**, *210*, 223–234. [CrossRef]
27. Hansen, J.D.; Wang, S.J.; Tang, J.M. A cumulated lethal time model to evaluate efficacy of heat treatments for codling moth *Cydia pomonella* (L.) (Lepidoptera: Tortricidae) in cherries. *Postharvest Biol. Technol.* **2004**, *33*, 309–317. [CrossRef]
28. Cheng, T.; Li, R.; Kou, X.X.; Wang, S.J. Influence of controlled atmosphere on thermal inactivation of *Escherichia coli* ATCC 25922 in almond powder. *Food Microbiol.* **2017**, *64*, 186–194. [CrossRef]
29. Karagöz, I.; Moreira, R.G.; Castell-Perez, M.E. Radiation D_{10} values for *Salmonella* Typhimurium LT2 and an *Escherichia coli* cocktail in pecan nuts (Kanza cultivar) exposed to different atmospheres. *Food Control.* **2014**, *39*, 146–153. [CrossRef]
30. Blessington, T.; Theofel, C.G.; Mitcham, E.J.; Harris, L.J. Survival of foodborne pathogens on inshell walnuts. *Int. J. Food Microbiol.* **2013**, *166*, 341–348. [CrossRef]
31. Zheng, A.J.; Zhang, L.H.; Wang, S.J. Verification of radio frequency pasteurization treatment for controlling *Aspergillus parasiticus* on corn grains. *Int. J. Food Microbiol.* **2017**, *249*, 27–34. [CrossRef]
32. Li, R.; Kou, X.X.; Hou, L.X.; Ling, B.; Wang, S.J. Developing and validating radio frequency pasteurisation processes for almond kernels. *Biosyst. Eng.* **2018**, *169*, 217–225. [CrossRef]
33. Wang, S.; Tiwari, G.; Jiao, S.; Johnson, J.A.; Tang, J. Developing postharvest disinfestation treatments for legumes using radio frequency energy. *Biosyst. Eng.* **2010**, *105*, 341–349. [CrossRef]
34. Hou, L.X.; Ling, B.; Wang, S.J. Development of thermal treatment protocol for disinfesting chestnuts using radio frequency energy. *Postharvest Biol. Technol.* **2014**, *98*, 65–71. [CrossRef]

35. Ling, B.; Hou, L.X.; Li, R.; Wang, S.J. Storage stability of pistachios as influenced by radio frequency treatments for postharvest disinfestations. *Innov. Food Sci. Emerg.* **2016**, *33*, 357–364. [CrossRef]
36. Kaur, B.P.; Rao, P.S. Modeling the combined effect of pressure and mild heat on the inactivation kinetics of *Escherichia coli, Listeria innocua*, and *Staphylococcus aureus* in black tiger shrimp (*Penaeus monodon*). *Front. Microbiol.* **2017**, *8*, 1311. [CrossRef]
37. Dong, Q.L. Modeling the thermal resistance of *Clostridium Sporogenes* spores under different temperature, pH and NaCl concentrations. *J. Food Process. Eng.* **2011**, *34*, 1965–1981. [CrossRef]
38. Kou, X.X.; Li, R.; Zhang, L.H.; Ramaswamy, H.; Wang, S.J. Effect of heating rates on thermal destruction kinetics of *Escherichia coli* ATCC25922 in mashed potato and the associated changes in product color. *Food Control.* **2019**, *97*, 39–49. [CrossRef]
39. Acosta, O.; Usaga, J.; Churey, J.J.; Worobo, R.W.; Padilla-Zakour, O.I. Effect of water activity on the thermal tolerance and survival of *Salmonella enterica* serovars Tennessee and Senftenberg in goat's milk caramel. *J. Food Prot.* **2017**, *80*, 922–927. [CrossRef]
40. Zhang, L.H.; Kou, X.X.; Zhang, S.; Cheng, T.; Wang, S.J. Effect of water activity and heating rate on *Staphylococcus aureus* heat resistance in walnut shells. *Int. J. Food Microbiol.* **2017**, *266*, 282–288. [CrossRef]
41. Garcia, M.V.; da Pia, A.K.R.; Freire, L.; Copetti, M.V.; Sant'Ana, A.S. Effect of temperature on inactivation kinetics of three strains of *Penicillium paneum* and *P. roqueforti* during bread baking. *Food Control.* **2019**, *96*, 456–462. [CrossRef]
42. Possas, A.; Valero, A.; García-Gimeno, R.M.; Pérez-Rodríguez, F.; de Souza, P.M. Influence of temperature on the inactivation kinetics of *Salmonella* Enteritidis by the application of UV-C technology in soymilk. *Food Control.* **2018**, *94*, 132–139. [CrossRef]
43. Li, R.; Kou, X.X.; Cheng, T.; Zheng, A.J.; Wang, S.J. Verification of radio frequency pasteurization process for in-shell almonds. *J. Food Eng.* **2017**, *192*, 103–110. [CrossRef]
44. Song, X.Y.; Ma, B.; Kou, X.X.; Li, R.; Wang, S.J. Developing radio frequency heating treatments to control insects in mung beans. *J. Stored Prod. Res.* **2020**, *88*, 101651. [CrossRef]
45. Fine, F.; Ferret, E.; Gervais, P. Thermal properties and granulometry of dried powders strongly influence the effectiveness of heat treatment for microbial destruction. *J. Food Prot.* **2005**, *68*, 1041–1046. [CrossRef] [PubMed]
46. Chen, L.; Jung, J.Y.; Chaves, B.D.; Jones, D.; Negahban, M.; Zhao, Y.Y.; Subbiah, J. Challenges of dry hazelnut shell surface for radio frequency pasteurization of inshell hazelnuts. *Food Control.* **2021**, *125*, 107948. [CrossRef]

Article

Influence of Radio Frequency Heating on the Pasteurization and Drying of Solid-State Fermented *Wolfiporia cocos* Products

Yu-Fen Yen and Su-Der Chen *

Department of Food Science, National Ilan University, Number 1, Section 1, Shen-Lung Road, Yilan City 26041, Taiwan; abcz550068@gmail.com
* Correspondence: sdchen@niu.edu.tw; Tel.: +886-920518028; Fax: +886-39351892

Abstract: Rice bran and soybean residue are high in nutrients and active ingredients. They are used as media in the solid-state fermentation of *Wolfiporia cocos*. They not only reduce raw material costs, but also raise the economic value and applications of soybean residues and rice bran. After 30 days of fermentation, the moisture content (w.b.) of the *W. cocos* product was approximately 40%, requiring it to be pasteurized and dried later. The objective of this research is to use radio frequency (RF) rapid heating technology to pasteurize and dry the solid-state fermented product. A 500 g bag of solid-state fermented *W. cocos* product took only 30 and 200 s at the RF electrode gap of 15 cm to pasteurize and reduce the moisture content (w.b.) below 15%, respectively; therefore, the methods can be used instead of the traditional 60 min autoclave sterilization and 100 min hot air drying at 45 °C. After RF treatment, the fermented *W. cocos* product was white, indicating that browning was prevented; the product contained 5.03% mycelium, 9.83% crude polysaccharide, 4.43% crude triterpene, 3.54 mg gallic acid equivalent/g dry weight (DW) of total polyphenols, and 0.38 mg quercetin equivalent/g DW of flavonoid contents and showed a good antioxidant capacity.

Keywords: radio frequency (RF); pasteurization; drying; *Wolfiporia cocos*; solid-state fermentation

Citation: Yen, Y.-F.; Chen, S.-D. Influence of Radio Frequency Heating on the Pasteurization and Drying of Solid-State Fermented *Wolfiporia cocos* Products. *Foods* **2022**, *11*, 1766. https://doi.org/10.3390/foods11121766

Academic Editors: Shaojin Wang, Rui Li and Susana Casal

Received: 14 March 2022
Accepted: 10 June 2022
Published: 15 June 2022

Publisher's Note: MDPI stays neutral with regard to jurisdictional claims in published maps and institutional affiliations.

Copyright: © 2022 by the authors. Licensee MDPI, Basel, Switzerland. This article is an open access article distributed under the terms and conditions of the Creative Commons Attribution (CC BY) license (https://creativecommons.org/licenses/by/4.0/).

1. Introduction

Wolfiporia cocos is a medicinal and edible fungi that is mainly harvested in China. Polysaccharides, triterpenoids, and bioactive compounds are abundant in *W. cocos*. The current demand for fungal health foods is high, with their market growing each year. If grains can be used as solid-state fermented media to provide suitable carbon and nitrogen sources for *W. cocos*, the mycelial growth and production of biologically active metabolites can be promoted [1]. The biofunctions of *W. cocos* include tumor inhibition [2], immunity improvement [3], and anti-inflammatory, antiaging, hypoglycemic [4], hypolipidemic, antibacterial [5], and antioxidant capacities. Soybean residues from soybean milk production can be rapidly dried by radio frequency (RF) energy [6], and rice bran from milled brown rice can be stabilized by RF heating [7]. When soybean residues and rice bran are mixed in a 1:1 ratio to form a solid-state medium for *W. cocos* fermentation, the cost of the process is reduced and a mycelial product rich in polysaccharides and triterpenoids is produced.

RF treatment is a type of dielectric heating. When food is placed between parallel top and bottom electrodes, the polar water molecules and charged ions in food can absorb electromagnetic radiation to generate heat via dipolar polarization and ionic movement. As a result, RF heating can overcome the issues caused by heat conduction and the convection of hot air heated from the outside to the inside, and RF energy provides volumetric and more uniform overall heating, deep penetration, and a moisture self-balance effect [8–10]. Because RF heating is an emerging processing technology, RF equipment has been used to study the rapid sterilization and drying of agricultural products or food [11].

The dielectric properties of vegetable powders, such as onion, chill, broccoli, tapioca flour, and potato starch, decrease with frequency and compacted density and increase with moisture content or temperature. The RF heating rates of vegetable powders, which range from 0.56 to 2.12 °C/s, are linearly related to moisture and the dielectric loss factor. RF technology has a fast-heating rate and a deep penetration depth, indicating that it may be an effective method for quickly pasteurizing dried vegetable powder, while maintaining a high product quality [12]. By intermittently rearranging layers during hot-air-assisted RF drying, carrot quality and heat uniformity are improved.

RF heating has been applied in food pasteurization. After 90 s of RF treatment, the reductions in *Salmonella* Typhimurium and *Escherichia coli* O157:H7 in creamy or chunky peanut butter were greater than 4 log CFU/g, and food quality was not affected [13]. The pathogens in black and red peppers were significantly reduced by RF heating for 50 and 40 s, respectively [14]. Heating prepackaged white bread to 58 °C or higher using combined RF and hot air treatment resulted in 4 log reductions in *Penicillium citrinum* spores and extended its storage time [15]. RF assisted the thermal processing pasteurization of low moisture content food, such as egg-white powder [16], powder infant formula milk [17], and walnut shells [18]. RF heating selectively killed the pathogens without damaging the food product due to the larger difference in the dielectric loss factor between target microorganisms and host foods [10,19].

Zhou and Wang [9] thoroughly overviewed recent advances in the RF drying of food and agricultural products. Fresh macadamia nuts [20] and walnuts [21] were dried using hot-air-assisted RF drying to significantly reduce drying time compared with drying only with hot air. Furthermore, the layer arrangement of carrot slices improved heat uniformity and quality produced by hot-air-assisted RF drying [22]. In addition, the RF vacuum system was controlled at 0.02 MPa, the final temperature of kiwi fruits was 60 °C, the drying time was reduced by 65% compared with hot air drying at 60 °C, and the quality of the RF vacuum-dried kiwi fruits was higher [23]. The total drying time of chicken powders was reduced, and the umami flavor of chicken powder was improved by vacuum RF drying [24].

After 30 days of solid-state fermentation, *W. cocos* has to be pasteurized to stop the fermentation reaction, and then it can be dried for storage. Traditional sterilization is achieved by autoclaving and then drying by hot or cold air [25], which are time-and energy-intensive. Furthermore, long high-temperature treatments may destroy the active ingredients in solid-state fermented *W. cocos* products. Therefore, the objectives of this study are to investigate the suitability of RF heating for the pasteurization and drying of solid-state fermented *W. cocos* products. The quality attributes of the RF-treated products were analyzed with bioactive components and according to the color of the products, and then the results were compared with those produced by traditional autoclaving and hot air drying.

2. Materials and Methods

2.1. Materials

Soybean residues were obtained from Kuang Chuan Dairy Co., Ltd. (Taoyuan, Taiwan). Rice bran was purchased from Jiyuan Farm (Yilan, Taiwan). *Wolfiporia cocos* (BCRC 36022) was purchased from Bioresource Collection and Research Center (Hsinchu, Taiwan). Potato dextrose agar (PDA) and potato liquid broth (PDB) were purchased from Difco Co., Ltd. (Sparks, MD, USA). Gallic acid, quercetin, ascorbic acid, butylated hydroxyanisole (BHA), ethylenediaminetetraacetic acid (EDTA), 1,1-diphenyl-2-picryl hydrazyl (DPPH), ferrozine, ferrous chloride ($FeCl_2 \cdot 4H_2O$), trichloroacetic acid (TCA), Folin–Ciocalteu phenol reagent, ergosterol standard, vanillin ($C_8H_8O_3$), and perchloric acid were purchased from Sigma Chemical Company (St. Louis, MO, USA). We obtained 99% methanol, 95% ethanol, sodium carbonate (Na_2CO_3), potassium dihydrogen phosphate (KH_2PO_4), ferric chloride ($FeCl_3$), glucose standard ($C_6H_{12}O_6$), sodium hydroxide (NaOH), and a phenol solution from WAKO Pure Chemical Industries, Ltd. (Osaka, Japan). Acetate and sulfuric acids were purchased from Union Chemical Works (Taipei, Taiwan).

2.2. Equipment

A radio frequency with hot air equipment (40.68 MHz, 5 kW, Yh-Da Biotech Co., Ltd., Yilan, Taiwan), electric oven (Channel DCM-45, Yilan, Taiwan), high-speed grinder (RT-40, Sci-Mistry Co., Ltd., Yilan, Taiwan), horizontal laminar flow hood (4HT-24, Sage Vision Co., Ltd., New Taipei, Taiwan), constant temperature incubator (LM-600R, Yihder Co., Ltd., New Taipei, Taiwan), high-temperature steam vertical autoclave (Tommy SS-325, Tokyo, Japan), vacuum concentrator (Eyela Oil Bath Osb-2000, Tokyo, Japan), centrifuge (Hsiangtai Centrifuge, Yihua Company, New Taipei, Taiwan), high-speed batch top centrifuge (Hermle Z300, Wehingen, Germany), vortex mixer (Vortex Genie 2, Scientific Industries, Inc., New York, NY, USA), spectrophotometer (Model U-200l, Hitachi Co., Tokyo, Japan), electronic precision scale, infrared thermometer (TM-300, Tenmars Electronics Co., Ltd., Taipei, Taiwan), colorimeter (Hunter Lab, Color Flex, Hunter Associates Laboratory Inc., Reston, VA, USA), ultrasonic cleaner (DC-600H, Delta, Yuantuo Technology Ltd., Taichung, Taiwan), fiber optic thermometer (FOB100, Omega Engineering, Norwalk, CT, USA), thermal imaging camera (TIM03, Zytemp, Hsinchu, Taiwan), multifunctional infrared thermometer (Testo104-IR, Hot Instruments Co., Ltd., New Taipei, Taiwan), microwave extraction system (Bio-Promotion Co., Ltd., Taoyuan, Taiwan), and HPLC equipment (Waters Co., Milford, MA, USA), which included a Waters^{TM}510 pump, Waters^{TM}717 autosampler, Athena C18 column (4.6 × 250 mm, 5 μm), and Waters^{TM}486 UV–Vis detector, were used in this study.

2.3. Sample Preparation

2.3.1. Maintenance and Pre-Activation of W. cocos

Wolfiporia cocos (BCRC 36022) was inoculated on a PDA plate medium and cultivated in a 25 °C incubator for 7 days. The strain was cut into 1 cm^2 pieces and the pieces were inoculated in a 500 mL flask with 150 mL of presterilized PDA medium at 25 °C and 150 rpm shaking for 7 d of pre-activation.

2.3.2. Solid-State Fermentation of W. cocos

The RF-dried soybean residue [6] and RF-stabilized rice bran [7] were mixed as a 1:1 ratio to form a solid-state medium (40% moisture content) in a plastic bag, and then sterilized in a 121 °C autoclave. After cooling, 10 mL of the pre-activated *W. cocos* solution was inoculated into 500 g of the solid-state medium and cultured for 30 d at 25 °C in an incubator.

2.4. Pasteurizing and Drying Solid-State Fermented W. cocos Products

2.4.1. RF Output Power Measurement

The polypropylene (PP) plastic bag (8 × 9 × 29 cm) with 500 g solid-state fermented *W. cocos* product was horizontally placed between two parallel electrode plates in the hot-air-assisted RF equipment, and the RF output power was measured for different electrode plate gaps. Because the maximum current and output power of the 5 kW, 40.68 MHz RF equipment were 1.6 A and 5 kW, respectively, the output current (I, A) was measured three times and the output RF power was calculated using the following formula:

$$\text{power output (kW)} = (I/1.6) \times 5 \tag{1}$$

2.4.2. RF Pasteurization

The solid-state fermented *W. cocos* products were heated by RF energy and samples were taken out every 10 s, and the surface temperature at three locations was measured with an infrared thermometer. The fermented products of various RF pasteurization times were inoculated on a PDA plate and cultured for 7 days in a 25 °C incubator to observe the lethal situation of *W. cocos*.

2.4.3. Heating and Drying Curves of Solid-State Fermented *W. cocos* Products during Hot-Air-Assisted RF Drying

During RF drying, the polypropylene (pp) plastic bag of the solid-state fermented *W. cocos* product was left open to allow water vapor to evaporate. The sample was removed at periodic intervals (20 s) to measure the weight change with an electronic balance and surface temperature with an infrared thermometer to determine the drying curve and temperature profile during RF drying. Moreover, the center and internal (2 cm from the edge) temperatures of the sample were determined by inserting a fiber optic thermometer.

2.4.4. Autoclaving and Hot-Air-Drying of the Solid-State Fermented *W. Cocos* Products

The PP plastic bag containing 500 g of solid-state fermented *W. cocos* product was heated in a 121 °C autoclave for 60 min, and then it was dried in a 45 °C hot-air-drying apparatus for 180 min. To determine the drying curve and temperature profile during hot air drying, the weight and temperature changes of the sample were measured at fixed time intervals.

2.5. Extraction of Solid-State Fermented *W. cocos* Products

The solid-state fermented *W. cocos* product was weighed to obtain a 2.5 g sample, and 50 mL of water or ethanol was added for 5 min microwave extraction, as described by Chen and Chen [25]. The hot water or ethanol extract was freeze-dried and dissolved to prepare the 20 mg/mL hot water or ethanol extract.

2.6. Analytical Methods

2.6.1. Moisture Content

Weigh 5 g of ground almonds in an aluminum dish and dry them in an oven at 105 °C for 12 h; then, remove and weigh them after reaching a constant.

$$\text{Moisture content (wet basis)} = (W_i - W_f)/W_i \times 100\%. \tag{2}$$

$$\text{Solid content (wet basis)} = W_f/W_i \times 100\%. \tag{3}$$

where W_f is the weight (g) of the dried sample and W_i is the initial weight (g) of the sample.

The weight (W_t) of sample during drying was measured, and then the dry basis moisture content of the sample during drying was calculated with the following equation:

$$\text{Moisture content (dry basis)} = (W_t - W_o)/W_o \text{ (g water/g dry material)} \tag{4}$$

where W_t is the weight (g) of the sample at drying time t and W_o is calculated by $W_i \times$ solid content.

2.6.2. Color Measurement

The color of sample was measured according to Chen et al. [7] with a color difference meter and standardized against a calibration white plate (X = 82.48, Y = 84.23, Z = 99.61; L* = 92.93, a* = − 1.26, b* = 1.17). The parameters determined were the degrees of lightness (L*), redness (+a*) or greenness (−a*), and yellowness (+b*) or blueness (−b*). All experiments were performed in six repetitions.

2.6.3. Mycelium Determination of Solid-State Fermented *W. Cocos* Products

The ergosterol content represents mycelium in the mycelial fermentation product, following the process described by Chang et al. [26], with some modifications. The methanol supernatant from extract was filtered with a 0.22 μm membrane after centrifugation; its ergosterol content was determined by HPLC using a C18 column (250 × 4.6 mm) using 100% methanol as the mobile phase, a flow rate of 1 mL/min, an injection volume of 20 μL, and a UV detector at 282 nm.

2.6.4. Crude Polysaccharide Analysis

With some modifications, the crude polysaccharide content was determined using the method described by Dubois et al. [27]. The hot water extract of the dried sample was mixed with four volumes of 95% ethanol and stirred the mixture vigorously, which was then collected by centrifugation at 5000× g for 20 min. The crude polysaccharide precipitate was washed twice with 95% (v/v) ethanol and then dried at 80 °C to remove residual ethanol. Then, the crude polysaccharides were dissolved in 1 mL of 1 N NaOH and determined the reducing sugar content of the supernatant using the phenol–sulfuric acid method.

2.6.5. Crude Triterpenoid Analysis

The crude triterpenoid content was measured according to Sun et al. [28] with some modifications. In an 80 °C dry bath incubator, the ethanol extract (0.1 mL) was evaporated to dryness. The dried extract was redissolved in 0.4 mL of 5% vanillin–acetic acid solution and 1 mL of perchloric acid solution at 60 °C. After a 15 min reaction, the extract was cooled to room temperature in an ice bath, and 5 mL of acetic acid was added. After 15 min of reaction, the absorbance at 548 nm was measured using a spectrophotometer to determine the triterpenoid content.

2.6.6. Total Phenol Analysis

The concentration of total phenolic compounds was determined using the method described by Antolovich et al. [29], with some modifications. A 0.2 mL sample of ethanol extract was mixed with 1 mL of Folin–Ciocalteu phenol reagent and 0.8 mL of 7.5% Na_2CO_3. Following the addition, the solution was incubated in the dark for 30 min at room temperature. The absorbance of the solution was measured at 765 nm and compared to a gallic acid calibration curve (0–500 ppm) using a linear regression equation of y = 0.0046x + 0.22 (R^2 = 0.993).

2.6.7. Total Flavonoid Analysis

Total flavonoids were analyzed using a modified version of the method described by Christel et al. [30]. A 1 mL extract sample was mixed with 1 mL of 2% methanolic $AlCl_3$. The solution was incubated at room temperature for 10 min. The absorbance of the solution was measured at 430 nm and the result was compared to a quercetin calibration curve (0–100 ppm) using a linear regression equation of y = 0.0206x + 0.0241 (R^2 = 0.999).

2.6.8. DPPH Radical Scavenging Activity

The DPPH free-radical-scavenging capacity of the extracts of the mycelial products obtained by solid-state fermentation was determined by following the method of Xu and Chang [31], with minor modifications. Briefly, 2 mL of the ethanol extract sample was mixed with a 2 mL ethanol solution of DPPH radical (final concentration was 0.2 mM). The mixture was vortexed vigorously for 1 min and then left at room temperature in the dark for 30 min. The absorbance of sample was then measured using a spectrophotometer at 517 nm against an ethanol blank. Ascorbic acid and BHA were used as controls. DPPH scavenging activity (%) = [1 − ($ABS_{sample}/ABS_{control}$)] × 100%.

2.6.9. Ferric-Reducing Antioxidant Power (FRAP) Assay

The FRAP assay was carried out by following the method of Xu and Chang [31]. A 2 mL ethanol extract sample was added to a 1 mM $FeCl_2·4H_2O$ (0.1 mL). The reaction was started by adding 0.25 mM ferrozine (0.2 mL). The mixture was vigorously shaken and allowed to stand for 10 min at room temperature. The absorbance was taken at 562 nm using a visible spectrophotometer. Ascorbic acid and EDTA were used as controls. Ferric reducing antioxidant power (%) = [1 − ($ABS_{sample}/ABS_{control}$)] × 100%.

2.6.10. Reducing Power

A 2.5 mL ethanol extract sample was mixed with 0.2 mL of 0.2 M phosphate buffer and 2.5 mL of 1% potassium ferricyanide. The mixture was incubated at 50 °C for 20 min. Approximately 2.5 mL of 10% trichloroacetic acid was added to the mixture. The mixture was then centrifuged for 10 min at 3000 rpm and the supernatant (5 mL) was mixed with 5 mL of distilled water and 1 mL of 0.1% ferric chloride. The absorbance was monitored at 700 nm using a spectrophotometer. Ascorbic acid and BHA were used as controls [32].

2.7. Statistical Analysis

The experimental results were presented as mean ± standard deviation (SD). A one-way analysis of variance (ANOVA) was performed and subsequently subjected to Duncan's multiple range tests of treatment mean by using Statistical Analysis System (SAS 9.4, SAS Institute, Cary, NC, USA), and the significant level was set at 0.05.

3. Results and Discussion

3.1. RF Pasteurization Conditions of Solid-State Fermented W. cocos Products

Figure 1 shows the RF output power of 500 g solid-state fermented *W. cocos* product placed between of two parallel electrode plates with different electrode gaps. The smaller the gap, the larger the RF output power; therefore, a gap for the electrode plate of 14, 15, and 16 cm was chosen for comparison of the temperature profile.

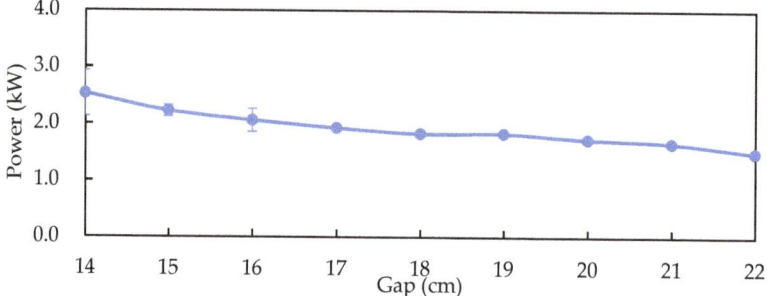

Figure 1. RF power output for different electrode gaps of 500 g solid-state fermented *W. cocos* product. Data are expressed as the mean ± SD (n = 3).

The average temperature profile of RF pasteurization (Figure 2) shows that the fermented product reached temperatures above 80 °C in less than 30 s with electrode plate gaps of 14, 15, and 16 cm. The temperature began to stabilize above 95 °C after 40 s of RF pasteurization. Because of the current output stability, we chose an electrode plate gap of 15 cm as the RF pasteurization condition for the solid-state fermented *W. cocos* products.

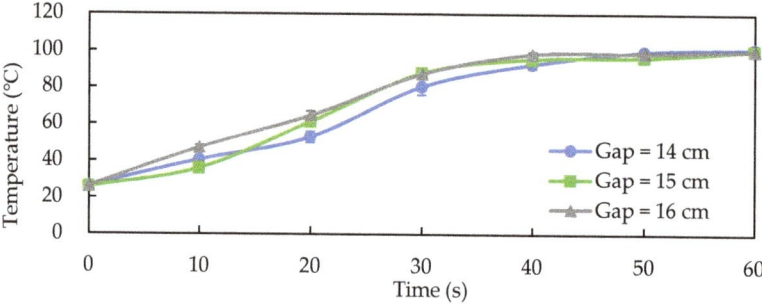

Figure 2. The temperature profile of 500 g solid-state fermented *W. cocos* mycelia product for different electrode gaps during RF pasteurization. Data are expressed as the mean ± SD (n = 3).

The central and internal temperatures of the samples were substantially higher than the surface temperature. After 1 min of RF pasteurization and 10 min of room-temperature cooling, the central and internal temperatures of the fermented product remained above 80 °C (Figure 3), which exceeded the conditions for inactivating *W. cocos* (*W. cocos* died above 80 °C in previous experiments; data not shown in this paper).

0 s   10 s   20 s   30 s   40 s   50 s   60 s

Figure 3. Center and internal temperature profiles of 500 g of solid-state fermented *W. cocos* product during RF with a gap of 15 cm, pasteurization of 1 min, and cooling.

As a test of the pasteurization effect, solid-state fermented *W. cocos* products were RF pasteurized for 0 to 60 s and then incubated in PDA for 7 d (Figure 4). We observed that the fermented products only required 30 s of RF heating to achieve the pasteurization requirements. Because RF generates heat by making the polar water molecules in the sample rapidly rotate and oscillate, the heating rate of the samples during RF heating depended on the moisture content, loading, dielectric constant, and RF electrode gap.

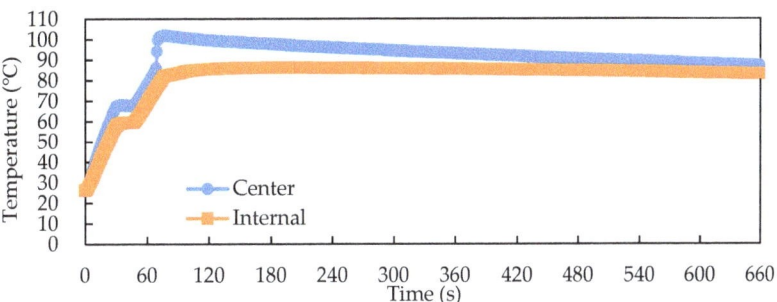

Figure 4. RF pasteurization with an electrode gap of 15 cm for solid-state fermented *W. cocos* products after 7 d of cultivation.

In previous studies, white bread (37.1% moisture content) was treated for less than 5 min with HARF to reach 58 °C or higher, which reduced the fungi spores and thereby extended the storage time [15]. Walnut shells (9% moisture content) were heated to over 60 °C by RF pasteurization for 10 min to kill *Staphylococcus aureus*. RF heating inhibited nucleic acid metabolism, translation, cell membrane transport, and cell wall biosynthesis, which eventually led to the cell death of *Staphylococcus aureus* [18]. Therefore, RF pasteurization is suitable for low-moisture-content foods due to the large difference in the dielectric loss between the target microorganisms and host foods. The method selectively heated and killed the microorganisms without damaging the food product [19].

Figure 5 presents the temperature profile of the solid-state fermented *W. cocos* products produced by an autoclave at 121 °C for 60 min and a cooling process. The figure shows that the temperature rose very slowly from 0 to 30 min. At 100 min, the highest central temperature of 114.1 °C was reached, which then gradually decreased. This indicated that the heat transfer of the solid-state fermented *W. cocos* products was poor when using an autoclave.

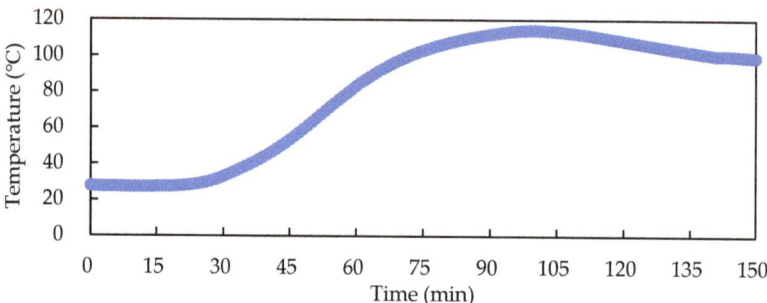

Figure 5. Center temperature profile of 500 g of solid-state fermented W. *cocos* product during 121 °C autoclaving for 60 min and cooling.

After 60 min of pasteurization in the traditional 121 °C autoclave, the solid-state fermented W. *cocos* products blackened due to the Maillard reaction (Figure 6). We observed no notable color change between the control group and the RF pasteurized solid-state fermented W. *cocos* product. This finding indicated that RF pasteurization can replace traditional autoclave treatment, reduce operation time, and improve product color preservation.

Figure 6. Effect of pasteurization method on the appearance of the solid-state fermented W. *cocos* products: (**a**) control, (**b**) RF heating for 30 s, and (**c**) 121 °C autoclaving for 60 min.

3.2. RF Drying Conditions of Solid-State Fermented W. cocos Products

After pasteurization, the solid-state fermented W. *cocos* products contained about 40% moisture content and required drying steps for storage. A 500 g packet of solid-state fermented W. *cocos* product was laid flat on a plate and dried under an electrode gap of 15 cm under the same conditions as the RF pasteurization treatment. The internal temperature of the sample increased from 45 °C to more than 100 °C after 50 s of RF drying. Due to water evaporation, the internal temperature remained around 100 °C, while the surface temperature increased from 25 °C to more than 60 °C. The surface temperature increased to more than 70 °C after 80 s of RF drying, and it also showed a flat phenomenon, indicating that it entered the latent heat stage of moisture evaporation (Figure 7).

Figure 8 shows the average temperature profile and drying curve of the solid-state fermented W. *cocos* product during RF drying. The temperature of the fermented product gradually increased and the moisture content slowly decreased during the first 50 s of RF drying. As a result, the sensible heat condition appeared in this first stage. However, after 50 s of RF drying, the average temperature of the fermented product gradually leveled off at approximately 85 °C and entered the latent heat moisture evaporation condition. The dry base moisture content of the fermented product linearly decreased with the drying time. The linear regression equation was $y = -0.0027x + 0.6806$, $R^2 = 0.9982$, indicating that the drying rate was constant in this period, with the drying rate at this time being 0.8419 g water/min, and only 200 s of RF drying time reduced the moisture content of the solid-state fermented W. *cocos* product from 40% to 15% or less.

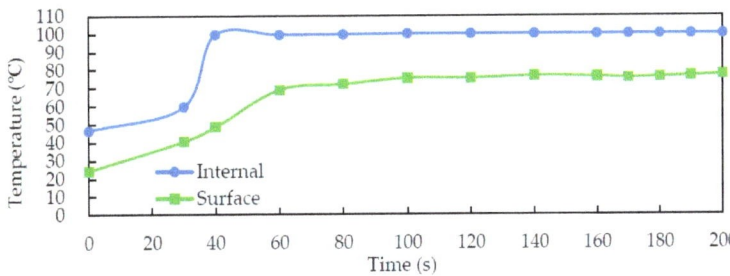

Figure 7. Internal and surface temperature profiles of the solid-state fermented *W. cocos* products during RF drying with an electrode gap of 15 cm. Data are expressed as mean ± SD (*n* = 3).

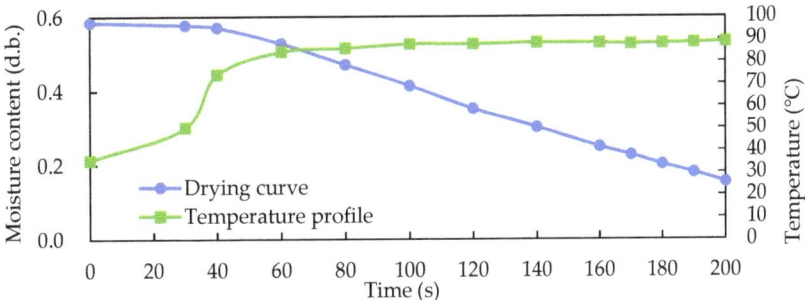

Figure 8. Average temperature profile and drying curve of the solid-state fermented *W. cocos* products during RF drying with an electrode gap of 15 cm. Data are expressed as the mean ± SD (*n* = 3).

If the solid-state fermented *W. cocos* product was used for continuous RF pasteurization and drying treatment, the temperature rising time during the drying process would be reduced by 50 s, thereby reducing the time required for the pasteurization and drying treatment of the fermented product to 180 s, which was extremely fast (Figure 9).

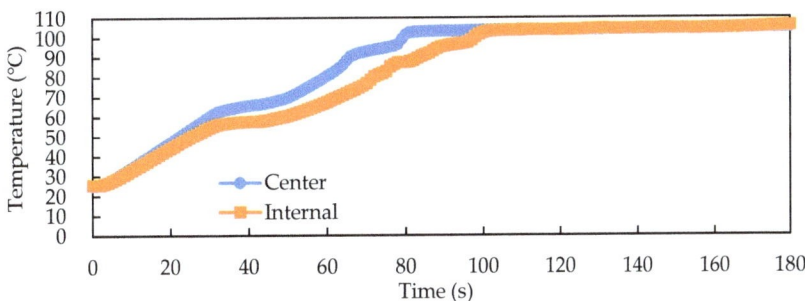

Figure 9. Center and internal temperature profiles of the solid-state fermented *W. cocos* products during continuous RF drying with an electrode gap of 15 cm, pasteurization of 30 s, and drying of 150 s.

Compared with a 45 °C hot-air-drying treatment, increasing the surface temperature of the solid-state fermented *W. cocos* products to more than 40 °C took 60 min. The constant drying rate period lasted from 0 to 60 min, the dry base moisture content gradually decreased, and then the drying rate decreased. A total of 100 min was required to reduce the moisture content of the fermented products to less than 15% in 45 °C with the hot-air-drying treatment (Figure 10). Therefore, the rapid overall effect of RF heating was able to quickly dry the solid-state fermented *W. cocos* products, demonstrating a highly efficient drying technology.

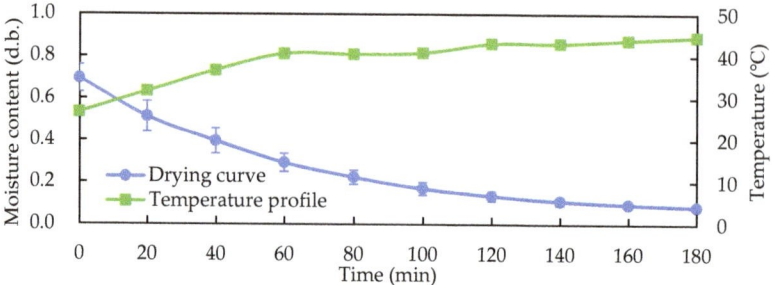

Figure 10. Temperature profile and drying curve during 45 °C hot-air drying of the solid-state fermented *W. cocos* products. Data are expressed as the mean ± SD (n = 3).

Comparing the total drying time required for in-shell walnuts using RF drying, vacuum drying, and hot-air drying, the former was the fastest (138 min), followed by vacuum drying (185 min) and hot-air drying (300 min) [9]. To improve drying efficiency, a combination of RF and hot-air drying may overcome the resistance of both heat transfer and mass transfer.

3.3. Active Component and Antioxidant Activity of the Solid-State Fermented W. cocos Mycelia Products

The solid-state fermented *W. cocos* product was subjected to RF pasteurization and drying, 121 °C autoclave and 45 °C hot-air-drying treatment; there was no significant difference in the mycelium, crude polysaccharides, crude triterpenes, total polyphenols, flavonoids, or antioxidant activity of the microwave extract. The solid-state fermented *W. cocos* mycelia product after RF pasteurization and drying contained 5.03% mycelium, 9.83% crude polysaccharides, 4.43% crude triterpenoids, 3.54 mg gallic acid equivalent/g DW of total polyphenols, and 0.38 mg quercetin equivalent/g DW of flavonoids. The supernatant of 50 mg/mL ethanol extract was assayed for antioxidant activity. The results show that the scavenging ability of DPPH free radicals and the chelating ability of ferrous iron were as high as 93.15% and 91.35%, respectively and had similar antioxidant activities to 5 mg/mL of ascorbic acid (93.44% and 91.40%, respectively), and the reducing power was as high as 0.71 (Table 1).

Table 1. Effect of different pasteurization and drying methods on the active components and antioxidant activity of the solid-state fermented *W. cocos* products.

Treatment	RF Pasteurization and RF Drying	121 °C Autoclave and 45 °C Hot-Air Drying
Mycelium (%)	5.03 ± 0.12	4.94 ± 0.05
Crude polysaccharides (%)	9.83 ± 0.24	9.35 ± 0.30
Crude triterpenoids (%)	4.43 ± 0.02	4.32 ± 0.01
Total polyphenols (mg gallic acid equivalent/g DW)	3.54 ± 0.21	3.55 ± 0.18
Flavonoids (mg quercetin equivalent/g DW)	0.38 ± 0.03	0.33 ± 0.02
Scavenging DPPH free radicals (%)	93.15 ± 2.46	92.11 ± 0.06
Chelating ferrous ion capacity (%)	91.35 ± 0.33	90.98 ± 0.29
Reducing power	0.71 ± 0.02	0.69 ± 0.03

Data are expressed as the mean ± SD (n = 4). A 50 mg/mL sample concentration was used for the antioxidant assay. Means in the same row are not significantly different (p > 0.05).

However, in the color analysis, the L* and b* values and whiteness of the solid-state fermented W. *cocos* product by RF pasteurization and drying treatment were significantly higher than those produced by 121 °C autoclaving and the 45 °C hot-air-drying treatment. It showed that the color of the solid-state fermented W. *cocos* product was better preserved through the RF pasteurization and drying process compared with the autoclaving and hot-air-drying process (Table 2).

Table 2. Effect of different pasteurization and drying methods on the color of the solid-state fermented W. *cocos* product.

Sample	L*	a*	b*	Whiteness (%)
RF pasteurization and RF drying	53.30 ± 0.31 [a]	9.64 ± 0.03 [a]	27.70 ± 0.18 [a]	44.86 ± 0.21 [a]
121 °C autoclave and hot-air drying	41.74 ± 0.03 [b]	9.68 ± 0.06 [a]	22.65 ± 0.15 [b]	36.75 ± 0.07 [b]

* Data are expressed as the mean ± SD ($n = 6$). [a,b] Means with different superscripts in the same column are significantly different ($p < 0.05$).

4. Conclusions

In this study, RF heating required only 30 s to stop fermentation and 200 s to dry 500 g of solid-state fermented W. *cocos* product, so it is superior to the traditional sterilization treatment requiring 60 min of autoclaving at 120 °C and followed by the 180 min of 45 °C hot-air-drying method. There was no significant difference in the active components or antioxidant activity between RF and traditional process. RF heating offers a novel method for pasteurizing and drying solid-state fermented fungi products.

Author Contributions: S.-D.C.: supervision, writing—review and editing, and project administration; Y.-F.Y.: investigation, analysis of the data, and writing of the original draft manuscript. All authors have read and agreed to the published version of the manuscript.

Funding: This research received no external funding.

Data Availability Statement: The data presented in this study are available in this article.

Acknowledgments: We thank Kuang Chuan Dairy Co. Ltd. for providing soybean residues that were a byproduct of soybean milk. We thank Yen-Hui Chen who helped with statistical analysis and discussion.

Conflicts of Interest: The authors declare no conflict of interest.

References

1. Chen, B.-H.; Yang, Y.-C.; Chen, S.-D. Study on hypoglycemic activities of *Poria cocos* solid-state fermented products. *Taiwan J. Agric. Chem. Food Sci.* **2017**, *53*, 34–42.
2. Chu, B.F.; Lin, H.C.; Huang, X.W.; Huang, H.Y.; Wu, C.P.; Kao, M.C. An ethanol extract of *Poria cocos* inhibits the proliferation of non-small cell lung cancer A549 cells via the mitochondria-mediated caspase activation pathway. *J. Funct. Foods* **2016**, *23*, 614–627. [CrossRef]
3. Zhang, J.H.; Tatsumi, E.; Ding, C.H.; Li, L.T. Angiotensin I-converting enzyme inhibitory peptides in douchi, a Chinese traditional fermented soybean product. *Food Chem.* **2006**, *98*, 551–557. [CrossRef]
4. Li, T.H.; Hou, C.C.; Chang, C.L.T.; Yang, W.C. Anti-hyperglycemic properties of crude extract and triterpenes from *Poria cocos*. *Evid. Based Complementary Altern. Med.* **2011**, *2011*, 128402. [CrossRef]
5. Wang, Y.; Xu, W.; Chen, Y. Surface modification on polyurethanes by using bioactive carboxymethylated fungal glucan from *Poria cocos*. *Colloids Surf. B* **2010**, *81*, 629–633. [CrossRef]
6. Chen, Y.-H.; Yen, Y.-F.; Chen, S.-D. Study of radio frequency drying on soybean residue. *Taiwan. J. Agric. Chem. Food Sci.* **2017**, *55*, 128–291.
7. Chen, Y.-H.; Yen, Y.-F.; Chen, S.-D. Effects of radio frequency heating on the stability and antioxidant properties of rice bran. *Foods* **2021**, *10*, 810. [CrossRef]
8. Marra, F.; Zhang, L.; Lyng, J.G. Radio frequency treatment of foods: Review of recent advances. *J. Food Eng.* **2009**, *91*, 497–508. [CrossRef]

9. Zhou, X.; Wang, S. Recent developments in radio frequency drying of food and agricultural products: A review. *Dry. Technol.* **2018**, *37*, 271–286. [CrossRef]
10. Dag, D.; Singh, R.K.; Kong, F. Developments in radio frequency pasteurization of food powders. *Food Rev. Int.* **2020**, *38*, 1197–1214. [CrossRef]
11. Wang, Y.; Li, Y.; Wang, S.; Zhang, L.; Gao, M.; Tang, J. Review of dielectric drying of foods and agricultural products. *Int. J. Agric. Biol. Eng.* **2011**, *4*, 1–19.
12. Ozturk, S.; Kong, F.; Trabelsi, S.; Singh, R.K. Dielectric properties of dried vegetable powders and their temperature profile during radio frequency heating. *J. Food Eng.* **2016**, *169*, 91–100. [CrossRef]
13. Ha, J.W.; Kim, S.Y.; Ryu, S.R.; Kang, D.H. Inactivation of *Salmonella enterica* serovar Typhimurium and Escherichia coli O157: H7 in peanut butter cracker sandwiches by radio-frequency heating. *Food Microbiol.* **2013**, *34*, 145–150. [CrossRef] [PubMed]
14. Kim, S.Y.; Sagong, H.G.; Choi, S.H.; Ryu, S.; Kang, D.H. Radio-frequency heating to inactivate *Salmonella* Typhimurium and *Escherichia coli* O157: H7 on black and red pepper spice. *Int. J. Food Microbiol.* **2012**, *153*, 171–175. [CrossRef]
15. Liu, Y.; Tang, J.; Mao, Z.; Mah, J.H.; Jiao, S.; Wang, S. Quality and mold control of enriched white bread by combined radio frequency and hot air treatment. *J. Food Eng.* **2011**, *104*, 492–498. [CrossRef]
16. Wei, X.; Lau, S.K.; Reddy, B.S.; Subbiah, J. A microbial challenge study for validating continuous radio-frequency assisted thermal processing pasteurization of egg white powder. *Food Microbiol.* **2020**, *85*, 103306. [CrossRef]
17. Lin, Y.; Subbiah, J.; Chen, L.; Verma, T.; Liu, Y. Validation of radio frequency assisted traditional thermal processing for pasteurization of powdered infant formula milk. *Food Control* **2020**, *109*, 106897. [CrossRef]
18. Jiang, H.; Gu, Y.; Gou, M.; Xia, T.; Wang, S. Radio frequency pasteurization and disinfestation techniques applied on low-moisture foods. *Crit. Rev. Food Sci. Nutr.* **2020**, *60*, 1417–1430. [CrossRef]
19. Zhang, L.; Ma, H.; Wang, S. Pasteurization mechanism of S. aureus ATCC 25923 in walnut shells using radio frequency energy at lab level. *LWT-Food Sci. Technol.* **2021**, *143*, 111129. [CrossRef]
20. Wang, Y.; Zhang, L.; Johnson, J.; Gao, M.; Tang, J.; Powers, J.R.; Wang, S. Developing hot air-assisted radio frequency drying for in-shell macadamia nuts. *Food Technol. Biotechnol.* **2014**, *7*, 278–288. [CrossRef]
21. Zhang, B.; Zheng, A.; Zhou, L.; Huang, Z.; Wang, S. Developing hot air-assisted radio frequency drying protocols for in-shell walnuts. *Emir. J. Food Agric.* **2016**, *28*, 459–467. [CrossRef]
22. Wang, C.; Kou, X.; Zhou, X.; Li, R.; Wang, S. Effects of layer arrangement on heating uniformity and product quality after hot air assisted radio frequency drying of carrot. *Innov. Food Sci. Emerg. Technol.* **2021**, *69*, 102667. [CrossRef]
23. Zhou, X.; Xu, R.; Zhang, B.; Pei, S.; Liu, Q.; Ramaswamy, H.S.; Wang, S. Radio frequency-vacuum drying of kiwifruits: Kinetics, uniformity, and product quality. *Food Bioprocess Technol.* **2018**, *11*, 2094–2109. [CrossRef]
24. Ran, X.L.; Zhang, M.; Wang, Y.; Liu, Y. Vacuum radio frequency drying: A novel method to improve the main qualities of chicken powders. *J. Food Sci. Technol.* **2019**, *56*, 4482–4491. [CrossRef]
25. Chen, B.-H.; Chen, S.-D. Microwave extraction of polysaccharides and triterpenoids from solid-state fermented products of Poria cocos. *Taiwan. J. Agric. Chem. Food Sci.* **2013**, *51*, 188–194.
26. Chang, C.Y.; Lue, M.Y.; Pan, T.M. Determination of adenosine, cordycepin and ergosterol contents in cultivated *Antrodia camphorata* by HPLC method. *J. Food Drug Anal.* **2005**, *13*, 338–342. [CrossRef]
27. DuBois, M.; Gilles, K.A.; Hamilton, J.K.; Rebers, P.T.; Smith, F. Colorimetric method for determination of sugars and related substances. *Anal. Chem.* **1956**, *28*, 350–356. [CrossRef]
28. Sun, Z.; Liu, H.; Huang, Y.; Ju, J. Determination of total triterpenoid and oleanolic acid contents in Tibetan medicine Indian Swertia. *Chin. J. Ethnomed. Ethnopharmacy* **2010**, *1*, 21.
29. Antolovich, M.; Prenzler, P.; Robards, K.; Ryan, D. Sample preparation in the determination of phenolic compounds in fruits. *Analyst* **2000**, *125*, 989–1009. [CrossRef]
30. Christel, Q.D. Phenolic compounds and antioxidant activities of buckwheat (*Fagopyrum esculentum* Moench) hulls and flour. *J. Ethnopharmacol.* **2000**, *72*, 35–42.
31. Xu, B.J.; Chang, S.K.C. A comparative study on phenolic profiles and antioxidant activities of legumes as affected by extraction solvents. *J. Food Sci.* **2007**, *72*, 159–166. [CrossRef] [PubMed]
32. Hsu, J.Y.; Chen, M.H.; Lai, Y.S.; Chen, S.D. Antioxidant profile and biosafety of white truffle mycelial products obtained by solid-state fermentation. *Molecules* **2021**, *27*, 109. [CrossRef] [PubMed]

Article

Sterilizing Ready-to-Eat Poached Spicy Pork Slices Using a New Device: Combined Radio Frequency Energy and Superheated Water

Ke Wang, Chuanyang Ran, Baozhong Cui, Yanan Sun, Hongfei Fu, Xiangwei Chen, Yequn Wang and Yunyang Wang *

College of Food Science and Engineering, Northwest A&F University, Yangling 712100, China
* Correspondence: wyy10421@nwafu.edu.cn; Tel.: +86-135-7241-2298

Abstract: In this study, a new device was used to inactivate *G. stearothermophilus* spores in ready-to-eat (RTE) poached spicy pork slices (PSPS) applying radio frequency (RF) energy (27.12 MHz, 6 kW) and superheated water (SW) simultaneously. The cold spot in the PSPS sample was determined. The effects of electrode gap and SW temperature on heating rate, spore inactivation, physiochemical properties (water loss, texture, and oxidation), sensory properties, and SEM of samples were investigated. The cold spot lies in the geometric center of the soup. The heating rate increased with increasing electrode gap and hit a peak under 190 mm. Radio frequency combined superheated water (RFSW) sterilization greatly decreased the come-up time (CUT) compared with SW sterilization, and a 5 log reduction in *G. stearothermophilus* spores was achieved. RFSW sterilization under 170 mm electrode gap reduced the water loss, thermal damage of texture, oxidation, and tissues and cells of the sample, and kept a better sensory evaluation. RFSW sterilization has great potential in solid or semisolid food processing engineering.

Keywords: RF heating; superheated water; poached spicy pork slices; sterilization; quality; RTE food; *G. stearothermophilus* spores

1. Introduction

Ready-to-eat (RTE) food, also known as convenient and precooked food, is earning the attention of people now with increasing income, for it is time-saving, hassle-free, and convenient [1]. Numerous research works have focused on safety control and new processing technologies of RTE food, but the products purchased by consumers are still at risk of contamination by microbes, such as *Listeria*, *Escherichia coli*, *Enterobacter sakazakii*, etc. [2–4].

Poached spicy pork slices (PSPS), a traditional Chinese cuisine available in any Sichuan restaurant worldwide, might have a considerable market if it was developed into an RTE food. Like other RTE meat products, RTE PSPS need to be sterilized as fore-cooking contamination may occur when PSPS are exposed to air, equipment, or food handlers [5]. Theoretically, heating of meat or meat product would induce protein degradation, myofibrillar shrinkage, and consequently water loss, causing tremendous changes in texture, color, sensory properties, and so on [6,7]. To ensure that the internal temperature of the product reaches the required temperature, high-pressure steam/water sterilization at 121 °C for 30 min is commonly used in industry, which will also cause the above-mentioned changes. The intrinsic quality of a meat product is an essential factor affecting consumer preference and the degree of acceptability for consumers [8]. Therefore, it is necessary to develop an innovative sterilization method to decrease the sterilization time while maximizing the preservation of its texture, color, and nutrition.

Radio frequency (RF) is an electromagnetic wave with a frequency ranging from 3 kHz to 300 MHz. RF heats food material by driving the polar molecules' rotation and movement of charged ions. RF heating has been studied in the drying, blanching, and sterilization of food materials because of its advantages of large penetration depth and high heating efficiency [9–12]. Nonuniform heating is the main obstacle to the industrialization of RF. Various kinds of food materials have been studied, such as meat products [13], grain products [14,15], and vegetable products [16], but reports about RTE food are rare.

Radio frequency combined hot water (RFHW) heating is not a new idea. Some previous studies carried out RFHW with two steps of RF preheating and a hot water bath, discovering it can greatly shorten the heating time and improve the quality of the product. Yang and Geveke's [17] study showed that RFHW heating can shorten the sterilization time by 78% while reducing the *salmonella* in eggs by at least 5 log and causing unobservable quality change. Xu et al. [14] successively applied a hot water bath and RF to *Nostoc sphaeroides*, observing that its temperature uniformity, nutrient content, and phycocyanin stability were different compared with samples subjected to high-pressure steam sterilization.

Although there have been many studies on RFHW heating, the literature on how the superposition of the two heating methods simultaneously affect the sterilization results and quality is scarce. Our laboratory designed a device that can carry out radio frequency combined superheated water (RFSW) sterilization, which overcomes this obstacle in terms of equipment. This study aimed to do the following: (1) determine the cold spot in PSPS during RFSW heating; (2) investigate the effect of superheated water (SW) temperature and electrode gap on heating rate and inactivation of spores of *G. stearothermophilus*; and (3) compare the weight loss, texture, and sensory and morphological structure differences between sterilization methods of RFSW and SW alone.

2. Materials and Methods

2.1. PSPS Preparation

The cooking method of PSPS in this article referred to the traditional Chinese method with moderate modification. Cooking-related material such as seasonings and pork tenderloin were purchased from the Hao&Duo supermarket in Yangling, China. Spring onion, ginger, and garlic were cut into pieces. In a wok, edible oil was heated to 150 °C.

Then, spring onion, ginger, garlic, and bean paste were added and stirred constantly for 40 s. Water was poured in, and the prepared seasonings were added. After boiling, the samples were removed, cooled, and stored at 4 °C for later use within 2 days. Pork tenderloin was cut into 1.0 cm × 3.0 cm × 1.0 cm slices (length × width × thickness) and placed into a vessel. Salt, vinegar, egg white, and potato starch were added successively and mixed adequately, and then the meat was left to stand for 30 min. The pork slices were uncrumpled, and then boiled for 4 min. They were removed with a colander, placed into a sterile vessel, and stored at 4 °C for later use within 2 days. The amount of all materials is shown in Table 1.

The sample (51.00 ± 0.50 g; soup 38.50 ± 0.04 g and pork slices 12.80 ± 0.03 g) was weighed and placed in a Teflon container (Figure 1). The Teflon container was sterilized by ultraviolet radiation prior to use. The ratio of height to diameter of the material was kept at 2.2. To keep the same initial temperature, the sample was held at 25 °C for 2–4 h before heating.

A waterproof and high-temperature-resistant adhesive was used to stick a plastic nut to the selected position outside the Teflon container. The plastic nut provided a cone thread, so that the fiber optic could be squeezed to isolate water when the plastic screw was tightened inward. The PSPS was inevitably stratified due to gravity. The pork slices gathered in the bottom of the container, and edible oil floated on the soup (Figure 1).

Table 1. Materials and dosage involved in PSPS cooking (listed in the order of use).

	Soup			Pork Slices	
	Dosage (g)	Information		Dosage (g)	Information
Spring onion	15.0 ± 0.1	Hao&Duo supermarket (Used stem only)	Salt	1.50 ± 0.05	Hao&Duo supermarket
Ginger	5.0 ± 0.1	Hao&Duo supermarket (Removed the peel)	Chinese rice wine	2.0 ± 0.1	Jingzhiliaojiu, Beijing Ershang Wangzhihe Food Co., Ltd., Beijing, China
Garlic	7.0 ± 0.2	Hao&Duo supermarket (Removed the peel)	Egg white	9.0 ± 0.5	Hao&Duo supermarket
Bean paste	15.0 ± 0.1	Pixian Bean paste, Pixian Star Seasoning Co., Ltd., Pixian, China	Potato starch	9.0 ± 0.5	Xi'an Panfeng Shunda Trading Co., Ltd., Xi'an, China
Salt	4.0 ± 0.1	Hao&Duo supermarket			
Chicken extract	0.40 ± 0.05	Hao&Duo supermarket			
Sugar	4.0 ± 0.1	Hao&Duo supermarket			
Light soy sauce	2.0 ± 0.1	Haday soy sauce, Foshan Haitian Condiment Food Co., Ltd., Foshan, China			
Chinese rice wine	2.0 ± 0.1	Jingzhiliaojiu, Beijing Ershang Wangzhihe Food Co., Ltd., Beijing, China			
Water	400 ± 1	Laboratory drinking water (Food grade)			

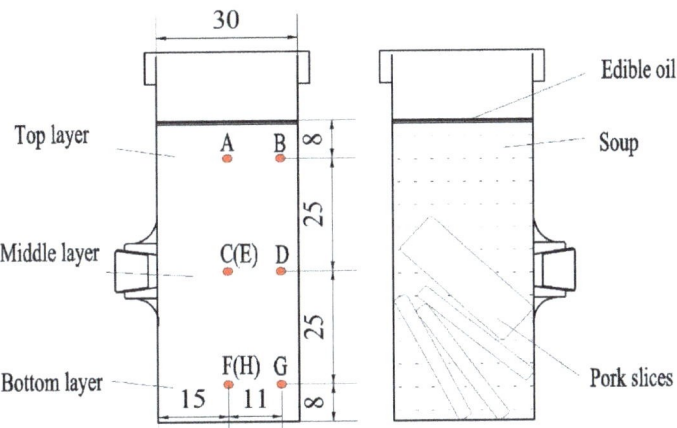

Figure 1. Position of fiber optic sensor in the Teflon container and distribution of the poached spicy pork slice sample in the Teflon container (all dimensions are in mm). Points A, B, C, D, F, and G were used to measure the soup temperature, whereas E and H were used for the pork slices' internal region.

2.2. Bacterial Culture and Spore Crop Preparation

The bacterial culture and spore crop preparation of this study was based on the method of Ahn et al. [18] with moderate modification. A strain of *G. stearothermophilus* (ATCC 7953, Shanghai Fuxiang Biotechnology Co., Ltd., Shanghai, China) was cultivated aerobically in a nutrient broth (NB, AOBOX biotechnology, Beijing, China) at 55 °C for 48 h. NB (200 μL) was evenly plated on the surface of NA (Land bridge, Beijing, China) containing 50 ppm $MnSO_4$ (Ghtech, Guangdong, China) and cultured at 55 °C for 5–10 days.

For bacteria suspension collection, 10 mL distilled water was added to each culture medium prior to scraping and pouring into a centrifuge tube. The bacteria suspension was heated at 90 °C for 30 min to inactivate the vegetative cells. Then, they were washed thrice by 6000× g centrifugation for 5 min. The collected spore suspension was resuspended in deionized water to 10^7–10^8 (CFU mL^{-1}) and stored at 4 °C for later use within 2 months.

2.3. RFSW Heating System

The RFSW heating system consisted of four parts of the SW generator (Figure 2a), an RF system (Figure 2b), sterilization kettle, and fiber optic temperature measurement system (Figure 2c). The SW generator with a power of 10 KW (TW-SP-25; Shanghai Triowin Automation Machinery Co., Ltd., Shanghai, China) could produce SW up to 130 °C and could automatically control the pressure in the sterilization kettle. The SW generator provided on-line cooling and it stopped when the internal temperature of the samples in the kettle reached the preset. A 6 KW, 27.12 MHz free running RF system (GJG-2.1-10AJY; Hebei Huashijiyuan High Frequency Equipment Co., Ltd., Hebei, China) with two parallel electrodes was used in this study. The electrode gap of the RF system ranged from 100 mm to 300 mm. The schematic of the sterilization kettle designed by our lab is shown in Figure 3. The sterilization kettle was composed of a hollow cylindrical cavity, two water pipes, and one pipe used for the fiber optic senor. For sealing, the top and bottom lids were fixed with the cavity by screws. The fiber optic temperature measurement system (HQ-FTS-D1F00, Herch Opto Electronic Technology Co., Ltd., Xi'an, China) consisted of fiber optic sensors, signal converter, and computer software. Detailed information is described in the studies by Wang et al. [19], Cui et al. [20], and Sun et al. [21].

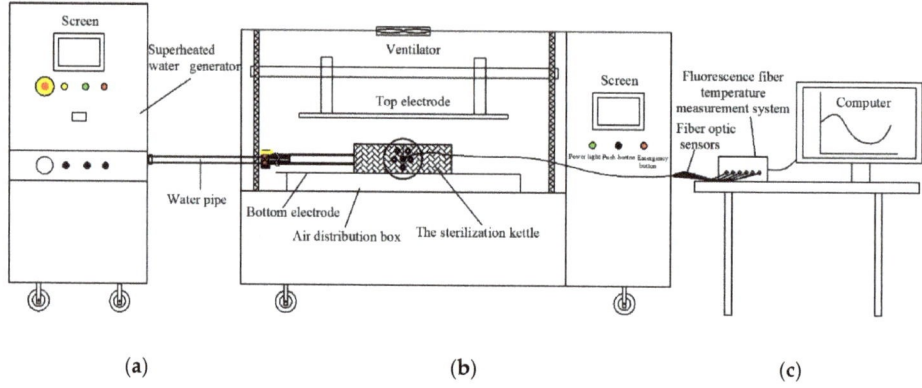

Figure 2. Simplified schematic diagram of RFSW heating system including superheated water generator (**a**), radio frequency heating system (**b**), and fiber optic temperature measurement system (**c**).

In preparation of the experiment, the SW generator was connected to the sterilization kettle with a thermal insulation water pipe. The Teflon container containing samples was positioned at the center of the bottom lid of the sterilization kettle. The fiber optic sensors were inserted into the sample at designated points (points A to H in Figure 1). Then, the top lid was tightened with bolts. The sterilization kettle was placed on the bottom electrode in the RF cavity (Figure 2b). The electrode gap was adjusted according to the test arrangement. Then, SW was pumped into the kettle, and the RF heating system was started. For safety reasons, the pressure in the kettle was maintained between 0.2 and 0.3 bar automatically. In particular, SW flowed in from the bottom of the kettle and flowed out from the upper part to fill the kettle with SW. The Teflon container containing samples was immersed in SW. The on-line cooling procedure was started immediately when the designated sterilization time was reached. Then, the SW was closed when the temperature in the kettle went down to 40 °C. This is because meat products should be cooled rapidly after heating treatment, since overheating destroys the quality of meat. By this structure, the sample can be heated by SW and RF simultaneously.

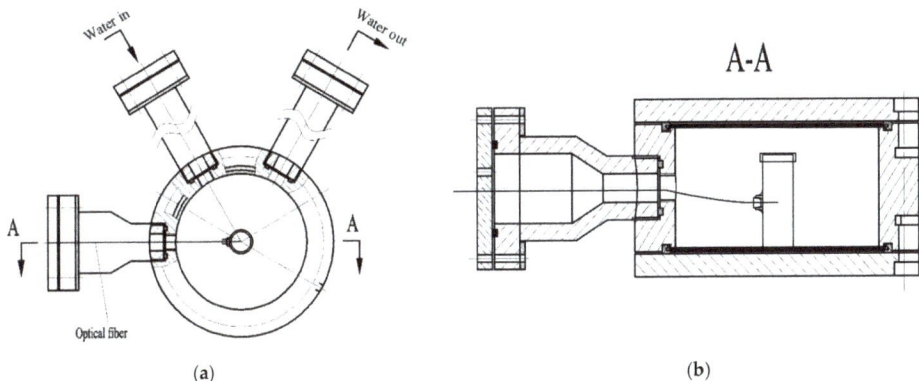

Figure 3. Schematic of the sterilization kettle (**a**) and partial section view of the sterilization kettle (**b**).

2.4. The Cold Spot and Heating Rate

To explore the cold spot of the sample in the Teflon container under RFHW heating, the fiber optic sensor was inserted at eight points distributed in three layers of PSPS. Two points (E and H) were inside the pork slices (Figure 1). The device was assembled as described in Section 2.3. The electrode gap was kept at 180 mm. Hot water set at 60 °C was pumped. Then, the RF heating system was turned on. Time was recorded when the temperature of the test point changed from 30 to 60 °C (come up time-1, CUT-1) to find the cold spot.

To evaluate the effect of SW temperature and electrode gaps on the heating rate of the samples, different SW temperatures (116 °C, 121 °C, and 124 °C) and electrode gaps (160, 170, 180, 190, and 200 mm) were chosen. The fiber optic sensor was inserted into the cold spot of the sample in the above-mentioned experiment. The time needed for the cold spot temperature of the sample to increase from 30 °C to SW temperature (come up time-2, CUT-2) was measured. Three electrode gaps in each temperature representing the slowest, the fastest, and the moderate heating rate of the cold spot were selected to obtain the temperature-time profile by measuring every minute or less in comparison with the control (SW sterilization).

2.5. Inactivation of G. stearothermophilus Spores

2.5.1. Spore Inoculation into PSPS

Before inoculation, the PSPS was maintained at 25 °C for 2–4 h. A spore suspension of 100 µL was injected into the PSPS sample using a 1 mL sterile injector (Shengguang medical instrument Co., Ltd., Henan, China) in a biosafety clean bench (YT-CJ-1N; Beijing Yataikelong Instrument Technology Co., Ltd., Beijing, China), and then, it was shaken for 30 min. The lid of the container was covered tightly anywhere outside the clean bench to prevent contamination. The initial concentration of *G. stearothermophilus* spores in the PSPS was approximately 10^6 CFU/g.

2.5.2. The Sterilization of RFSW

Three electrode gaps (chosen in Section 2.4) and three SW temperatures (116 °C, 121 °C, and 124 °C) were used to find appropriate parameters in sterilization that could reduce *G. stearothermophilus* spores by at least 5 log and maximize the protection of PSPS quality. The PSPS samples sterilized by SW alone served as the control (CON-SW).

2.5.3. Enumeration of Surviving Cells

PSPS samples were cooled after sterilization and were placed into a sterile homogeneous bag. They were serially (1:10) diluted with 0.8% saline after blending for 2 min in a laboratory mixer (LC-08, Ningbo Licheng Instrument Co., Ltd., Ningbo, China). Then, 0.1 mL of each dilution (1 ml of blended PSPS sample) was poured into NA, spread evenly, and incubated reversely at 55 °C for 24–48 h. The minimum detection limit of this method was 1 CFU g^{-1}.

2.6. Quality Analysis

Quality analysis was performed immediately after the sterilized sample was cooled to room temperature.

2.6.1. Water Loss

Before sterilization, the boiled pork slices were weighted (W_0). After sterilization, the PSPS sample was cooled to room temperature. Then, the pork slices were removed from the PSPS mixture. The surface moisture was wiped off slightly with a tissue. The sterilized pork slices (W_1) were weighed. The water loss was calculated using the following Formula (1):

$$\text{Weight loss (\%)} = (W_0 - W_1)/W_0 \times 100\% \tag{1}$$

Each treatment was replicated three times.

2.6.2. Texture Measurement

The textural characteristics of pork slices sterilized by RFSW and CON-SW were measured by a texture analyzer (TA.XT Plus, Stable Micro system, Ltd., Godalming, UK). A 5 mm-diameter probe was inserted twice into the pork slices at 2 mm depth to obtain the hardness, chewiness, and resilience data. The apparatus parameters of pretest, test, and post-test speeds were set at 1 mm/s [21]. Each treatment was replicated three times.

2.6.3. Measurement of Thiobarbituric Acid Reactants (TBARS)

The TBARS values of PSPS sample unsterilized and sterilized by RFSW/CON-SW, including pork slices and soup, were determined by using a method described by Liu, et al. [22] with slight modification. Specifically, 10 g sample (pork slices or soup) was mixed with 50 mL 7.5% aqueous solution of trichloroacetic acid (TCA with 0.1% of ethylene diamine tetraacetic acid) for malondialdehyde extraction. Then, the mixture was fully homogenized, shaken for 5 min, and centrifuged at $6000\times g$ for 5 min. Next, 2 mL supernatant was mixed with 2 mL 0.02 mol/L thiobarbituric acid (TBA) and reacted at 100 °C in a water bath for 30 min. It was cooled quickly and centrifuged at $1600\times g$ for 5 min. Finally, the supernatant was mixed with 2 mL trichloromethane (analytical reagent), shaken for 10 s, and left to stand for delamination. The absorbance of the supernatant was measured at 532 and 600 nm by a spectrophotometer (P7, Shanghai MAPADA instruments Co., Ltd., Shanghai, China). The blank solution of mixed 2 mL TBA and 2 mL TCA was used as the control. The TBARS value was expressed as mg/kg of malondialdehyde and calculated by using the following Formula (2) [22]:

$$\text{TBARS (mg/kg)} = (A_{532} - A_{600})/150 \times (1/10) \times 72.6 \times 100 \tag{2}$$

Each treatment was replicated three times.

2.6.4. Sensory Evaluation

Sensory evaluation of PSPS samples was carried out using quantitative descriptive analysis (QDA). Before the analysis, 10 panelists met weekly for a twelve-week period to establish descriptor attributes and terminology around attributes of PSPS. The analytical content was developed and improved through consensus and voting methods, and the evaluation process and descriptive terms with specific meanings were established through group discussions [23]. A descriptor list was developed to evaluate the attributes: appearance, odor, tenderness, juiciness, soup taste, and overall acceptability. Panelists may have given a low evaluation to the odor and taste of the sample because of its unpleasant overcooked and rancid flavor. The score of the items ranged from 1 to 5 with 1 being the lowest (inferior) and 5 being the highest (superior). The samples used for testing were reheated by a microwave before evaluation, and the microwave oven was turned off when the temperature at the cold spot reached 60 °C. Each sample was made in duplicate, and the order of the samples' placement was randomized. Moreover, PSPS samples (50 mL soup with three pieces of meat slices) were put in special edible plastic cups. Panelists assessed the random coded samples individually in a quiet, odorless, and enclosed environment. Each sample was evaluated three times by a total of 10 panelists of sensory evaluators.

2.7. Microstructure Analysis

A scanning electron microscope (SEM) (Nano SEM-450, USA) was used to intuitively reflect the effects on the fiber structure of meat slices in the processing of RFSW or SW alone and observe the changes in the cross-sections in cell morphology before and after sterilization.

The samples were prepared by soaking sections ($5 \times 5 \times 3$ mm^3) in 1.5 mL of 2% (v/v) glutaraldehyde solution at 4 °C for more than 10 h fixation. Then, the samples were rinsed with phosphate buffer (pH 7.2, 100 mmol/L) three times for 10–15 min each, dehydrated in different gradient concentration ethanol (30%, 50%, 70%, 80%, and 90% (v/v)), in turn, three times each for 30 min. After dehydration, the sample was taken out and dried in supercritical CO_2 fluid overnight [20,21]. Finally, samples were coated in a sputter coater and photographed in SEM.

2.8. Statistical Analysis

All experiments were performed in triplicate and the results were expressed as mean ± standard deviation. Significant differences ($p < 0.05$) were analyzed by one-way analysis of variance (ANOVA) using SPSS software (Version 20.0, IBM Corp., Armonk, NY, USA).

3. Results and Discussion

3.1. Determination of Cold Spot

It is important to reveal how the time–temperature relationship impacts the inactivation of microorganism and the quality of a meat product [8]. Eight points in the PSPS sample were measured to discuss RFSW heating uniformity. The cold spot representing the heat limitation is meaningful for the food sterilization process.

The CUT-1 in eight points of the PSPS sample is shown in Figure 4. The central point in the middle layer (point C) was the cold spot, but it showed no significant difference ($p > 0.05$) from the two other points (points D and E) in the middle layer. The overall heating rate in the bottom layer was slower than that in the top layer. The nonuniform heating of PSPS samples resulted from the nonuniform distribution of RF energy and samples. The top, middle, and bottom layers of the samples consisted of oil, soup (salt solution), and pork slices, respectively. The RFSW heating of PSPS manifested as a margin heating pattern in this study, which was contrary to the core heating pattern identified by Jiang et al. [10]. Possibly, the PSPS sample was surrounded by SW, which transfers heat from the surface to the interior. The fastest heating rate was observed in the top layer. That was because the contact surface between edible oil and soup had a better heating efficacy base due to the fact that the electric intensification happens where two contact dielectric materials have a notable difference in dielectric constant [24]. The dielectric constant of edible oil was the smallest among all food items, whose result was similar to the research by Valantina et al. [25]. In the top layer, point B was heated faster than point A ($p < 0.05$), which possibly led to the outcome that the electric field was deflected by the edges in a sample with a cylindrical shape. Thus, point C was selected as the cold spot for the subsequent experiments.

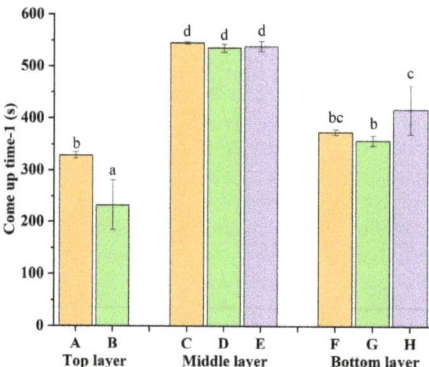

Figure 4. Come-up time-1 from 30 °C to 60 °C under 180 mm electrode gap at different points of the PSPS sample. Six points (A, B, C, D, F, and G) were in soup and two points (E and H) were in pork slices. (Different lowercase letters (a–d) indicate a significant difference ($p < 0.05$)).

3.2. The Heating Rate and Curve

The inactivation efficiency of microorganisms of a meat product depends on the temperature achieved during heat treatment, whereas the quality change depends on the time [26]. Therefore, the sterilization process needs to be built in a mild but time-saving way.

The effect of SW and electrode gap on heating rate is shown in Table 2. The average heating rate of 116 °C increased continuously with the increasing electrode gap from 160 mm to 190 mm and decreased sharply from 190 to 200 mm; the same results were found for 121 °C and 124 °C. Many studies indicated that a narrower electrode gap strengthens an electric field [11,27]. Also, a narrower electrode gap leads to worse heating uniformity when food materials were relatively smaller-sized and placed on the center of the bottom electrode of the RF cavity [28]. Consequently, the heating rate of the cold spot in the PSPS sample increased, possibly because it absorbed more RF energy as the electrode gap increased (from 160 to 190 mm). It is hard to conclude about the effect of SW temperature on heating rate. To find the appropriate condition, the electrode gaps of 190, 200, and 170 mm (representing the fastest, slowest, and moderate heating rates, respectively) in each SW temperature were selected to complete the inactivation test of *G. stearothermophilus* spores.

Table 2. Come-up time and heating rate from 30 °C to superheated water set temperature at different electrode gaps.

Superheated Water Temperature (°C)	Electrode Gaps (mm)	Come-Up Time-2 (min)	Heating Rate (°C/min)
116	160	12.09 ± 0.52	7.12
	170	11.35 ± 0.03	7.58
	180	10.26 ± 0.05	8.38
	190	9.57 ± 0.87	8.99
	200	12.70 ± 0.81	6.77
121	160	13.25 ± 0.07	6.87
	170	11.54 ± 0.48	7.89
	180	10.63 ± 0.88	8.56
	190	10.19 ± 0.51	8.93
	200	13.61 ± 0.56	6.69
124	160	13.54 ± 0.13	6.94
	170	13.05 ± 0.41	7.23
	180	11.62 ± 0.14	8.10
	190	11.12 ± 0.14	8.47
	200	16.43 ± 0.11	5.72

Temperature–time curve of PSPS at different electrode gaps at 116 °C, 121 °C, and 124 °C is shown in Figure 5.

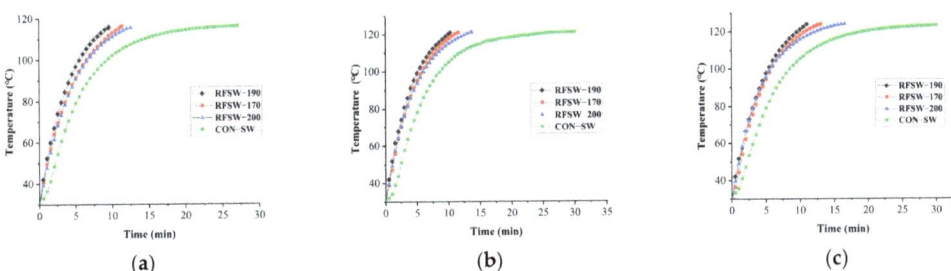

Figure 5. Time–temperature profiles of point C in poached spicy pork slices in different electrode gaps at 116 °C (a), 121 °C (b), and 124 °C (c). RFSW means radio frequency combined superheated water at an electrode gap; CON-SW means superheated-water sterilization without radio frequency energy.

RFSW sterilization reduced at least half of the sterilization time, thereby reaching the target temperature compared with the CON-SW treatment. A greater slope difference between two curves of RFSW and CON-SW in a temperature was observed in the back part of the curve than in the front part, thereby showing that the electrode gap significantly increased the heating rate of the latter part of the heating process. As the temperature difference between PSPS and SW decreased, the mainstay of heat transfer impetus changed from temperature difference to RF energy. This conclusion can be supported by another phenomenon that the heating rate difference between RFSW-190 and RFSW-170 or RFSW-200 was more obvious at 116 °C and 121 °C than at 124 °C in the front part of the heating process. At 124 °C, a greater temperature difference was found between SW and PSPS samples. We interestingly found that 121 °C-30 min SW sterilization was not enough for the PSPS sample to reach an internal temperature of 121 °C, even though this is widely used in the food industry.

3.3. Inactivation of G. stearothermophilus spores

G. stearothermophilus spores, which have extremely strong heat resistance, are known as an indicator species in thermal food processing and used as a substitute for *Clostridium botulinum* to address biosafety concerns [18]. The inactivation of *G. stearothermophilus* spores in PSPS by RFSW sterilization at different temperatures and electrode gaps is shown in Figure 6. The SW temperature and electrode gap significantly affected the inactivation of *G. stearothermophilus* spores. The shortest time of 5 log reduction in *G. stearothermophilus* spores was observed in RFSW-190 /124 °C sterilization for 12 min, whereas the CON-SW/116 sterilization had the longest time of 36 min. Under each SW temperature, RFSW-190 sterilization showed the fastest spore reduction rate, whereas RFSW-200 had the slowest. In accordance with the heating rate results in Section 3.2, it could be concluded that the efficiency of *G. stearothermophilus* spores inactivation was positively correlated with heating rate. When the SW temperature increased from 116 to 121 °C and 124 °C, less time was needed to reach the 5 log reduction in *G. stearothermophilus* spores under a certain electrode gap, indicating that temperature was an important factor in RFSW sterilization. Whether higher temperature of thermal treatment and shorter time were more desirable for food sterilization has not been determined. To protect the edible quality and nutrients, conducting a physicochemical property test was necessary. Therefore, a sample at the end of each sterilization process was selected for subsequent quality measurement.

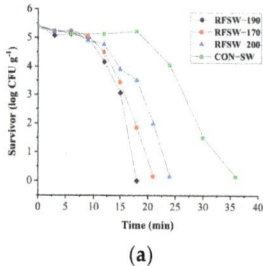 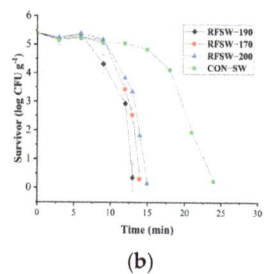 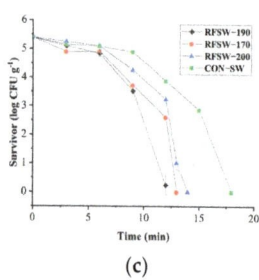

Figure 6. Log survivor of *G. stearothermophilus* spores in PSPS under the different electrode gaps and SW temperatures of 116 °C (**a**), 121 °C (**b**), and 124 °C (**c**).

From the comparison and detailed analysis of the time–temperature profile and *G. stearothermophilus* spores inactivation curve, more valuable information could be obtained. Under 121 °C, the number of surviving spores sterilized by RFSW increased at 6 min compared with 3 min, indicating that *G. stearothermophilus* spores may germinate in soup and pork slices from 3 min to 6 min. At 6 min, according to Figure 5, the temperature of the cold spot reached around 105 °C, which seems insufficient for *G. stearothermophilus* spore activation, according to the research of Finley et al. [29]. Some nutrients with small molecular weight in the PSPS (e.g., amino acids and sugars) help accelerate spore germination [30]. Moreover, a higher temperature was reached in some points except for the cold spot. The time–temperature profile and the inactivation curve showed great correspondence.

The holding times (time above the target temperature) at 124 °C (0.9 min at 190 mm, 0 min at 170 mm, and −2.25 min at 200 mm) and 121 °C (2.81 min at 190 mm, 2.46 min at 170 mm, and 1.39 min at 200 mm) showed a slower heating rate corresponding to shorter holding time, but at 116 °C (8.44 min at 190 mm, 9.65 min at 170 mm, and 11.3 min at 200 mm), the opposite result was observed. SW temperature influenced the sterilization efficiency in some way, perhaps in relation to the heating cumulative effect [29,30]. Moreover, a larger slope can be observed in the tail of most curves, indicating that a critical temperature varied the sterilization efficiency.

3.4. Quality Evaluation

3.4.1. Water Loss

The mixture of potato starch and egg white gelatinized and formed a gel covering the surface of the pork slices at high temperature, thereby somehow preventing water loss. Nonetheless, the electrode gap in each SW temperature still affected the water loss markedly according to Table 3. Pork slices sterilized by RFSW-200 and RFSW-170 had more significant ($p < 0.05$) smaller water loss than those sterilized by RFSW-190 and CON-SW at 116 °C. At 121 °C, a significant difference was observed between CON-SW and RFSW sterilization ($p > 0.05$), whereas no significant difference was found at 124 °C. Prolonged heating time and elevated temperature are the two main factors affecting water loss [26], in which a too-fast heating rate as well as longer heating time both could lead to severe moisture drop, because water expulsion happens when muscle fiber shrinks and sarcomere length decreases [26,31]. This finding possibly accounted for the fact that CON-SW and RFSW-190 sterilization resulted in higher water loss. CON-SW sterilization more rapidly reached 5 log *G. stearothermophilus* spores reduction at 124 °C than at 121 °C and 116 °C (18, 24, and 36 min, respectively), so that less time was allowed to accumulate the water loss difference between CON-SW and RFSW treatment, thereby causing a smaller structural change. RFSW-170 in each temperature was shown to have a better balance between heating time and temperature and consequently led to the best condition for RFSW to reduce water loss.

Table 3. Water loss and texture of PSPS sample sterilized by radio frequency combined superheated water in different electrode gaps and at different superheated water temperatures. RFSW means radio frequency combined superheated water at an electrode gap; CON-SW means superheated water sterilization alone.

Sterilization		Water Loss (%)	Texture		
			Hardness (g)	Chewiness	Resilience
116 °C	RFSW-200	8.47 ± 0.88 ab	157.63 ± 15.22 ab	41.29 ± 2.82 a	0.181 ± 0.003 ab
	RFSW-190	18.15 ± 2.14 f	324.97 ± 3.68 h	158.85 ± 2.12 h	0.162 ± 0.003 a
	RFSW-170	7.91 ± 0.37 a	161.97 ± 9.23 ab	43.18 ± 6.13 ab	0.181 ± 0.001 ab
	CON-SW	16.51 ± 0.62 f	264.51 ± 5.11 f	70.94 ± 3.21 c	0.161 ± 0.002 a
121 °C	RFSW-200	11.80 ± 0.97 cde	213.06 ± 11.26 e	50.32 ± 1.83 ab	0.174 ± 0.001 a
	RFSW-190	13.87 ± 0.67 e	249.26 ± 0.54 f	99.85 ± 5.59 e	0.162 ± 0.001 a
	RFSW-170	10.45 ± 1.12 bc	181.13 ± 5.79 cd	53.86 ± 2.47 b	0.274 ± 0.004 d
	CON-SW	17.71 ± 0.84 f	307.32 ± 8.84 g	102.05 ± 5.93 e	0.163 ± 0.003 a
124 °C	RFSW-200	11.74 ± 0.34 cde	168.79 ± 1.01 bc	107.99 ± 12.39 e	0.232 ± 0.033 c
	RFSW-190	12.85 ± 0.84 de	190.41 ± 10.1 d	133.25 ± 8.92 f	0.204 ± 0.001 b
	RFSW-170	9.56 ± 0.82 ab	145.61 ± 7.21 a	83.89 ± 4.19 d	0.307 ± 0.032 e
	CON-SW	10.51 ± 2.62 bcd	228.37 ± 5.28 e	144.24 ± 4.35 g	0.242 ± 0.006 c

Different lowercase letters (a–f) indicate a significant difference ($p < 0.05$) within the same column. RFSW-radio frequency combined superheated water in an electrode gap; CON-SW-superheated water treatment without radio frequency energy.

3.4.2. Texture Measurement

The texture of samples sterilized by RFSW and CON-SW under each temperature is shown in Table 3. The hardness and chewiness of samples sterilized by RFSW-170 and RFSW-200 were significantly smaller ($p < 0.05$) than those sterilized by RFSW-190 and CON-SW, possibly because RFSW-190 reached a higher end temperature for its fastest heating rate, and CON-SW needed a long duration of time to achieve a 5 log reduction of in *G. stearothermophilus* spores. Three components primarily determined the hardness of the meat, namely, myofibrillar proteins, connective tissues, and collagen. The secondary, tertiary, and quaternary structures of these protein-based substances can be affected differently by different heat intensities. A too-fast heating rate of the sample induced denaturing, the connective tissue and collagen dissolving, and myofiber shrinkage, and longer heating time of sample led to severe protein denaturation as well, with both increasing the hardness of the meat [31]. Moreover, according to the results of water loss, RFSW-190 caused greater water loss than RFSW-170 and RFSW-200 and gave results closer to CON-SW, which also led to severe protein denaturation. During the heating, structure-related protein starts to denature, the connective tissue and collagen dissolve, and myofibers shrink, thereby increasing the toughness of the meat [7]. Palka and Daun [32] indicated that compared with lower temperature, a closer myofiber fiber gap was observed at 121 °C, thereby showing that a higher temperature leads to more severe myofiber shrinkage. In terms of resilience, sample resilience under sterilization by RFSW-170 was significantly higher ($p < 0.05$) than that under sterilization by RFSW-190, RFSW-200, and CON-SW at 121 °C and 124 °C, whereas no significant difference ($p > 0.05$) was found at 116 °C. The resilience of meat might be related to the swelling of fiber, as reflected by the diameter of the fiber. RFSW-170 reached 5 log *G. stearothermophilus* spores reduction under moderate heating, thereby preventing the excessive increase in myofiber diameter.

3.4.3. Oxidation Analysis

Under the required cooking or sterilization intensity, a lower temperature and a shorter time are desired to reduce oxidation of meat products in industry. Strong heat would induce the oxidization of polyunsaturated fatty acids, thereby yielding aldehydes, free fatty acids, and other substances. This would cause quality failure, i.e., off flavor, which may lead to safety problems and reduce consumer acceptability [33,34].

The influence of SW temperature and electrode gap on the TBARS value of pork slices and soup is shown in Figure 7. TBARS value reflects the secondary oxidation degree of food. Apparently, oxidation in soup sterilized by RFSW or CON-SW was greater than in pork slices. Soup in the top layer of the PSPS sample reached a higher temperature according to the result in Figure 4. Autoxidation became more intense at higher temperature [35]. Soup contains more fatty acids and transfers heat faster compared with pork slices. The PSPS sample sterilized by RFSW-170 produced significantly ($p < 0.05$) less malonaldehyde than that sterilized by RFSW-190 at 116 °C and 121 °C. At 124 °C, no significant difference was observed among samples under all sterilization conditions. These phenomena indicated that electrode gap influenced the extent of oxidation less under high temperature or over a long time.

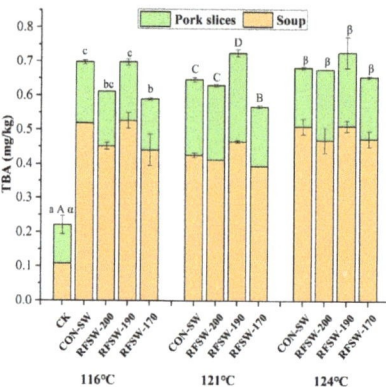

Figure 7. Thiobarbituric reactive substances of poached spicy pork slices sample sterilized by RFSW under different electrode gaps and superheated water temperatures. Different lowercase letters (a–f, A–D, and α,β) indicate significant differences ($p < 0.05$) within the different temperatures of 116 °C, 121 °C, and 124 °C. CK means unsterilized poached spicy pork slices; RFSW means radio frequency combined superheated water at a certain electrode gap; CON-SW means superheated water sterilization alone.

To summarize, the 170 mm electrode gap was the best condition to reduce heating time and protect PSPS quality. Thus, RFSW-170 was selected to complete the sensory evaluation.

3.4.4. Sensory Evaluation

The sensory-evaluation results of samples sterilized by RFSW-170 in each temperature and CON-SW are presented in Figure 8. Significant differences ($p < 0.05$) were observed in appearance, odor, soup taste, and overall acceptability between RFSW and CON-SW, whereas no significant differences ($p > 0.05$) were found in tenderness and juiciness. The scores of each item under RFSW at three temperatures showed no significant differences ($p > 0.05$), indicating that the panelists could hardly tell the distinction among PSPSs.

Continuous heating promotes oxidation in pork slices and lipid oxidation and hydrolysis in the edible oil of the soup, leading to off-flavor generation and the loss of sensory quality [36,37]. The panelists preferred the odor and soup taste of RFSW-sterilized samples compared with the CON-SW group, which might be due to the lower level of TBARS (Section 3.4.3) in meat sterilized by RFSW-170.

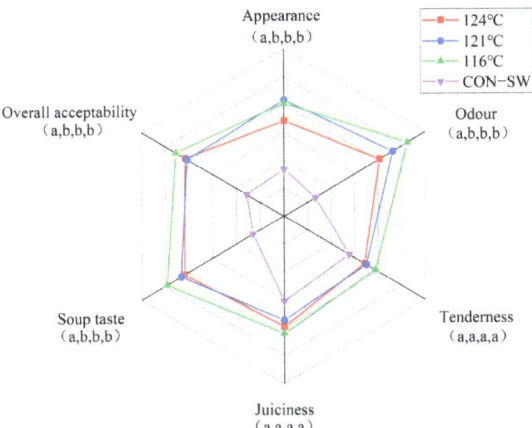

Figure 8. Sensory evaluation of poached spicy pork slice samples sterilized by radio frequency combined superheated water in three temperatures and superheated water alone in 116 °C. The four letters in brackets belong to the four conditions from the center to the outside. Different lowercase letters (a–b) indicate a significant difference ($p < 0.05$) within the same object.

3.4.5. Microstructure Analysis on SEM Imaging

Changes in the morphological myofibrillar structure of sliced pork in SEM images subjected to RFSW and conventional SW sterilization are shown in Figure 9. Figure 9a demonstrates the internal microstructure of raw pork slices, which clearly showed the sparse arrangement of myofibrils with clearly visible gaps. When samples were heated to 116 °C by SW (Figure 9b) alone, the myofibrils of the pork slices were arranged more compactly, contrasting with control. Similar phenomena can be found in Figure 9c–e when samples were treated by RFSW (116 °C) at the fixed electrodes of 170 mm, 190 mm, and 200 mm, respectively. This phenomenon may be attributed to denaturation of the proteins that make up the myofibrillar and drainage of interfiber water [38].

When the temperature rose to 121 °C, the peptide bonds were hydrolyzed and denatured, and collagen cross-links were broken; meanwhile, the hardness of the pork slices decreased (Figure 9f–i). Similarly, more significant results can be found in Figure 9j–m when samples were treated at the SW temperature of 124 °C. Li et al. [39] revealed that the structure of chicken muscle fibers was packed more tightly after the chicken was heated from 45 to 95 °C at the center temperature. As a consequence, the myofibrillar structure of sliced pork heated by RFSW heating at a fixed electrode of 170 mm showed less variation compared to the control because the samples were heated at a relatively suitable heating rate under this condition, which made its heating rate faster than that with SW heating without reaching excessive temperature in a short period of time. Moreover, a faster heating rate resulted in less structural change in myofibrils, reducing the water loss of the pork slices and decreasing the hardness of the pork slices, which corresponded to the results above in Sections 3.4.1 and 3.4.2. Thus, revealing the microstructural changes in meat during the process of sterilization is necessary.

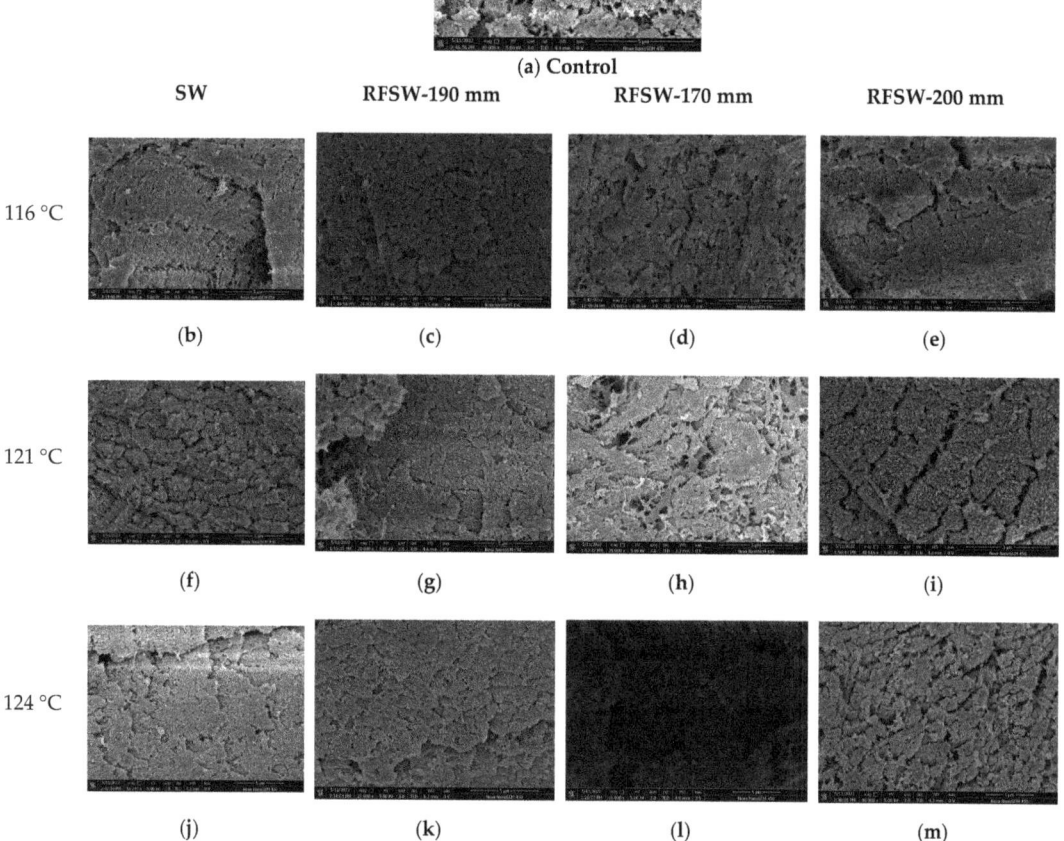

Figure 9. SEM micrographs of myofibrils in pork slices after sterilization with different electrode gaps and superheated-water temperature (control group means starching pork slice without sterilization).

4. Conclusions

A new device, which can apply RF energy and SW simultaneously, was employed in this study to eliminate microorganisms inoculated in RTE PSPS. The availability of the test results proved that the application was stable and serviceable. The cold spot lay in the geometric center of the Teflon container, but we were surprised by the difference in heating rate between the upper and bottom layers of the PSPS sample. Such a difference shows that the shape and properties of the material had a great impact on the electric field distribution and heating rate. A deeper study of RF heating complex food materials will pose a greater challenge. Less time was needed for the cold spot in PSPS to reach the SW temperature when RFSW sterilization was applied compared with when SW sterilization was applied. Also, less time was required for the PSPS sample to reach a 5 log reduction in *G. stearothermophilus* spores under RFSW sterilization. It was found that RFSW-190, despite its fastest heating rate, resulted in even worse quality compared to the conventional method. On the contrary, RFSW-170 sterilization overall reduced the water loss and thermal damage to texture, oxidation, tissues, and cells of the PSPS sample and kept better sensory properties. Thus, RFSW-170 was more suitable for sterilization in this study. It is hard to reach a conclusion about the effect of SW temperature on the quality of PSPS samples, but RF energy affected it less at 124 °C.

Although we achieved some desirable results in terms of heating rate and quality protection, it is still unclear whether this new RTE food sterilization technology saves more energy, which is an important issue to be explored in subsequent research. This study suggested considerable promise for the application of RFSW to RTE food sterilization.

Author Contributions: Conceptualization, Y.W. (Yunyang Wang), C.R. and K.W.; methodology, K.W., C.R. and Y.W. (Yunyang Wang); software, B.C., Y.S. and K.W.; validation, K.W. and C.R.; formal analysis, Y.W. (Yunyang Wang), Y.S., C.R. and K.W.; investigation, Y.W. (Yunyang Wang) and K.W.; resources, Y.W. (Yequn Wang), X.C., H.F. and C.R.; data curation, B.C., K.W. and Y.S.; writing-original draft preparation, K.W. and C.R.; writing-review and editing, Y.W. (Yunyang Wang) and K.W.; visualization, K.W. and B.C.; supervision, Y.W. (Yunyang Wang); project administration, Y.W. (Yunyang Wang); funding acquisition, Y.W. (Yunyang Wang), All authors have read and agreed to the published version of the manuscript.

Funding: This research was funded by the general program (Grant No. 31371854) of the National Nature Science Foundation of China, the Key Research Project of Shaanxi Province (2017ZDXM-SF-104).

Institutional Review Board Statement: Not applicable.

Informed Consent Statement: Not applicable.

Data Availability Statement: Data on this study are available in the article.

Acknowledgments: The authors would like to thank the instrument-sharing platform of the College of Food Science & Engineering (Northwest A&F University, China).

Conflicts of Interest: The authors declare no conflict of interest.

References

1. Osman, I.; Osman, S.; Mokhtar, I.; Setapa, F.; Shukor, S.A.M.; Temyati, Z. Family Food Consumption: Desire towards Convenient Food Products. *Procedia-Soc. Behav. Sci.* **2014**, *121*, 223–231. [CrossRef]
2. Tirloni, E.; Nauta, M.; Vasconi, M.; Di Pietro, V.; Bernardi, C.; Stella, S. Growth of Listeria monocytogenes in ready-to-eat "shrimp cocktail": Risk assessment and possible preventive interventions. *Int. J. Food Microbiol.* **2020**, *334*, 108800. [CrossRef] [PubMed]
3. Carballo, D.; Moltó, J.C.; Berrada, H.; Ferrer, E. Presence of mycotoxins in ready-to-eat food and subsequent risk assessment. *Food Chem. Toxicol.* **2018**, *121*, 558–565. [CrossRef] [PubMed]
4. Vasconcellos, L.; Carvalho, C.T.; Tavares, R.O.; de Mello Medeiros, V.; de Oliveira Rosas, C.; Silva, J.N.; dos Reis Lopes, S.M.; Forsythe, S.J.; Brandão, M.L.L. Isolation, molecular and phenotypic characterization of *Cronobacter* spp. in ready-to-eat salads and foods from Japanese cuisine commercialized in Brazil. *Food Res. Int.* **2018**, *107*, 353–359. [CrossRef]
5. Horita, C.N.; Baptista, R.C.; Caturla, M.Y.R.; Lorenzo, J.M.; Barba, F.J.; Sant'Ana, A.S. Combining reformulation, active packaging and non-thermal post-packaging decontamination technologies to increase the microbiological quality and safety of cooked ready-to-eat meat products. *Trends Food Sci. Technol.* **2018**, *72*, 45–61. [CrossRef]
6. Bedane, T.F.; Pedrós-Garrido, S.; Quinn, G.; Lyng, J.G. The impact of emerging domestic and commercial electro-heating technologies on energy consumption and quality parameters of cooked beef. *Meat Sci.* **2021**, *179*, 108550. [CrossRef]
7. Yildiz-Turp, G.; Sengun, I.Y.; Kendirci, P.; Icier, F. Effect of ohmic treatment on quality characteristic of meat: A review. *Meat Sci.* **2013**, *93*, 441–448. [CrossRef]
8. Jiménez-Colmenero, F.; Herrero, A.M.; Cofrades, S.; Ruiz-Capillas, C. Meat: Eating Quality and Preservation. In *Encyclopedia of Food and Health*; Caballero, B., Finglas, P.M., Toldrá, F., Eds.; Academic Press: Oxford, UK, 2016; pp. 685–692. [CrossRef]
9. Liu, Y.; Zhang, Y.; Wei, X.; Wu, D.; Dai, J.; Liu, S.; Qin, W. Effect of radio frequency-assisted hot-air drying on drying kinetics and quality of Sichuan pepper (*Zanthoxylum bungeanum* Maxim.). *LWT* **2021**, *147*, 111572. [CrossRef]
10. Jiang, H.; Ling, B.; Zhou, X.; Wang, S. Effects of combined radio frequency with hot water blanching on enzyme inactivation, color and texture of sweet potato. *Innov. Food Sci. Emerg. Technol.* **2020**, *66*, 102513. [CrossRef]
11. Yao, Y.; Sun, Y.; Cui, B.; Fu, H.; Chen, X.; Wang, Y. Radio frequency energy inactivates peroxidase in stem lettuce at different heating rates and associate changes in physiochemical properties and cell morphology. *Food Chem.* **2021**, *342*, 128360. [CrossRef]
12. Zhang, C.; Hu, C.; Sun, Y.; Zhang, X.; Wang, Y.; Fu, H.; Chen, X.; Wang, Y. Blanching effects of radio frequency heating on enzyme inactivation, physiochemical properties of green peas (*Pisum sativum* L.) and the underlying mechanism in relation to cellular microstructure. *Food Chem.* **2021**, *345*, 128756. [CrossRef] [PubMed]

13. Dong, J.; Kou, X.; Liu, L.; Hou, L.; Li, R.; Wang, S. Effect of water, fat, and salt contents on heating uniformity and color of ground beef subjected to radio frequency thawing process. *Innov. Food Sci. Emerg. Technol.* **2021**, *68*, 102604. [CrossRef]
14. Xu, J.; Zhu, S.; Zhang, M.; Cao, P.; Adhikari, B. Combined radio frequency and hot water pasteurization of Nostoc sphaeroides: Effect on temperature uniformity, nutrients content, and phycocyanin stability. *LWT* **2021**, *141*, 110880. [CrossRef]
15. Dag, D.; Singh, R.K.; Kong, F. Effect of surrounding medium on radio frequency (RF) heating uniformity of corn flour. *J. Food Eng.* **2021**, *307*, 110645. [CrossRef]
16. Wang, C.; Kou, X.; Zhou, X.; Li, R.; Wang, S. Effects of layer arrangement on heating uniformity and product quality after hot air assisted radio frequency drying of carrot. *Innov. Food Sci. Emerg. Technol.* **2021**, *69*, 102667. [CrossRef]
17. Yang, Y.; Geveke, D.J. Shell egg pasteurization using radio frequency in combination with hot air or hot water. *Food Microbiol.* **2020**, *85*, 103281. [CrossRef]
18. Ahn, J.; Lee, H.Y.; Balasubramaniam, V.M. Inactivation of Geobacillus stearothermophilus spores in low-acid foods by pressure-assisted thermal processing. *J. Sci. Food Agric.* **2015**, *95*, 174–178. [CrossRef]
19. Wang, K.; Huang, L.; Xu, Y.; Cui, B.; Sun, Y.; Ran, C.; Fu, H.; Chen, X.; Wang, Y.; Wang, Y. Evaluation of Pilot-Scale Radio Frequency Heating Uniformity for Beef Sausage Pasteurization Process. *Foods* **2022**, *11*, 1317. [CrossRef]
20. Cui, B.; Sun, Y.; Wang, K.; Liu, Y.; Fu, H.; Wang, Y.; Wang, Y. Pasteurization mechanism on the cellular level of radio frequency heating and its possible non-thermal effect. *Innov. Food Sci. Emerg. Technol.* **2022**, *78*, 103026. [CrossRef]
21. Sun, Y.; Wang, K.; Dong, Y.; Li, K.; Liu, H.; Cui, B.; Fu, H.; Chen, X.; Wang, Y.; Wang, Y. Effects of radiofrequency blanching on lipoxygenase inactivation, physicochemical properties of sweet corn (*Zea mays* L.), and its correlation with cell morphology. *Food Chem.* **2022**, *394*, 133498. [CrossRef]
22. Liu, F.; Zhu, Z.; Dai, R. Effects of carnosine on color stability and lipid oxidation of beef patties. *Sci. Technol. Food Ind.* **2009**, *30*, 140–143. [CrossRef]
23. Gillespie, R.; Ahlborn, G.J. Mechanical, sensory, and consumer evaluation of ketogenic, gluten-free breads. *Food Sci. Nutr.* **2021**, *9*, 3327–3335. [CrossRef] [PubMed]
24. Huang, Z.; Marra, F.; Wang, S. A novel strategy for improving radio frequency heating uniformity of dry food products using computational modeling. *Innov. Food Sci. Emerg. Technol.* **2016**, *34*, 100–111. [CrossRef]
25. Rubalya Valantina, S. Measurement of dielectric constant: A recent trend in quality analysis of vegetable oil—A review. *Trends Food Sci. Technol.* **2021**, *113*, 1–11. [CrossRef]
26. Zell, M.; Lyng, J.G.; Cronin, D.A.; Morgan, D.J. Ohmic cooking of whole beef muscle—Evaluation of the impact of a novel rapid ohmic cooking method on product quality. *Meat Sci.* **2010**, *86*, 258–263. [CrossRef]
27. Gao, M.; Tang, J.; Wang, Y.; Powers, J.; Wang, S. Almond quality as influenced by radio frequency heat treatments for disinfestation. *Postharvest Biol. Technol.* **2010**, *58*, 225–231. [CrossRef]
28. Tiwari, G.; Wang, S.; Tang, J.; Birla, S.L. Analysis of radio frequency (RF) power distribution in dry food materials. *J. Food Eng.* **2011**, *104*, 548–556. [CrossRef]
29. Finley, N.; Fields, M.L. Heat activation and heat-induced dormancy of Bacillus stearothermophilus spores. *Appl. Microbiol.* **1962**, *10*, 231–236. [CrossRef]
30. Georget, E.; Kushman, A.; Callanan, M.; Ananta, E.; Heinz, V.; Mathys, A. Geobacillus stearothermophilus ATCC 7953 spore chemical germination mechanisms in model systems. *Food Control* **2015**, *50*, 141–149. [CrossRef]
31. N'Gatta, K.C.; Kondjoyan, A.; Favier, R.; Sicard, J.; Rouel, J.; Gruffat, D.; Mirade, P.-S. Impact of Combining Tumbling and Sous-Vide Cooking Processes on the Tenderness, Cooking Losses and Colour of Bovine Meat. *Processes* **2022**, *10*, 1229. [CrossRef]
32. Palka, K.; Daun, H. Changes in texture, cooking losses, and myofibrillar structure of bovine M. semitendinosus during heating. *Meat Sci.* **1999**, *51*, 237–243. [CrossRef]
33. Sun, X.; Wang, Y.; Li, H.; Zhou, J.; Han, J.; Wei, C. Changes in the volatile profile, fatty acid composition and oxidative stability of flaxseed oil during heating at different temperatures. *LWT* **2021**, *151*, 112137. [CrossRef]
34. Addis, P.B. Occurrence of lipid oxidation products in foods. *Food Chem. Toxicol. Int. J. Publ. Br. Ind. Biol. Res. Assoc.* **1986**, *24*, 1021–1030. [CrossRef]
35. Choe, E.; Min, D.B. Mechanisms and factors for edible oil oxidation. *Compr. Rev. Food Sci. Food Saf.* **2006**, *5*, 169–186. [CrossRef]
36. Barbosa-Cánovas, G.V.; Medina-Meza, I.; Candoğan, K.; Bermúdez-Aguirre, D. Advanced retorting, microwave assisted thermal sterilization (MATS), and pressure assisted thermal sterilization (PATS) to process meat products. *Meat Sci.* **2014**, *98*, 420–434. [CrossRef]
37. Traore, S.; Aubry, L.; Gatellier, P.; Przybylski, W.; Jaworska, D.; Kajak-Siemaszko, K.; Santé-Lhoutellier, V. Effect of heat treatment on protein oxidation in pig meat. *Meat Sci.* **2012**, *91*, 14–21. [CrossRef]
38. Li, Z.; Li, M.; Du, M.; Shen, Q.W.; Zhang, D. Dephosphorylation enhances postmortem degradation of myofibrillar proteins. *Food Chem.* **2018**, *245*, 233–239. [CrossRef] [PubMed]
39. Li, C.; Wang, D.; Xu, W.; Gao, F.; Zhou, G. Effect of final cooked temperature on tenderness, protein solubility and microstructure of duck breast muscle. *LWT-Food Sci. Technol.* **2013**, *51*, 266–274. [CrossRef]

Article

Microstructure, Digestibility and Physicochemical Properties of Rice Grains after Radio Frequency Treatment

Zhenna Zhang [1], Bin Zhang [1], Lin Zhu [2] and Wei Zhao [1,*]

[1] School of Food Science and Technology, Jiangnan University, Wuxi 214122, China; zhangzhenna321@163.com (Z.Z.); 15617571473@163.com (B.Z.)
[2] Key Laboratory of Preservation Engineering of Agricultural Products, Institute of Agricultural Products Processing, Ningbo Academy of Agricultural Sciences, Ningbo 315040, China; zhulin0822@163.com
* Correspondence: zhaow@jiangnan.edu.cn

Abstract: Radio frequency (RF) energy has been successfully applied to rice drying, sterilization, and controlling pests. However, the effects of RF treatment on the microstructure, physicochemical properties, and digestibility of rice have rarely been studied. This study investigated the alteration of a multiscale structure, pasting, rheology, and digestibility of rice grains after the RF treatment. A microstructure analysis demonstrated that the RF treatment caused starch gelatinization and protein denaturation in rice grains with an increasing treatment time. After the RF treatment, indica and japonica rice (IR and JR) remained as A-type crystals, with the formation of an amylose–lipid complex. In contrast, the crystalline structure of waxy rice (WR) was disrupted. The RF treatment led to a decrease in crystallinity and short-range ordered structures. However, the DSC results indicated that the RF treatment enhanced the T_o, T_p, and T_c of IR and JR. The RF treatment resulted in an increase in the resistant starch (RS) of IR and JR, thereby reducing the digestibility. In addition, the pasting profiles of IR and JR after RF treatment were reduced with the increase in treatment time, while the RF-treated WR showed an opposite trend. The storage modulus (G') and loss modulus (G") of all samples after the RF treatment obviously increased compared to the control.

Keywords: rice; radio frequency; microstructure; physicochemical properties; digestibility

Citation: Zhang, Z.; Zhang, B.; Zhu, L.; Zhao, W. Microstructure, Digestibility and Physicochemical Properties of Rice Grains after Radio Frequency Treatment. *Foods* **2022**, *11*, 1723. https://doi.org/10.3390/foods11121723

Academic Editors: Shaojin Wang and Rui Li

Received: 10 May 2022
Accepted: 7 June 2022
Published: 13 June 2022

Publisher's Note: MDPI stays neutral with regard to jurisdictional claims in published maps and institutional affiliations.

Copyright: © 2022 by the authors. Licensee MDPI, Basel, Switzerland. This article is an open access article distributed under the terms and conditions of the Creative Commons Attribution (CC BY) license (https:// creativecommons.org/licenses/by/ 4.0/).

1. Introduction

Rice (*Oryza sativa* L.) is widely cultivated and consumed by humans worldwide, as it plays a vital role in human nutrition, energy supply, and food security [1]. Compared with other cereals, rice is already classified as a high-glycemic-index (GI) food, owing to its higher digestible energy [2]. The high-GI food is correlated with diet-associated diseases such as type II diabetes, obesity, and cardiovascular disorders [3], which pose a threat to human health. Therefore, there is a need to develop low-GI rice to improve human health.

Various methods have been applied to modify rice GI by manipulating the starch structure to reduce the rate of starch digestion. Depending on the rate of digestion, starch is generally divided into rapidly digestible starch (RDS), slowly digestible starch (SDS), and resistant starch (RS) [4]. SDS and RS have been considered beneficial for the control of postprandial blood glucose, thereby preventing the occurrence of these metabolic diseases [5,6], and eventually displaying positive health benefits. Hence, increasing the SDS or RS contents of rice to satisfy the requirements of consumers has attracted more attention. Compared to the chemical and enzymatic modification of starch, physical methods are preferred by consumers, as it is natural and highly safe [7].

Radio frequency (RF) heating has been successfully used in food processing, such as drying, thawing, disinfestations, pasteurization, blanching, and roasting [8]. It is regarded as a promising alternative to replace conventional thermal methods due to faster heating and a higher penetration depth. During RF heating, electromagnetic energy is converted into thermal energy by dipole rotation and ionic migration in food [8]. The application of

RF heating in rice mainly focused on insect control and storage [9–11]. However, as far as we know, the effects of RF treatment on the microstructure, physicochemical properties, and digestibility of rice grain have rarely been investigated. Our previous study showed that the RS content of rice flour was enhanced after RF treatment, leading to the strengthening of its resistance to enzyme hydrolysis [12]. White rice is composed of starch, protein, lipids, and non-starch polysaccharides. The starch properties and starch–protein/lipid interactions in rice grains play a critical role in improving digestibility. For example, Gong et al. [13] reported that rice starches with high short- and medium-amylose chains accelerated short-term retrogradation, thereby reducing the digestibility of starch. Li et al. [14] showed that the perfect arrangement of amylopectin double helices was conducive for the starch to resist the digestion of amylase. Moreover, it has been demonstrated that the interaction between starch and protein/lipids in rice remarkably affects starch digestibility [15,16]. Therefore, the microstructure of rice grains during RF treatment should be investigated to provide sufficient knowledge for industry application.

Hence, this work investigates the effect of the RF treatment on the microstructure, physicochemical, and digestive characteristics of different rice varieties. The results suggest that RF technology can be applied to reduce the digestibility of rice as an alternative to conventional methods and provide a theoretical basis for industry applications.

2. Materials and Methods

2.1. Materials

Three rice samples, including indica (Simiao), japonica (Daohuaxiang 2), and waxy (lvnuo 619) rice (IR, JR, and WR), used in this study were obtained from a supermarket in Wuxi (Jiangsu, China). Pepsin (P7000), pancreatin (P7545), and amyloglucosidase (A7095) were provided by Sigma-Aldrich (Shanghai, China). The Glucose oxidase-peroxidase kit was obtained from the Nanjing Jiancheng Bioengineering Institute (Nanjing, China). The other chemical reagents used in this work were of analytical grade.

2.2. Sample Preparation

Rice grains were soaked in distilled water (1:3 w/v) and equilibrated at room temperature for 6 h. After soaking, the water was decanted, and the sample was drained. The moisture content of the sample was approximately 30%. Subsequently, these samples were subjected to RF system.

2.3. RF Treatment

An RF system with 6 kW, 27.12 MHz (GJG-2.1-10A-JY; Hebei Huashijiyuan High Frequency Equipment Co., Ltd., Shijiazhuang, China) was applied in this study for the modification of rice, as shown in Figure 1. According to our previous studies, the electrode gap was fixed at 120 mm for the whole experiment. The preprepared rice (250 g) was put into a cylindrical container, which was treated by the RF system at 120 mm for various times (10, 20, and 30 min). The temperature of samples was detected and collected by an optical fiber sensor. The RF-treated samples were cooled to 25 °C and then equilibrated at 4 °C for 48 h. After that, samples were dried, ground, and passed through a 100-mesh sieve for further analysis. The rice with different RF treatment times was tagged as IR-RF-10, IR-RF-20, IR-RF-30, JR-RF-10, JR-RF-20, JR-RF-30, WR-RF-10, WR-RF-20, and WR-RF-30.

2.4. Scanning Electron Microscopy (SEM)

SEM (SU8100, Hitachi, Ltd., Tokyo, Japan) was performed to analyze the microstructure of rice grains before and after RF treatment. According to the method of Zhong et al. [17], a razor blade was used to apply pressure to the cross-sectional axis of the rice grain to naturally fracture the rice grain. Then, the fractured rice grain was mounted on a stub using double-sided carbon-coated tape and was coated with gold. The micrographs of the fractured surface were collected at an accelerating voltage of 3.0 kV with different magnifications.

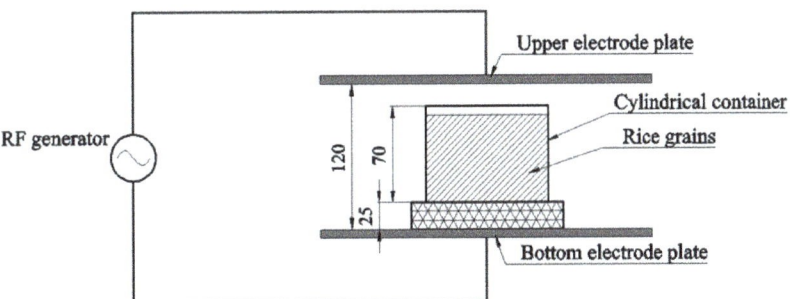

Figure 1. Schematic of the sample placed in the RF heating system (27.12 MHz, 6 kW). All dimensions are in mm.

2.5. X-ray Diffraction (XRD)

The XRD patterns of samples were measured with an X-ray diffractometer (D2 PHASER, Bruker AXS Inc., Karlsruhe, Germany). The experimental conditions were as follows: scanning range of 5–40°, step size of 0.05°, and step duration of 0.5 s. Jade 6 software was used for analyzing the crystallinity of samples.

2.6. Fourier-Transform Infrared Spectroscopy (FTIR)

An FTIR spectrometer (IS10, Nicolet, WI, USA) was applied to detect the chemical structures of samples. Generally, samples were fully ground with KBr (1:100, w/w) and then compressed into tablets. The prepared sample was scanned from 4000 to 400 cm^{-1} with a resolution of 4 cm^{-1}.

2.7. Differential Scanning Calorimetry (DSC)

The thermal properties of native and RF-treated rice were determined with DSC3 (Mettler-Toledo, Switzerland) following the method of Zhang et al. [18]. Approximately 3 mg samples and 6 µL deionized water were placed in a 40 µL aluminum pan, which was tested from 25 to 100 °C at a heating rate of 10 °C/min. The relative parameters, including onset temperature (T_o), peak temperature (T_p), conclusion temperature (T_c), and gelatinization enthalpy (ΔH), were obtained.

2.8. Pasting Properties

Rapid viscosity analyzer (Newport Scientific Pty. Ltd., Sidney, Australia) was used to characterize the pasting properties of samples. First, 3 g samples and 25 g deionized water were placed in an aluminum canister. After that, the prepared sample was measured using the STD 2 procedure based on the method of Ma et al. [19]. The relative pasting parameters, including peak viscosity (PV), trough viscosity (TV), breakdown (BD), final viscosity (FV), and setback (SB), were collected.

2.9. Rheological Properties

Discovery Hybrid Rheometer-3 (TA instruments, New Castle, DE, USA) was applied to evaluate the rheological properties of samples, following the method of Zhang et al. [18]. After pasting properties measurement, the gelatinized samples were cooled to 25 °C and then placed on the testing platform equipped with a 40 mm parallel plate at a gap of 1 mm for rheological measurements. The dynamic frequency sweeps were tested from 0.1 to 100 rad/s at 0.5% strain. Storage modulus (G′), loss modulus (G″), and loss tangent (tan δ = G″/G′) of samples were determined.

2.10. In Vitro Digestibility

The in vitro digestibility of native and RF-treated rice grains was carried out as described by Englyst et al. [4] with certain modifications. Firstly, 200 mg samples were fully mixed with 10 mL sodium acetate buffer (pH 5.2, 0.1 mol/L) at room temperature. Then, they were heated for 30 min in a water bath at 95 °C with magnetic stirring. After that, the mixture was cooled down to 37 °C and equilibrated for 10 min. Afterwards, it was mixed with pepsin solution (10 mL, 5 mg/mL) and incubated for 30 min in a shaking water bath (180 rpm) at 37 °C. Next, 5 mL of sodium acetate buffer was added to the reaction mixtures and incubated for another 30 min. Thereafter, 5 mL mixed enzyme solution containing pancreatin and amyloglucosidase was added to each sample and allowed to react in a shaking water bath at 37 °C for 120 min. At various time points (0, 20, and 120 min), the hydrolyzed solution (0.1 mL) was collected and put into a centrifuge tube containing anhydrous ethanol (0.9 mL). After centrifugation at 5000 r/min for 10 min, the hydrolyzed glucose content in the supernatant was analyzed with a glucose oxidase-peroxidase kit. The RDS, SDS, and RS contents were calculated using the following equation:

$$\text{RDS (\%)} = (G_{20} - G_0) \times 0.9 \times 100/\text{TS} \quad (1)$$

$$\text{SDS (\%)} = (G_{120} - G_{20}) \times 0.9 \times 100/\text{TS} \quad (2)$$

$$\text{RS (\%)} = (1 - \text{RDS} - \text{SDS}) \times 100 \quad (3)$$

where G_0, G_{20}, and G_{120} reflect the mass of glucose released after enzymatic hydrolysis for 0, 20, and 120 min, respectively, and TS is the total starch content.

2.11. Statistical Analysis

All tests were performed at least in triplicate. Statistical analysis was conducted with Minitab 18 software (State College, PA, USA). Data were subjected to one-way analysis of variance (ANOVA) and the Tukey test at the 95% confidence level. Results were exhibited as means with standard deviations.

3. Results and Discussion

3.1. Microstructure

Figure 2 exhibits the transverse-section SEM images of native and RF-treated rice grains. They showed that the cross-section of the rice grains exhibited some cracks due to intercellular cleavage, and intracellular cleavage simultaneously occurred when the grains were cracked open [17]. In native rice, the starch granules were polygonal and were closely packed (Figure 2(a1–c1)). The surface layer of native rice was rough, with small protein molecules scattered on the surface and embedded in gaps between starch granules [20]. After RF treatment, the starch granules lost their angular character with the increasing RF treatment time, especially for RF-30. This could be attributed to the RF treatment inducing a high temperature in rice grains, resulting in partial starch gelatinization (Figure 2(a3–c3)). It has been reported that changes in rice grain structure happen mainly due to the gelatinization of starch [21]. Moreover, the protein structure in rice could be disrupted by the RF treatment, which could reconstruct and interact with starch, thereby altering the internal structure of the rice [22]. Furthermore, the starch granules in rice grains after the RF treatment swelled compared with those in native rice (Figure 2(a2–c2)). This phenomenon may have resulted due to the increased interior temperature, resulting in an increased interior pressure, which led to starch swelling and even gelatinization [23,24]. In addition, a hole appeared in the center of the WR subjected to the RF treatment, implying that the core of the RF-treated WR was overheated, causing either endosperm shrinkage or thermal disintegration [24].

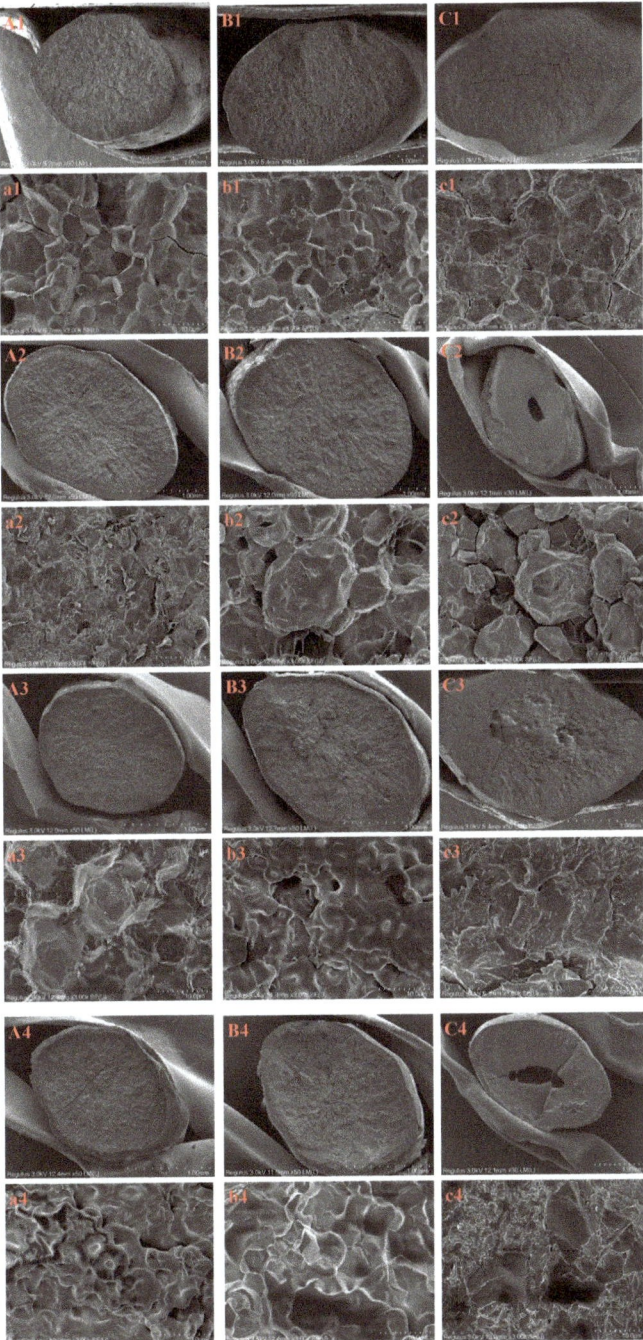

Figure 2. SEM images of different varieties of rice grains (indica, japonica, and waxy rice, coded "(**A**)/(**a**), (**B**)/(**b**), and (**C**)/(**c**)", respectively) subjected to RF treatment at different treatment times (0, 10, 20, and 30 min, coded "1, 2, 3, and 4", respectively). (**A**–**C**): the general profile of the cross-section; (**a**–**c**): partially enlarged view.

3.2. Crystalline Structure

The X-ray diffractograms of untreated and RF-treated IR, JR, and WR are exhibited in Figure 3. The corresponding relative crystallinity is summarized in Table 1, which is generally applied to assess the long-range crystalline structure of rice [17]. All the native rice exhibited a typical A-type crystalline pattern, with main peaks at 2θ of 15°, 17°, 18°, and 23° [18]. After the RF treatment, the XRD patterns of IR and JR remained with a weakened diffraction intensity compared to the control. Moreover, there was a peak at 20° of the RF-treated IR and JR, suggesting the formation of amylose–lipid complexes (V-type pattern). With the increase in RF treatment time, the decrease in diffraction intensity was consistent with the presence of gelatinized starch [18]. Especially for WR, the typical diffraction peaks almost disappeared with a single peak after the RF treatment, implying that starch granules were in gelatinized and fused states.

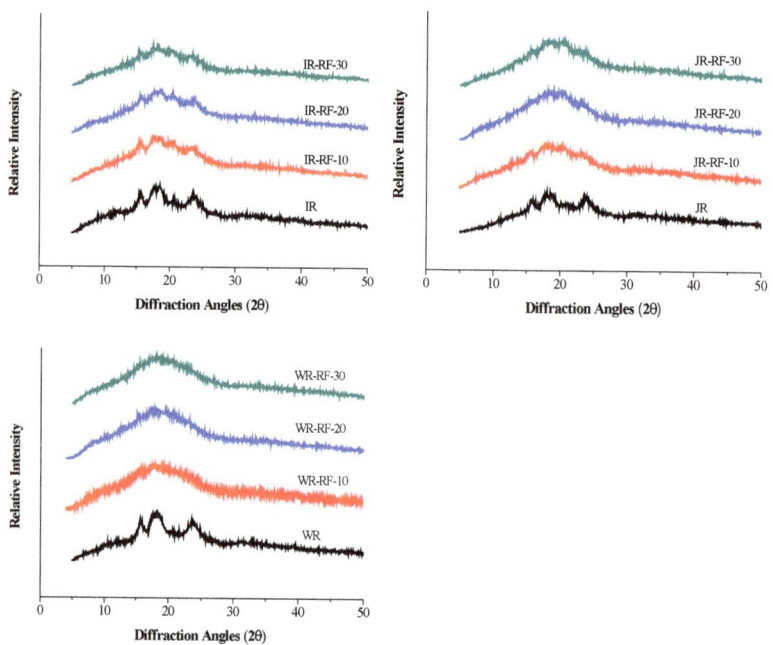

Figure 3. X-ray diffractograms of rice grains were submitted to RF treatment at different times (10, 20, and 30 min). IR: indica rice; JR: japonica rice; WR: waxy rice.

Table 1. The X-ray diffraction, FTIR, and thermal characteristics of native and RF-treated rice grains.

Samples	RC (%)	$R_{1022/995}$	$R_{1048/1022}$	T_o	T_p	T_c	ΔH
IR	25.7 ± 0.8 [bc]	0.98 ± 0.009 [d]	0.71 ± 0.006 [abc]	65.3 ± 0.07 [c]	71.8 ± 0.1 [d]	78.5 ± 0.5 [c]	3.2 ± 0.2 [b]
IR-RF-10	21.6 ± 0.6 [cd]	1.28 ± 0.009 [abc]	0.67 ± 0.001 [bc]	72.1 ± 0.8 [ab]	77.1 ± 0.5 [a]	81.5 ± 1.8 [abc]	0.8 ± 0.1 [cd]
IR-RF-20	22.1 ± 2.5 [cd]	1.35 ± 0.03 [a]	0.62 ± 0.01 [c]	71.9 ± 0.05 [ab]	77.4 ± 0.2 [a]	83.1 ± 0.7 [ab]	1.3 ± 0.2 [c]
IR-RF-30	20.5 ± 2.1 [d]	1.35 ± 0.03 [ab]	0.69 ± 0.003 [bc]	72.6 ± 0.3 [a]	78.1 ± 0.1 [a]	83.9 ± 0.3 [a]	1.3 ± 0.1 [c]
JR	28.3 ± 1.3 [b]	1.02 ± 0.004 [d]	0.71 ± 0.002 [abc]	59.6 ± 0.07 [d]	66.4 ± 0.1 [e]	73.1 ± 0.3 [d]	4.1 ± 0.5 [b]
JR-RF-10	20.9 ± 0.8 [cd]	1.25 ± 0.02 [bc]	0.69 ± 0.02 [bc]	68.1 ± 3.2 [bc]	74.4 ± 0.7 [bc]	80.9 ± 1.5 [abc]	1.1 ± 0.2 [c]
JR-RF-20	21.3 ± 1.6 [cd]	1.27 ± 0.02 [abc]	0.69 ± 0.006 [bc]	68.7 ± 1.1 [abc]	73.7 ± 0.2 [c]	79.1 ± 0.5 [c]	0.8 ± 0.2 [cd]
JR-RF-30	19.5 ± 0.5 [d]	1.29 ± 0.05 [abc]	0.65 ± 0.04 [bc]	69.9 ± 0.9 [ab]	75.3 ± 0.3 [b]	79.8 ± 1.1 [bc]	0.6 ± 0.3 [cd]
WR	37.0 ± 0.7 [a]	0.98 ± 0.01 [d]	0.74 ± 0.02 [ab]	58.5 ± 0.5 [d]	65.7 ± 0.6 [e]	73.2 ± 1.1 [d]	5.5 ± 0.4 [a]
WR-RF-10	7.38 ± 0.7 [e]	1.21 ± 0.07 [c]	0.79 ± 0.04 [a]	-	-	-	-
WR-RF-20	7.10 ± 0.1 [e]	1.22 ± 0.004 [c]	0.74 ± 0.01 [ab]	-	-	-	-
WR-RF-30	4.35 ± 0.3 [e]	1.24 ± 0.05 [c]	0.73 ± 0.02 [ab]	-	-	-	-

Different letters in the same column indicate a significant difference ($p < 0.05$). IR: indica rice; JR: japonica rice; WR: waxy rice; RF: radio frequency.

On the other hand, the relative crystallinity of RF-treated rice grains was reduced compared to the control. It further confirmed that the RF treatment resulted in starch gelatinization and crystallite destruction. The reduction in RC may be ascribed to the friction and collision that occurred among polar molecules, rapidly producing more heat in rice grains, causing the destruction of starch granules and the degradation of molecular structures. In addition, the increased mobility of starch chains during the RF treatment also induced the disruption of crystalline regions [7]. Different changes of RC among the three rice grains indicated that the amylose content played a critical role in crystallite formation [20].

3.3. Short-Range Ordered Structure

The short-range ordered structure (single- and double-helical structures) of all samples was analyzed by FTIR in the range of 4000 to 500 cm^{-1}. Generally, the absorption bands at 1022 and 1047 cm^{-1} are related to the amorphous and crystalline regions of starch, respectively, and the alteration of ordered structures can be evaluated by the values of 1047/1022 cm^{-1} [25]. Moreover, the absorption band at 995 cm^{-1} is associated with C-OH bending vibrations [26], and the 1022/995 cm^{-1} is used to reveal the degree of double helices [27]. As shown in Table 1, the RF-treated samples exhibited lower values of 1047/1022 cm^{-1} than that of the control, except for WR-RF-10 and WR-RF-20. It signified that the RF treatment reduced the degree of ordered structures of starch at the short-range level, which might be ascribed to the partial gelatinization of starch after the RF treatment. The high temperature produced by the RF system might have disrupted the amorphous and crystalline structures of starch, resulting in the destruction of the packing or winding of the helical structures [28]. In addition, the changes of 1022/995 cm^{-1} values were opposite to the 1047/1022 cm^{-1} values. The RF-treated samples showed higher values of 1022/995 cm^{-1} than the untreated samples. The increase in 1022/995 cm^{-1} values after the RF treatment suggested the destruction of single- and double-helical structures, which were in agreement with XRD results.

3.4. Thermal Properties

The thermal properties of native and RF-treated rice grains are summarized in Table 1. The thermal properties of all the samples were significantly affected by the RF treatment, especially the RF treatment time. After the RF treatment, the T_o, T_p, and T_c of IR and JR were transferred to higher temperatures compared to the untreated sample. The increased T_o, T_p, and T_c of the RF-treated samples indicated that the RF treatment reinforced the interactions between the starch chains or starch and other components. Similar results were observed by Zhong et al. [29], showing that microwave irradiation enhanced the T_o and T_p of rice. Moreover, the higher gelatinization temperature of RF-treated samples implied that more energy was needed when the initial gelatinization of samples occurred. Changes in gelatinization temperature were mainly attributed to the amylose content and the amylopectin fine structure [30]. The higher gelatinization temperature of RF-treated IR could be primarily due to the higher amylose content. Furthermore, the gelatinization temperature of IR and JR increased with the increasing RF treatment time, suggesting that the RF treatment could improve the interaction among starch chains. However, compared with the untreated samples, the ΔH significantly decreased for the RF-treated samples, implying that the content of double-helical structures was reduced. This could be due to the partial gelatinization of rice grains caused by the RF treatment, but to a different extent [31]. In addition, the T_o, T_p, T_c, and ΔH of the RF-treated WR decreased compared with the native sample. As revealed by the XRD and SEM results, the RF treatment seriously destroyed the microstructure of the WR.

3.5. Pasting Properties

Pasting properties are one of the most sensitive indicators in rice grains, responsible for the rice cooking and eating quality. The typical pasting curves of all samples are displayed in Figure 4, and the corresponding related parameters are presented in Table 2. Obviously, the pasting properties of the RF-treated rice grains were different from those of the untreated rice. This may be attributed to the degradation of starch and protein, the alteration of microstructure and intermolecular forces, and the enhanced hydrolytic enzyme activities of the RF-treated rice [22].

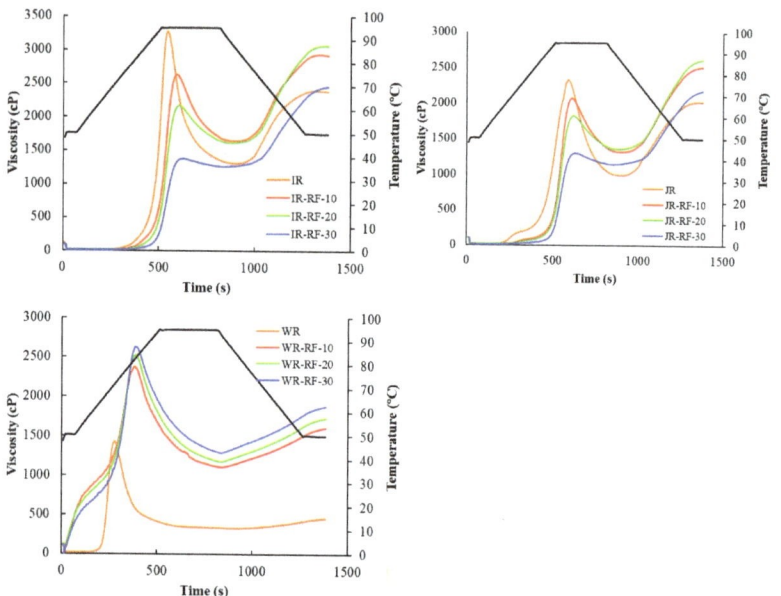

Figure 4. Pasting profile of native and RF-treated rice at different treatment times (10, 20, and 30 min). The black line corresponds to temperature. IR: indica rice; JR: japonica rice; WR: waxy rice.

Table 2. The pasting and in vitro digestibility properties of rice grains subjected to RF treatment.

Samples	PV (Pa·s)	BD (Pa·s)	T (Pa·s)	FV (Pa·s)	SB (Pa·s)	RDS (%)	SDS (%)	RS (%)
IR	3238 ± 4.9 a	1939 ± 9.9 a	1299 ± 4.9 bc	2371 ± 3.54 e	1072 ± 1.4 d	63.8 ± 0.8 abc	3.2 ± 1.2 a	32.8 ± 0.4 ab
IR-RF-10	2600 ± 30.4 b	979 ± 7.8 c	1621 ± 22.6 a	2897 ± 21.9 b	1276 ± 0.7 b	59.8 ± 1.0 bc	5.1 ± 0.2 a	34.9 ± 0.8 ab
IR-RF-20	2192 ± 46.7 de	564 ± 19.8 de	1628 ± 26.9 a	3071 ± 24.0 a	1443 ± 2.8 a	59.5 ± 3.5 bc	4.8 ± 1.5 a	35.5 ± 0.1 ab
IR-RF-30	1379 ± 1.4 g	112 ± 12.0 f	1266 ± 10.6 c	2435 ± 22.6 de	1168 ± 33.2 c	60.7 ± 1.4 abc	4.5 ± 1.0 a	34.6 ± 0.4 ab
JR	2354 ± 14.8 cd	1319 ± 27.6 b	1035 ± 12.7 e	2100 ± 14.8 g	1065 ± 2.1 d	64.3 ± 0.6 abc	1.9 ± 1.0 a	33.7 ± 1.6 ab
JR-RF-10	2035 ± 46.0 e	735 ± 27.6 d	1300 ± 18.4 bc	2485 ± 25.5 d	1185 ± 7.1 c	60.1 ± 2.3 abc	1.3 ± 0.6 a	38.4 ± 1.6 a
JR-RF-20	1809 ± 13.4 f	468 ± 4.2 e	1341 ± 17.7 b	2588 ± 24.0 c	1246 ± 6.4 b	57.6 ± 2.9 c	7.6 ± 2.3 a	34.6 ± 0.4 ab
JR-RF-30	1316 ± 24.0 g	164 ± 6.3 f	1151 ± 17.7 d	2198 ± 36.1 f	1047 ± 18.4 d	60.6 ± 0.4 abc	1.3 ± 0.6 a	38.0 ± 1.0 a
WR	1335 ± 28.7 g	1000 ± 44.0 c	335 ± 14.8 f	467 ± 20.5 k	132 ± 5.7 g	57.8 ± 1.4 c	4.4 ± 1.2 a	37.7 ± 0.2 a
WR-RF-10	2359 ± 20.5 cd	1243 ± 35.4 b	1116 ± 14.8 d	1612 ± 14.1 j	495 ± 0.7 f	65.0 ± 0.8 ab	3.2 ± 2.0 a	31.6 ± 1.2 b
WR-RF-20	2518 ± 7.1 bc	1347 ± 2.8 b	1171 ± 4.2 d	1714 ± 17.0 i	543 ± 12.7 ef	67.1 ± 1.2 a	2.3 ± 0.8 a	30.5 ± 2.0 b
WR-RF-30	2618 ± 10.6 b	1326 ± 14.1 b	1292 ± 3.5 bc	1879 ± 4.24 h	586 ± 0.7 e	62.1 ± 2.0 abc	2.6 ± 1.2 a	35.2 ± 3.3 ab

Different letters in the same column indicate a significant difference ($p < 0.05$). IR: indica rice; JR: japonica rice; WR: waxy rice; RF: radio frequency.

The pasting profiles of IR and JR after the RF treatment were similar to that of the untreated samples. As shown in Figure 4, the gelatinization time of the RF-treated IR and JR was delayed, meaning that the RF treatment had inhibiting effects on starch gelatinization. Moreover, the RF treatment caused a decrease in the PV and BD of IR and JR, whereas it enhanced the TV, FV, and SB. The PV of IR and JR significantly decreased with the increase in RF treatment time. The reduction in PV in IR and JR might have been due to the disruption of hydrogen bonds and glycosidic linkages [32] caused by the high temperature during the RF treatment. A similar phenomenon for the PV of rice after a microwave treatment was reported [33]. The high temperature affected the starch lamellar structure, resulting in gelatinization and a decreased paste viscosity. The decreased BD of the RF-treated IR and JR indicated that the RF treatment improved the thermal stability of starches with respect to the control sample. This result was in agreement with Sun et al. [34], suggesting that a heat moisture treatment reduced the BD of early indica rice. However, Zhong et al. [17] found that the microwave treatment enhanced the PV and BD of rice. The different changes in pasting properties may be due to the experiment conditions and rice varieties. Moreover, the higher SB of the RF-treated IR and JR implied that the recrystallization of amylose molecules increased the chance of starch retrogradation [35]. RF energy generated high internal thermal pressures in rice grains, which probably destroyed the cell wall and disintegrated the starch granules, leading to the leaching of compounds, consequently increasing the SB.

On the contrary, the pasting viscosity of the WR subjected to the RF treatment significantly increased compared to the control. As shown in Table 2, the PV, TV, SB, BD, and FV of the RF-treated WR were higher than untreated samples. The elevated PV may be associated with the destruction of starch granules and the damage of the microstructure (crystalline and helical structure), which favored the granules swelling and raised the stretching of starch molecules. Additionally, the increased PV may be due to the interactions between water and stretched molecules, which formed intra- and inter-hydrogen bonds and networked chains [28]. In addition, the BV represents the heat and shear resistance of starch granules at high temperatures. The greater BV implied that the RF treatment reduced the thermal stability of starch paste. The higher SB of the RF-treated WR may be due to the degradation of amylopectin, leading to the rearrangement of starch chains. The different changes in the RF-treated rice grains could be related to their varied compositions and structures.

3.6. Rheological Characterization

Understanding the rheological properties of rice gels is instructive for the production of gluten-free products. The rheological behavior of rice gels formed by the RF-treated samples is presented in Figure 5. The G' and G" of all samples displayed a raised trend with the increasing angular frequency. The G' values were higher than the G" values for all samples, except for the native WR, meaning that these gels exhibited weak gel behaviors. Conversely, the native WR displayed a dominant viscous property (G' < G"). Moreover, both the G' and G" of rice grains after the RF treatment significantly increased compared with the control, meaning that the RF treatment enhanced the strength of the cross-linked gel network. As confirmed by RVA, the setback values of the RF-treated samples increased. It showed that the RF treatment accelerated the retrogradation of rice gels and improved their gelling ability [12]. Similarly, Solaesa et al. [32] found that the microwave treatment markedly enhanced the G' and G" of rice flour. They confirmed that gels created from MV-treated flour were more stable. In addition, the tan δ of all RF-treated gels was obviously reduced compared to those of untreated gels, implying that the tightness and stability of the internal structure of rice grains were destroyed after the RF treatment [22]. With the increase in angular frequency, the loss tangent (tan δ) of IR and JR before and after the RF treatment firstly decreased (0.1–0.5 rad/s) and then increased. The relative tan δ values were less than one, implying that the IR and JR gels displayed a solid-like behavior. On the other hand, the tan δ of the untreated WR exhibited a downward trend with the increasing angular frequency, while the tan δ of the RF-treated WR first decreased and then slightly increased. The gels created from the untreated WR exhibited a higher tan δ value (tan δ > 1) at a lower frequency (0.1–20 rad/s), suggesting that the gels showed more viscous characteristics, whereas the gels presented a more solid-like behavior (tan δ < 1) at a higher frequency (20–100 rad/s). Moreover, the RF-treated WR gels differed from the control, which showed a more solid-like behavior (tan δ < 1). These differences among the three rice varieties could be associated with changes in other components of rice, such as protein, lipids, and phenolic acids [17].

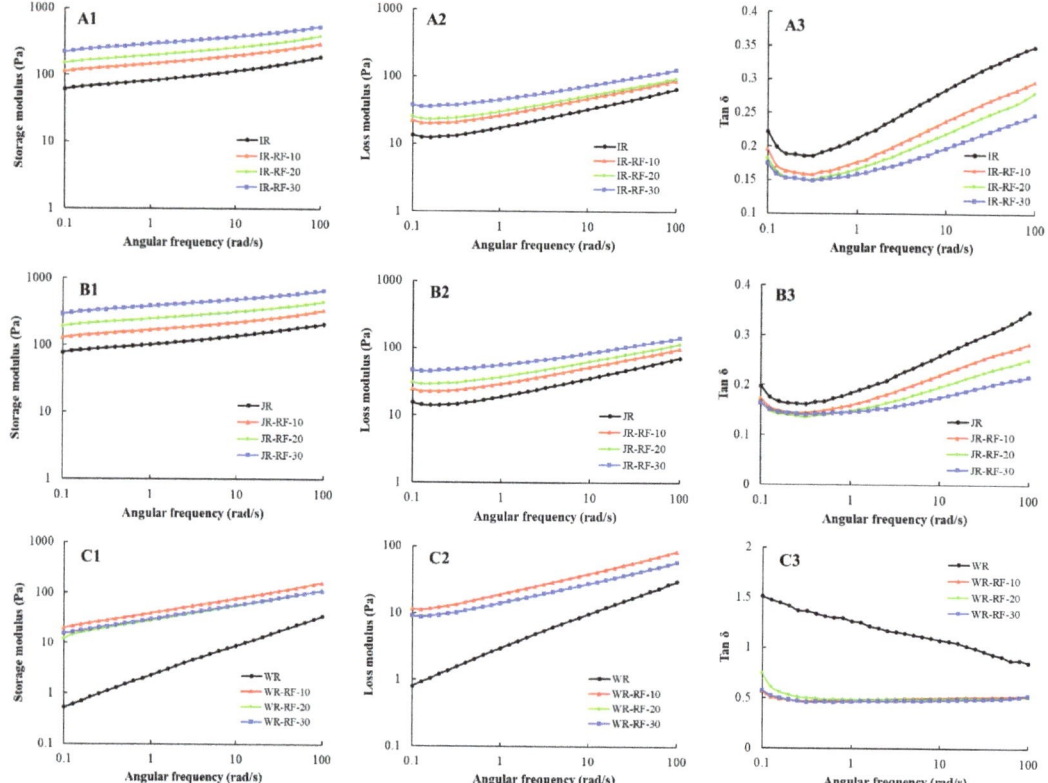

Figure 5. The dynamic viscoelastic properties of untreated and RF-treated rice. (**A1–A3**): storage modulus (G′); (**B1–B3**): loss modulus (G″); (**C1–C3**): tan δ; IR: indica rice; JR: japonica rice; WR: waxy rice.

3.7. In Vitro Digestibility of Rice

The digestibility of rice grains subjected to the RF treatment was assessed with the starch fractions. Thereby, the results of RDS, SDS, and RS are summarized in Table 2. The digestibility of rice grains after the RF treatment was altered to various degrees, especially for the RDS and RS contents. Compared with native IR and JR, the RF treatment led to a decrease in the RDS content and an increase in the RS fraction. The increased RS indicated that the RF treatment reinforced the interaction between the starch and lipid, thus, inhibiting the physical accessibility to enzymes. Cheng et al. [36] suggested that the formation of amylose–lipid complexes after the heat moisture treatment enhanced the RS content and reduced the swelling and digestibility of starch. Moreover, the protein could have been denatured during the RF treatment and adhered to the surface of starch granules [37], which restrained the susceptibility of starch to enzymes. However, the RF-treated WR had lower SDS and RS contents than those in the native sample, and this fraction could be transformed into RDS. This phenomenon could be due to the complete gelatinization of starch molecules, as confirmed by the SEM and DSC results. The differences in the digestibility of the three rice varieties depended on many factors, such as the amylose/amylopectin content, chemical components (proteins and lipids), starch granular structure, and experiment conditions [30,38]. It has been reported that the amylopectin chain length influences the formation of crystalline starch polymorphs, thereby affecting the digestibility of starch [39].

3.8. Mechanism of RF Treatment for Rice

In this study, we discovered that the RF treatment had the potential to improve the structural, physicochemical, and digestive properties of rice grains. During the RF treatment, the friction and collision among polar molecules rapidly produced more heat in rice grains, thereby facilitating the interactions between starch chains or starch and other components. The formation of starch–lipid complexes increased the RS content and reduced the starch digestibility. The RF treatment promoted the formation of protein gels network, which restrained the contact between starch and enzymes. Moreover, the RF treatment improved the stability of rice grains as reflected by the enhancement of the gelatinization temperature (T_o, T_p and T_c). Moreover, the RF treatment destroyed the crystalline and short-range ordered structures of starch granules, and changed the microstructure of endosperm cells. The chemical and structural changes in turn altered the physicochemical properties of rice grains.

4. Conclusions

The microstructures, physicochemical properties, and digestibility of rice grains were altered after the RF treatment. In terms of the RF-treated IR and JR, the formation of amylose–lipid complexes contributed to the increase in the RS content, thereby reducing the digestibility. Moreover, the enhanced T_o, T_p, and T_c values of the RF-treated IR and JR suggested that new aggregation structures formed during the RF treatment, which was also conducive to reducing the digestibility. In addition, with the increase in treatment time, the RF treatment resulted in the gelatinization of starch molecules, the destruction of the crystalline structure, and the destruction of short-range ordered structures to different degrees. Furthermore, the RF treatment improved the physicochemical properties (pasting and rheology) of rice gains. These results confirmed that the RF treatment exhibits the potential for modifying the digestibility of rice. However, the effects of the RF treatment for the cooking time and quality of rice compared to the conventional method remain unclear, and should be studied in future work.

Author Contributions: Conceptualization, Z.Z.; Data curation, Z.Z. and B.Z.; Funding acquisition, W.Z. and L.Z.; Investigation, Z.Z.; Methodology, Z.Z.; Software, B.Z.; Supervision, W.Z.; Writing—original draft, Z.Z.; Writing—review and editing, W.Z. All authors have read and agreed to the published version of the manuscript.

Funding: This research was funded by the Public welfare science and technology projects of Ningbo (202002N3087). And The APC was funded by School of Food Science and Technology, Jiangnan University.

Institutional Review Board Statement: Not applicable.

Informed Consent Statement: Not applicable.

Data Availability Statement: Research data are not shared.

Conflicts of Interest: The authors declare no conflict of interest.

References

1. Maldaner, V.; Coradi, P.C.; Nunes, M.T.; Müller, A.; Carneiro, L.O.; Teodoro, P.E.; Ribeiro Teodoro, L.P.; Bressiani, J.; Anschau, K.F.; Müller, E.I. Effects of intermittent drying on physicochemical and morphological quality of rice and endosperm of milled brown rice. *LWT—Food Sci. Technol.* **2021**, *152*, 112334. [CrossRef]
2. Azizi, R.; Capuano, E.; Nasirpour, A.; Pellegrini, N.; Golmakani, M.T.; Hosseini, S.M.H.; Farahnaky, A. Varietal differences in the effect of rice ageing on starch digestion. *Food Hydrocoll.* **2019**, *95*, 358–366. [CrossRef]
3. Butardo, V.M.; Sreenivasulu, N. Tailoring grain storage reserves for a healthier rice diet and its comparative status with other cereals. *Int. Rev. Cel. Mol. Bio.* **2016**, *323*, 31–70.
4. Englyst, H.N.; Kingman, S.M.; Cummings, J.H. Classification and measurement of nutritionally important starch fractions. *Eur. J. Clin. Nutr.* **1992**, *46*, S33. [PubMed]
5. Zheng, J.; Huang, S.; Zhao, R.; Wang, N.; Kan, J.; Zhang, F. Effect of four viscous soluble dietary fibers on the physicochemical, structural properties, and in vitro digestibility of rice starch: A comparison study. *Food Chem.* **2021**, *362*, 130181. [CrossRef]

6. Toutounji, M.R.; Farahnaky, A.; Santhakumar, A.B.; Oli, P.; Butardo, V.M., Jr.; Blanchard, C.L. Intrinsic and extrinsic factors affecting rice starch digestibility. *Trends Food Sci. Technol.* **2019**, *88*, 10–22. [CrossRef]
7. Silva, W.M.; Biduski, B.; Lima, K.O.; Pinto, V.Z.; Hoffmann, J.F.; Vanier, N.L.; Dias, A.R. Starch digestibility and molecular weight distribution of proteins in rice grains subjected to heat-moisture treatment. *Food Chem.* **2017**, *219*, 260–267. [CrossRef]
8. Ling, B.; Cheng, T.; Wang, S. Recent developments in applications of radio frequency heating for improving safety and quality of food grains and their products: A review. *Crit. Rev. Food Sci. Nutr.* **2020**, *60*, 2622–2642. [CrossRef]
9. Liu, Q.; Qu, Y.; Liu, J.; Wang, S. Effects of radio frequency heating on mortality of lesser grain borer, quality and storage stability of packaged milled rice. *LWT—Food Sci. Technol.* **2021**, *140*, 110813. [CrossRef]
10. Yang, C.; Zhao, Y.; Tang, Y.; Yang, R.; Yan, W.; Zhao, W. Radio frequency heating as a disinfestation method against Corcyra cephalonica and its effect on properties of milled rice. *J. Stored Prod. Res.* **2018**, *77*, 112–121. [CrossRef]
11. Zhou, L.Y.; Wang, S.J. Verification of radio frequency heating uniformity and Sitophilus oryzae control in rough, brown, and milled rice. *J. Stored Prod. Res.* **2016**, *65*, 40–47. [CrossRef]
12. Zhang, Z.N.; Wang, Y.Y.; Ling, J.G.; Yang, R.J.; Zhu, L.; Zhao, W. Radio frequency treatment improved the slowly digestive characteristics of rice flour. *LWT—Food Sci. Technol.* **2022**, *154*, 112862. [CrossRef]
13. Gong, B.; Cheng, L.; Gilbert, R.G.; Li, C. Distribution of short to medium amylose chains are major controllers of in vitro digestion of retrograded rice starch. *Food Hydrocoll.* **2019**, *96*, 634–643. [CrossRef]
14. Li, C.; Hu, Y.; Zhang, B. Plant cellular architecture and chemical composition as important regulator of starch functionality in whole foods. *Food Hydrocoll.* **2021**, *117*, 106744. [CrossRef]
15. Ye, J.P.; Hu, X.T.; Luo, S.J.; McClements, D.J.; Liang, L.; Liu, C.M. Effect of endogenous proteins and lipids on starch digestibility in rice flour. *Food Res. Int.* **2018**, *106*, 404–409. [CrossRef]
16. Amagliani, L.; O'Regan, J.; Kelly, A.L.; O'Mahony, J.A. The composition, extraction, functionality and applications of rice proteins: A review. *Trends Food Sci. Technol.* **2017**, *64*, 1–12. [CrossRef]
17. Zhong, Y.J.; Xiang, X.Y.; Chen, T.T.; Zou, P.; Liu, Y.F.; Ye, J.P.; Luo, S.J.; Wu, J.Y.; Liu, C.M. Accelerated aging of rice by controlled microwave treatment. *Food Chem.* **2020**, *323*, 126853. [CrossRef]
18. Zhang, Z.N.; Zhang, M.Q.; Zhang, B.; Wang, Y.Y.; Zhao, W. Radio frequency energy regulates the multi-scale structure, digestive and physicochemical properties of rice starch. *Food Biosci.* **2022**, *47*, 101616. [CrossRef]
19. Ma, M.; Zhang, Y.; Chen, X.; Li, H.; Sui, Z.; Corke, H. Microwave irradiation differentially affect the physicochemical properties of waxy and non-waxy hull-less barley starch. *J. Cereal Sci.* **2020**, *95*, 103072. [CrossRef]
20. Sittipod, S.; Shi, Y.C. Changes of starch during parboiling of rice kernels. *J. Cereal Sci.* **2016**, *69*, 238–244. [CrossRef]
21. Paiva, F.F.; Vanier, N.L.; Berrios, J.; Pinto, V.Z.; Wood, D.; Williams, T.; Pan, J.; Elias, M.C. Polishing and parboiling effect on the nutritional and technological properties of pigmented rice. *Food Chem.* **2016**, *191*, 105–112. [CrossRef] [PubMed]
22. Liu, Q.; Kong, Q.; Li, X.; Lin, J.; Chen, H.; Bao, Q.; Yuan, Y. Effect of mild-parboiling treatment on the structure, colour, pasting properties and rheology properties of germinated brown rice. *LWT—Food Sci. Technol.* **2020**, *130*, 109623. [CrossRef]
23. Li, J.; Han, W.; Xu, J.; Xiong, S.; Zhao, S. Comparison of morphological changes and in vitro starch digestibility of rice cooked by microwave and conductive heating. *Starch Stärke* **2014**, *66*, 549–557. [CrossRef]
24. Olatunde, G.A.; Atungulu, G.G. Milling behavior and microstructure of rice dried using microwave set at 915 MHz frequency. *J. Cereal Sci.* **2018**, *80*, 167–173. [CrossRef]
25. Vela, A.J.; Villanueva, M.; Solaesa, Á.G.; Ronda, F. Impact of high-intensity ultrasound waves on structural, functional, thermal and rheological properties of rice flour and its biopolymers structural features. *Food Hydrocoll.* **2021**, *113*, 106480. [CrossRef]
26. Vela, A.J.; Villanueva, M.; Ronda, F. Low-frequency ultrasonication modulates the impact of annealing on physicochemical and functional properties of rice flour. *Food Hydrocoll.* **2021**, *120*, 106933. [CrossRef]
27. Monroy, Y.; Rivero, S.; Garcia, M.A. Microstructural and techno-functional properties of cassava starch modified by ultrasound. *Ultrason. Sonochem.* **2018**, *42*, 795–804. [CrossRef]
28. Wang, H.; Wang, Y.; Xu, K.; Zhang, Y.; Shi, M.; Liu, X.; Chi, C.; Zhang, H. Causal relations among starch hierarchical structure and physicochemical characteristics after repeated freezing-thawing. *Food Hydrocoll.* **2022**, *122*, 107121. [CrossRef]
29. Zhong, Y.; Tu, Z.; Liu, C.; Liu, W.; Xu, X.; Ai, Y.; Liu, W.; Chen, J.; Wu, J. Effect of microwave irradiation on composition, structure and properties of rice (Oryza sativa L.) with different milling degrees. *J. Cereal Sci.* **2013**, *58*, 228–233. [CrossRef]
30. Zhu, D.; Fang, C.; Qian, Z.; Guo, B.; Huo, Z. Differences in starch structure, physicochemical properties and texture characteristics in superior and inferior grains of rice varieties with different amylose contents. *Food Hydrocoll.* **2021**, *110*, 106170. [CrossRef]
31. Villanueva, M.; Harasym, J.; Munoz, J.M.; Ronda, F. Microwave absorption capacity of rice flour. Impact of the radiation on rice flour microstructure, thermal and viscometric properties. *J. Food Eng.* **2018**, *224*, 156–164. [CrossRef]
32. Solaesa, Á.G.; Villanueva, M.; Muñoz, J.M.; Ronda, F. Dry-heat treatment vs. heat-moisture treatment assisted by microwave radiation: Techno-functional and rheological modifications of rice flour. *LWT—Food Sci. Technol.* **2021**, *141*, 110851. [CrossRef]
33. Rockembach, C.T.; Mello El Halal, S.L.; Mesko, M.F.; Gutkoski, L.C.; Elias, M.C.; de Oliveira, M. Morphological and physicochemical properties of rice grains submitted to rapid parboiling by microwave irradiation. *LWT—Food Sci. Technol.* **2019**, *103*, 44–52. [CrossRef]
34. Sun, Q.; Wang, T.; Xiong, L.; Zhao, Y. The effect of heat moisture treatment on physicochemical properties of early indica rice. *Food Chem.* **2013**, *141*, 853–857. [CrossRef] [PubMed]

35. Lang, G.H.; Timm, N.S.; Neutzling, H.P.; Ramos, A.H.; Ferreira, C.D.; de Oliveira, M. Infrared radiation heating: A novel technique for developing quick-cooking rice. *LWT—Food Sci. Technol.* **2022**, *154*, 112758. [CrossRef]
36. Cheng, K.C.; Chen, S.H.; Yeh, A.I. Physicochemical properties and in vitro digestibility of rice after parboiling with heat moisture treatment. *J. Cereal Sci.* **2019**, *85*, 98–104. [CrossRef]
37. Qadir, N.; Wani, I.A. In-vitro digestibility of rice starch and factors regulating its digestion process: A review. *Carbohyd. Polym.* **2022**, *291*, 119600. [CrossRef]
38. Bora, P.; Ragaee, S.; Marcone, M. Effect of parboiling on decortication yield of millet grains and phenolic acids and in vitro digestibility of selected millet products. *Food Chem.* **2019**, *274*, 718–725. [CrossRef]
39. Li, C.; Wu, A.; Yu, W.; Hu, Y.; Li, E.; Zhang, C.; Liu, Q. Parameterizing starch chain-length distributions for structure-property relations. *Carbohyd. Polym.* **2020**, *241*, 116390. [CrossRef]

Article

Developing Radio-Frequency Roasting Protocols for Almonds Based on Quality Evaluations

Ting-Yu Lian and Su-Der Chen *

Department of Food Science, National Ilan University, Number 1, Section 1, Shen-Lung Road, Yilan 26041, Taiwan; j6j61215@gmail.com
* Correspondence: sdchen@niu.edu.tw; Tel.: +886-920518028; Fax: +886-39351892

Abstract: Hot air roasting is a popular method for preparing almonds, but it takes a long time. We roasted almonds via dielectric heating using 5 kW, 40.68 MHz batch radio-frequency (RF) equipment and analyzed their quality and aroma using a gas chromatography/ion mobility spectrometer and sensory evaluation. Almonds with an initial moisture content of 8.47% (w.b.) were heated at an RF electrode gap of 10 cm; the target roasting temperature of 120 °C was achieved at weights of 0.5, 1, 1.5, and 2 kg for 4, 3.5, 7.5, and 11 min, respectively; and the moisture content was reduced to less than 2% (w.b.). For comparison, 1 kg of almonds was roasted in a 105 °C conventional oven for 120 min. The darker color and lower moisture content, water activity, and acid value of the RF-roasted almonds were favorable for preservation. The aroma analysis using gas chromatography/ion mobility spectroscopy (GC–IMS) revealed that the aroma signal after roasting was richer than that of raw almonds, and principal component analysis (PCA) demonstrated that the aromas of roasted and commercial almonds were similar. The RF-roasted almonds presented a better flavor, texture, and overall preferability compared to commercial almonds. RF heating could be used in the food industry to roast nuts.

Keywords: almonds; radio frequency (RF); roasting; aroma; quality

1. Introduction

Almonds have a high nutritional value and are high in unsaturated fatty acids and vitamins. They are consumed after being roasted, a process which increases their hardness and crispness, enhances their aroma, and gives them a roasted appearance. Roasting can also deactivate enzymes, eliminate pests, and kill pathogenic microbes [1]. The most common type of oven is a traditional hot air oven, and the roasting method comprises hot air heating at over 100 °C for 120 min, which is extremely time-consuming. This method may lack a pasteurization effect and requires a lengthy processing time, increasing the processing cost. Furthermore, prolonged roasting can easily produce an undesirable flavor [1]. Therefore, new processing strategies must be developed to achieve an increased roasting speed while maintaining superior product quality.

Yang et al. [2] used infrared radiation (IR) combined with hot air technology to roast almonds. Ten grams of almonds were heated to the final roasting temperature of 130 °C by IR within 1 min and were then roasted with hot air for 15 min. The initial roasting time was reduced by more than 80%; the bacterial reduction increased by 38%; and the sensory quality was not significantly changed (appearance, texture, taste, and overall acceptability). Agila et al. [3] roasted 50 g of almonds at 177 °C for 5 min before roasting them in a microwave oven at 135 °C for 5 min (the final temperature of the almonds was 108 °C). According to the results and sensory evaluations, the aroma of the microwave-roasted almonds was superior to that of the oven-roasted almonds. Kosoko et al. [4] roasted 2.5 kg of cashew kernels in a hot air and halogen oven at 200 °C for 40 min. The rapid temperature cycle of the halogen oven effectively reduced the moisture content of the nuts. The halogen-oven-roasted cashews were highly acceptable according to the entire sensory evaluation. However, infrared and microwave heating penetration depths are minimal, and the thickness and quantity of the treated samples are limited. In the two above mentioned tests, for example, the weight of roasted almonds was only 10 g and 50 g, which is insufficient for large-scale production. Although the heating effect of a halogen oven is superior to that of a standard hot air oven for industrial applications, the short heating distance prevents it from producing a significant heating effect and thus from effectively reducing the processing time [5].

In recent years, radio-frequency (RF) heating has been used to roast nuts such as almonds [1], peanuts [6], and cashews [5,7]. RF roasting is a fast dielectric heating method that operates at frequencies ranging from 10 MHz to 300 MHz. The electromagnetic field is rapidly transformed by the top and lower electrode plates, causing the polar molecules and charged ions in the sample to violently rotate and shake, resulting in frictional heat generation. Furthermore, due to the deep penetration of RF waves, the heating is fast and uniform, reducing the heating time and improving product quality [8]. RF treatment has great potential as a new nut-roasting method because it can achieve disinfestation, pasteurization, and drying effects when the target temperature of 120 °C is reached.

The objectives of this study were to develop an RF roasting protocol; compare the quality of almonds roasted using commercial, oven, and RF heating methods; and evaluate the aroma and sensory quality of RF-roasted almonds.

2. Materials and Methods

2.1. Materials

The almonds used in this study were purchased from Beans Group Foods Science and Technology Co. (Taoyuan, Taiwan), and their origin was California, USA. 1,1-Diphenyl-2-picryl hydrazyl (DPPH), ascorbic acid (vitamin C), synthetic glacial acetic acid, ethylene-diaminetetraacetic acid (EDTA), and antioxidant butylated hydroxyl anisole (BHA) were purchased from Sigma Chemical Co. (St. Louis, MO, USA). Ethanol (95%), diethyl ether, phenolphthalein indicator, potassium hydroxide (KOH), and potassium hydrogen phthalate (KHP) were purchased from WAKO Pure Chemical Industries, Ltd. (Osaka, Japan).

2.2. Roasting Methods

2.2.1. Determination of RF Roasting Conditions

The RF system with hot air equipment (Yh-Da Biotech Co., LTD., Yilan, Taiwan) (Figure 1) was designed to produce 5 kW and operate at 40.68 MHz, with a hot air temperature of 100 °C. The hot air was blown in from the right side. Samples of 0.5 and 1 kg were placed in a polypropylene (PP) plastic basket with holes 23 cm in diameter and 8 cm in height, and 1.5 and 2 kg specimens were placed in a larger PP plastic basket with holes 27 cm in diameter and 9.5 cm in height. RF treatment was performed with different electrode plate gaps. Because the RF equipment had a maximum current of 1.6 Amp and a maximum output power of 5 kW, the average output power of the RF system was calculated by reading the current (A) and using the following formula: power output (kW) = (5/1.6) A.

The surface temperature of the sample was measured at the center and 5 cm on either side of the center by a multifunctional infrared thermometer (Testo104-IR, Hot Instruments Co., LTD., New Taipei City, Taiwan), and the final surface temperature of the sample was also measured by an infrared thermometer (TIM-03, HILA International Inc., Taipei, Taiwan). To obtain the temperature profile, the increase in the temperature for different electrode plate gaps was recorded every 30 s until the surface temperature of the sample reached 120 °C. The dry-basis moisture content change was measured according to the weight change in the sample during the RF roasting.

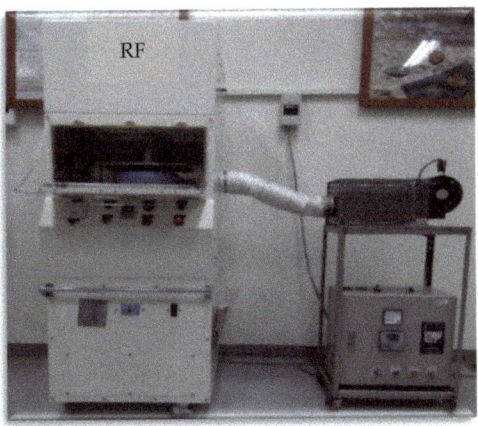

Figure 1. 5 kW, 40.68 MHz RF hot air equipment (**left**) and a sample basket placed between RF electrodes (**right**).

2.2.2. Conventional-Oven-Roasted Almonds

The 1 kg sample was placed in a single layer on a stainless-steel tray and roasted in an oven (Channel DCM-45, Taiwan) at 105 °C for 120 min; the temperature and weight of the sample were recorded.

2.3. Analytical Methods

2.3.1. Moisture Content

We weighed 5 g of ground almonds in an aluminum dish using an electronic precision scale (HDW-15L, Hengxin Metrology Technology Co., Ltd., Yilan, Taiwan), dried them in an oven at 105 °C for 12 h, and then removed and weighed them after reaching a constant weight.

2.3.2. Water Activity

The water activity (Aw) was measured using a water-activity analyzer (HC2-AW, Rotronic Instruments Ltd., Zwillikon, Switzerland). The ground almond samples were loaded into the analyzer and measured at 25 °C for 10 min each time, and the data were recorded.

2.3.3. Color Measurement

The color of the samples was measured with a color difference meter (Hunter LAB, Color Flex, Virginia, VA, USA) and standardized against a white calibration plate (X = 82.48, Y = 84.23, Z = 99.61, $L^* = 92.93$, $a^* = -1.26$, $b^* = 1.17$). The parameters determined were the degree of lightness (L^*), redness ($+a^*$), greenness ($-a^*$), yellowness ($+b^*$), and blueness ($-b^*$). All experiments were performed in six repetitions.

2.3.4. DPPH Radical Scavenging Ability Assays

We extracted 2 mL of the supernatant using a focused ultrasonic machine (20 k Hz, 1400 W, Ultrasonic Co., Ltd., New Taipei City, Taiwan) and subjected it to a DPPH assay according to [9]. We applied 20 mg/mL ascorbic acid, BHA, and EDTA as standards. All experiments were performed in three repetitions.

2.3.5. Acid Value

The acid value was determined according to the Chinese National Standard analytical methods for edible oil (CNS 3647 N6082) [10]. All experiments were performed in four repetitions.

2.3.6. GC–IMS Analysis

The aroma analysis by GC–IMS was based on the method of Thomas et al. [11]. One gram of each sample was placed in a 20 mL headspace vial. The sample was heated in a heater at 50 °C for 20 min by the autosampler system, which was equipped with a 1 mL syringe with an injection rate of 170 μL/s. The sample was then injected into the gas chromatography/ion mobility spectroscopy machine (GC–IMS, Flavour Spec®, GAS Dortmund, Germany) in the non-diversion mode using a 20 m long 0.53 nm ID non-polar capillary column (CC), model OV-5 (5%-diphenyl, 95%-dimethylpolysiloxane), operated at 45 °C. The volatile gases were separated by injecting them into the GC using nitrogen (purity $\geq$ 99.999%) as the carrier gas. The gradient of the nitrogen flow rate after injection was as follows: 0~5 min—increased from 2 mL/min to 15 mL/min; 5~5.5 min—maintained at 15 mL/min, 5.5~13 min—increased from 15 mL/min to 30 mL/min; and 13~20 min—decreased from 30 mL/min to 2 mL/min. Subsequently, the gas was separated into the IMS ionization zone chamber. Data analysis was carried out using IMS Control TFTP Server software provided by G.A.S (Dortmund, Germany).

2.3.7. Sensory Evaluation

A panel of 65 people conducted a nine-point hedonic sensory evaluation of the roasted almonds in regard to appearance (how much they liked the appearance of the samples in terms of size, thickness, and completeness); aroma (how much they liked the strong smell of the samples); flavor (how much they liked the taste of the samples in terms of sourness, sweetness, and bitterness); taste (how much they liked the texture, chewiness, and crispness of the samples); aftertaste (the aftertaste after swallowing); and overall performance. All samples were allocated a three-digit number and selected randomly, and a nine-point scale for sensory evaluation was applied, with a score of 1 indicating very much dislike, 5 indicating neither like nor dislike, and 9 indicating very much like.

2.4. Statistical Analysis

Experimental results are expressed as means ± standard deviation (SD). One-way analysis of variance (ANOVA) was performed and subsequently subjected to Duncan's multiple range tests of treatment mean using Statistical Analysis System (SAS 9.4, SAS Institute, Cary, NC, USA), and the significance level was set at 0.05.

3. Results and Discussion

3.1. Effect of Almond Loading on RF Heating

The effect of the RF electrode plate gap on the output power for almonds at varying loading capacities is illustrated in Figure 2. The RF energy output for 1, 1.5, and 2 kg almonds decreased as the RF electrode plate gap increased, similar to the results of Chen et al. [12], who used 1, 2, and 3 kg rice bran at an electrode plate gap of 6–16 cm, with the smaller gap demonstrating a higher RF output power for heavier rice bran. The same pattern was observed in our study. The RF output power was high for different loading levels of almonds at a lower electrode plate gap of 10 cm; therefore, a 10 cm gap was chosen to compare the RF roasting conditions for different loading levels of almonds.

The drying and heating curves of the almonds loaded at different amounts with a 10 cm RF electrode plate gap (Figure 3) showed that for 0.5 and 1 kg almonds, the temperature increased to 100 °C within 2 min of RF heating, before the rate of heating slowed down significantly until reaching 120 °C, but the moisture content of the almonds decreased more rapidly.

For 1.5 and 2 kg almonds, a rapid increase in temperature was observed at the beginning of roasting, followed by a steady increase. However, due to the increased loading of almonds, the RF roasting time also increased. The slowest heating rate for 2 kg almonds required only 10 min to reach 120 °C, and 1 kg almonds demonstrated the fastest heating rate of 33.37 °C/min, requiring only 3.5 min to reach 120 °C and achieve the roasting effect.

Figure 4 shows the heating and drying curves of the almonds roasted in a conventional hot air oven. The temperature of the almonds was affected by heat transfer resistance in the later stages. After the almond temperature rose to 80 °C at 40 min, it took twice as long to increase the temperature by less than 20 °C, and the maximum temperature of the almonds reached only 95 °C at the end of the baking. Although the dry-basis moisture content of the almonds showed a steady decreasing trend, the dry-basis moisture content of the final almond sample only decreased to 0.043 kg water/kg dry material, and it took nearly 40 times longer than the RF technique. The drying efficiency was poor, and the high moisture content made the almonds less crisp.

The moisture content of the RF-roasted almonds did not drop significantly due to the low temperature at the beginning, but a higher moisture content could provide more polar water molecules to quickly raise the temperature. At the beginning of RF heating, the almond temperature rose rapidly in a straight line, and in the later stages most of the energy was turned into moisture evaporation heat, which was accompanied by a lower moisture content and also caused the temperature rise to slow down. Therefore, the final RF-roasted almond temperature reached about 120 °C, and the moisture content was reduced to less than 2%, which was far from the initial moisture content of 8.47% in the dried almonds before roasting. This result is in agreement with that of Wang et al. [13], who observed that the moisture content of hazelnuts in shells with RF heating decreased significantly from 34% to 19% in the initial stage (1 min) and slowed down to reach 10% in the later stages (after 2 min), while it took 22 h to reach approximately 10% if the hazelnuts were dried only by hot air at 40 °C. During the later stages of the conventional roasting of the samples, as the moisture content was already very low, causing heat conduction resistance, the thicker samples in particular required a very long roasting and drying time, whereas the RF roasting of almonds demonstrated more efficient heating.

3.2. Processing Performance of RF Roasting

The electrode plate gap was negatively correlated with the RF output power, but the sample loading amount was positively correlated with the RF output power (Figure 2). However, considering the heating efficiency and uniformity of the sample, 1 kg of almonds was chosen as the loading amount to achieve the fastest roasting conditions and heating treatment with three different RF electrode plate gaps (9, 10, and 11 cm) (Figure 5). The temperature of the almonds increased linearly as the gap between the RF electrode plates increased. The almonds were heated faster at a gap of 9 cm between the electrode plates, but the heating uniformity was poor. The heating speed was slower at a gap of 11 cm between the electrode plates. An electrode plate gap of 10 cm was chosen as the operating conditions for the subsequent RF roasting of almonds.

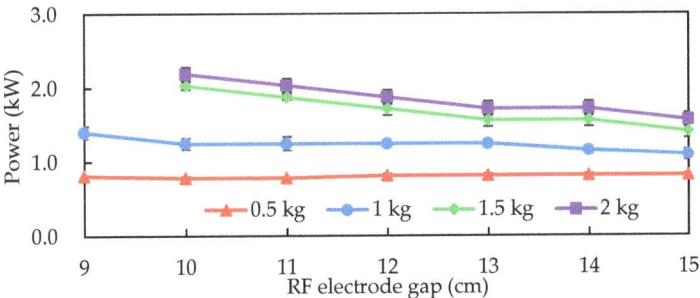

Figure 2. Effect of different loading levels of almonds and different 5 kW, 40.68 MHz RF electrode plate gaps on power. Data are expressed as mean ± S.D. (n = 3).

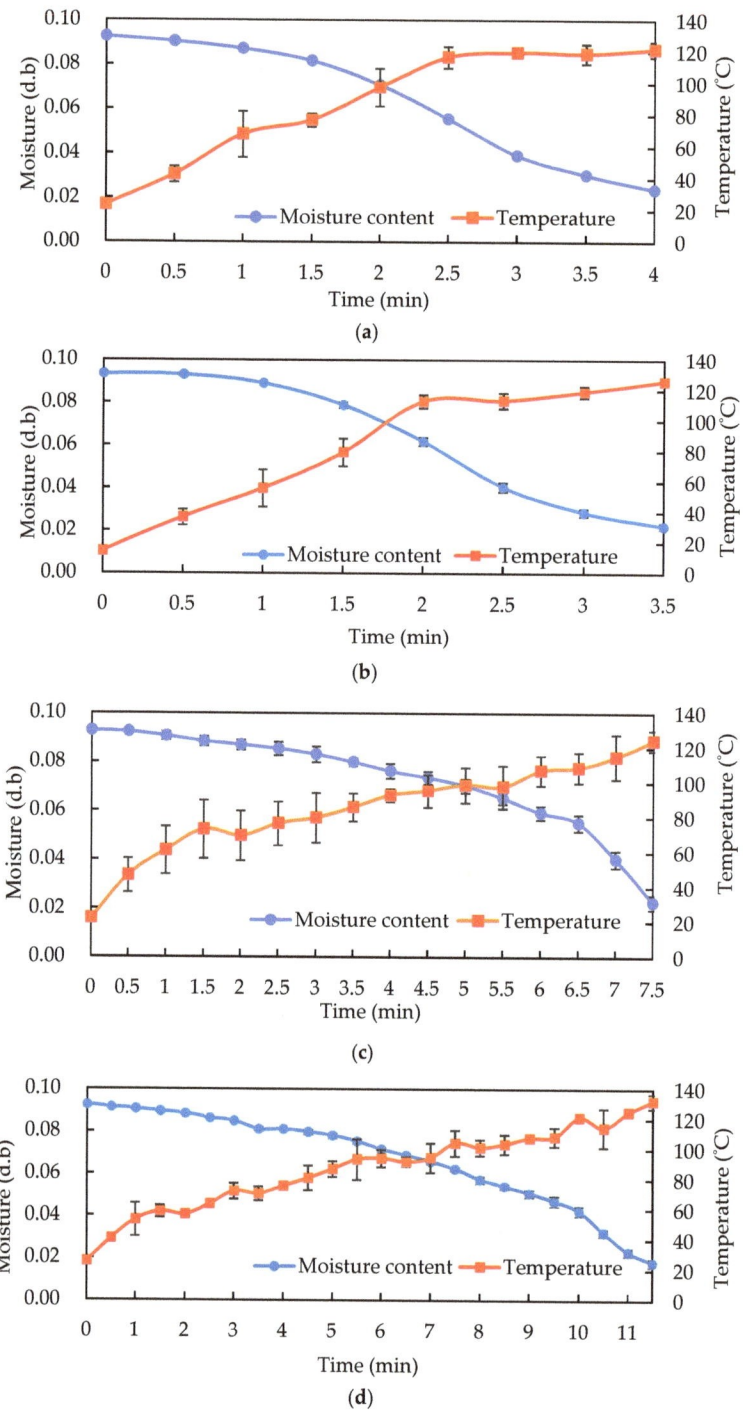

Figure 3. The drying and temperature curves of (**a**) 0.5, (**b**) 1, (**c**) 1.5, and (**d**) 2 kg almonds during 5 kW, 40.68 MHz batch RF roasting at 10 cm electrode gap. Data are expressed as mean ± S.D. (n = 3).

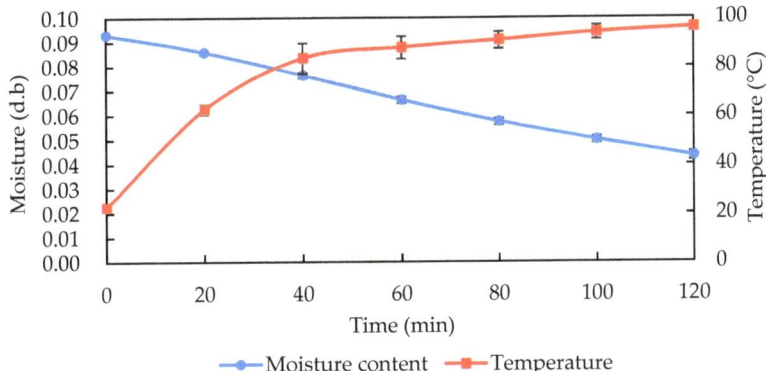

Figure 4. The drying curve and temperature profile of 1 kg almonds during 105 °C oven roasting. Data are expressed as mean ± S.D. (*n* = 3).

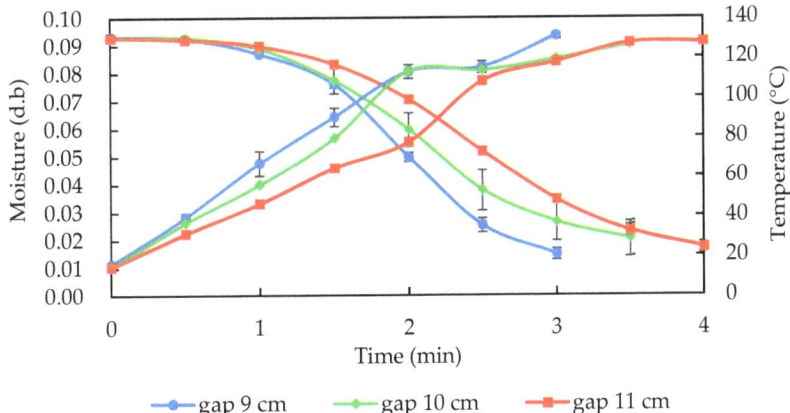

Figure 5. The drying and temperature curves of 1 kg almonds during 5 kW, 40.68 MHz RF roasting at different electrode gaps. Data are expressed as mean ± S.D. (*n* = 3).

The results were similar to those of Jiao et al. [6], who determined the appropriate electrode gap for peanuts by considering the heating uniformity: the peanuts reached 90 °C after 5.5, 11, and 18 min at gaps of 9, 10, and 11 cm, respectively. The smaller the RF electrode plate gap, the faster the heating, but considering the heating efficiency and uniformity, the electrode plate gap of 10 cm was selected as the subsequent operating conditions. Figure 3 also shows the drying curve of these three RF heating gaps: the dry-basis moisture content of the almonds at gaps of 9, 10, and 11 cm decreased from 0.092 kg water/kg dry material to 0.017, 0.026, and 0.035 kg water/kg dry material, respectively. The moisture evaporated rapidly, and the loss of moisture raised the hardness and brittleness of the almonds. The results of the texture profile analysis (TPA) conducted by Xu et al. [1] showed that after roasting, the moisture content of almonds decreased, while the hardness increased.

In addition, the surface temperature of the almonds was measured using an infrared thermometer (Figure 6). The average temperature for the three gaps was higher than the target temperature of 120 °C, and the temperature distribution was very uniform. This is also a characteristic of RF heating and roasting—because the sample heats up quickly and the heat distribution is very uniform, it is less likely to contain cold spots.

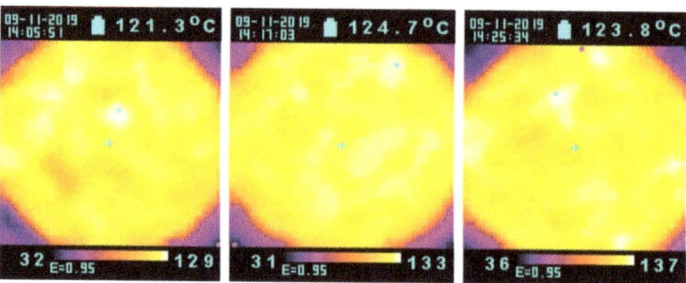

Figure 6. The surface temperature of 1 kg almonds after 5 kW, 40.68 MHz RF roasting (**left**: 9 cm gap, 3 min; **center**: 10 cm gap, 3.5 min; and **right**: 11 cm gap, 4 min).

3.3. Quality, Aroma, and Sensory Evaluation of RF-Roasted Almonds

The roasting process increases the crispness of nuts, gives them a roasted color, and produces a special roasted flavor. Table 1 showed that the moisture content and water activity of the almonds were significantly reduced after roasting. The moisture content (w.b.) of the almonds roasted using hot-air-assisted RF heating for 3.5 min decreased from 8.47% to 1.57%, and the water activity decreased from 0.74 to 0.34. The moisture content and water activity results were lower than those of commercial roasted almonds and almonds roasted in an oven for 120 min, which already had a moisture content below 5.80% for the safe storage of nuts [4]. While almonds with a water activity of 0.2 to 0.3 have a longer shelf life [1], less water is more favorable for nut storage. Oil quality changes are a very important indicator of nut shelf life. The acid value tended to decrease after roasting, with the lowest acid value of 0.34 mg KOH/g obtained for RF-roasted almonds, probably due to the rapidity of RF heating, which inactivated lipase and reduced free fatty acid production.

Table 1. Quality parameters of raw, RF-roasted, conventional-oven-roasted, and commercial almonds.

Sample	Raw	RF	Oven	Commercial
MC (%)	8.47 ± 0.12 [a]	1.57 ± 0.06 [d]	3.6 ± 0.10 [b]	2.3 ± 0.26 [c]
Aw	0.74 ± 0.00 [a]	0.34 ± 0.03 [c]	0.49 ± 0.01 [b]	0.56 ± 0.01 [b]
L*	44.43 ± 0.72 [a]	41.72 ± 0.84 [b]	41.10 ± 0.84 [b]	41.64 ± 0.59 [b]
a*	15.58 ± 0.17 [a]	15.56 ± 0.11 [a]	14.32 ± 26.10 [c]	15.08 ± 0.22 [b]
b*	29.36 ± 0.26 [a]	26.70 ± 0.54 [b]	26.10 ± 0.42 [c]	26.67 ± 0.51 [b]
AV (mg/g)	0.58 ± 0.09 [a]	0.34 ± 0.02 [c]	0.48 ± 0.06 [b]	0.47 ± 0.02 [b]
DPPH (%)	82.50 ± 0.62 [a]	64.60 ± 0.46 [c]	72.20 ± 0.36 [b]	52.10 ± 1.01 [d]

Data are expressed as mean ± S.D ($n = 5$). Means with different superscript letters in the same row are significantly different ($p < 0.05$).

In terms of color (Table 1), roasting reduced the reddish and yellowish color of the almonds, and the brightness was lowest in the oven-roasted almonds. Although the RF roasting temperature of 120 °C was higher than the oven roasting temperature of 105 °C, the color change in the almonds was closer to that of the untreated almonds, due to the short duration of the RF roasting.

In terms of the ability to scavenge DPPH free radicals, the antioxidant capacity of the untreated almonds was the highest, due to the fact that the roasting process breaks down the cells, followed by hot air roasting. RF roasting had a higher final temperature, and so the DPPH antioxidant effect was poorer, but it was better compared to the commercial roasted almonds. The results were similar to those of Liao et al. [5], who roasted cashew kernels with a thickness of 5 cm using RF (120–130 °C, 30 min) and hot air (140 °C, 33 min), determining that raw cashews had the best antioxidant capacity, while there was no significant difference between the RF and hot air treatments ($p > 0.05$).

Figure 7 shows the GC–IMS analysis of the volatile organic compounds in raw, RF-roasted, oven-roasted, and commercial roasted almonds. The red area indicates more volatile components, and the darker the color, the more components, while the blue area indicates the opposite. As shown in Figure 7, most of the signals appeared at the retention time of 0~150 s and the drift time of 1.0–1.5. The raw almonds showed fewer and weaker odor signals, while the roasted almonds clearly produced more odor signals, especially those roasted at a higher RF temperature (120 °C), which could be clearly seen at the retention time of 0–100 s and the drift time of 1.0–1.5. This was probably due to the fact that the RF roasting method provided heat to the almonds both internally and externally, and the almonds received heat from more sources, thus producing a higher volatile content, which conducted heat mainly on the surface of the almonds. This resulted in differences in aroma presentation, and it has been found that microwaves generate richer volatile compounds than ovens and frying [3].

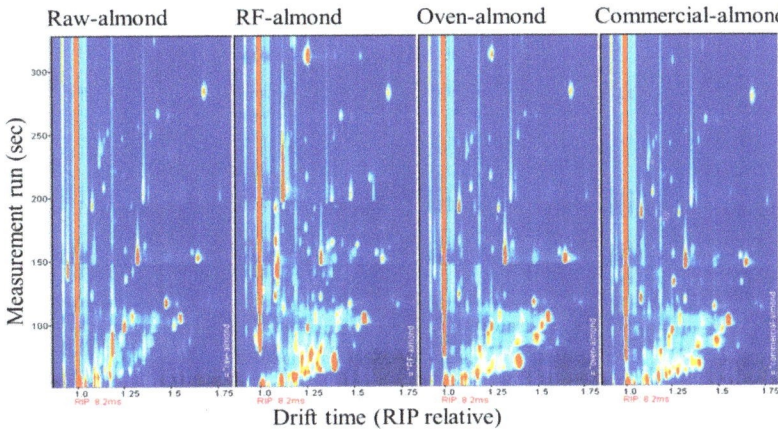

Figure 7. GC–IMS plot comparison of raw, RF-roasted, oven-roasted, and commercial almonds.

According to the differences in the volatile compound signals in the GC–IMS results, the fingerprint profiles of 32 characteristic compounds were further selected (Figure 8). The main volatile compounds of the raw almonds were found in the 1–15 and 28–32 fingerprint profiles. Most of the roasted almonds retained their original flavor, while in 16–26, the flavor signal had the highest intensity for the almonds roasted at an RF temperature of 120 °C. In 27–32, 100 °C oven roasting and commercially available 100 °C hot air roast produced strong flavor signals, indicating that almonds need to be roasted at high temperatures to produce a special aroma.

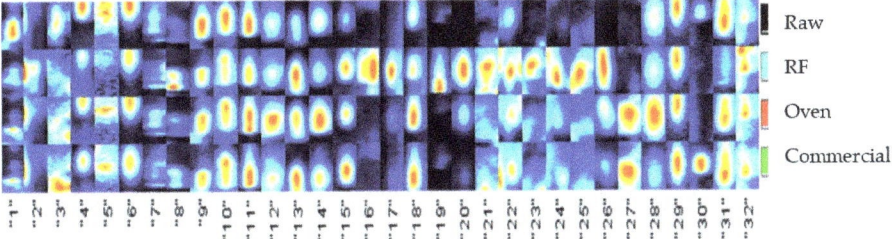

Figure 8. The characteristic aroma fingerprint of raw, RF-roasted, oven-roasted, and commercial almonds.

The principal component analysis (PCA) in Figure 9 shows that the aroma of the almonds before and after RF roasting was different, while oven roasting and commercially available hot air roasting were both conducted at 100 °C and with slow heat transfer, so the aroma was similar. The difference between the aroma of the commercially available almonds and that of the RF almonds was probably due to the fact that the almonds were roasted quickly by RF at a temperature of 120 °C.

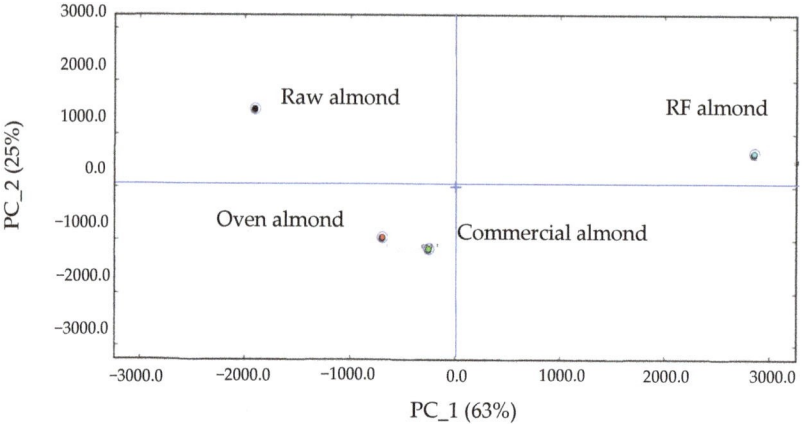

Figure 9. Principal component analysis of raw, RF-roasted, oven-roasted, and commercial almonds.

The results were similar to those of Liao et al. [7], who investigated the changes in the aroma content of cashew nuts during roasting and found that the aroma concentration increased with time and with higher temperatures. This may be related to the formation of volatiles in roasted nuts due to the Mena reaction, which cause aroma production and changes in flavor. Xu et al. [1] also analyzed the aroma of raw and hot-air-assisted-RF-roasted almonds using GC–MS and found that they contained 61 and 87 volatile components, respectively, and that the concentration of the roasted flavor components such as aldehydes, ketones, esters, alcohols, furans, pyrroles, and pyridine derivatives increased significantly after roasting, with most of the new volatile compounds commonly considered hot roasted flavors in oily nuts. Moreover, the sensory characteristics, nutritional quality, and oxidative stability of roasted macadamia nuts were greatly improved compared to raw nuts during storage [14].

Table 2 shows the 9-point-scale hedonic sensory evaluation by 65 tasters. Comparing the results of the commercial and RF-roasted almonds, there was no significant difference in appearance, aroma, or aftertaste; however, the flavor, texture, and overall acceptance of the RF-roasted almonds scored significantly higher than those of the commercial almonds, and the flavor score of 6.63 was significantly higher than the commercial almonds' 6.03 points. This may be due to the above principal component analysis, which demonstrated that the RF-roasted almonds produced several special aromas that were different to those produced by hot-air-roasted almonds. In addition, the moisture content of the almonds roasted by RF energy was lower, causing a crispier texture that scored 6.77 points, which was significantly higher than the 6.03 points garnered by the commercial almonds, while the other five sensory attributes eared scores greater than 6 points. The overall acceptance score of the RF-roasted almonds was 6.58 points higher than that of the commercial almonds (6.15 points).

Table 2. Consumer 9-point-scale hedonic sensory evaluation of RF-roasted and commercial roasted almonds.

Roasting	Appearance	Aroma	Flavor	Texture	Aftertaste	Overall
RF	6.57 ± 1.33	6.35 ± 1.78	6.63 ± 1.57 *	6.77 ± 1.49 *	6.75 ± 1.79	6.85 ± 1.58 *
Commercial	6.71 ± 1.28	5.85 ± 1.55	6.03 ± 1.58	6.03 ± 1.85	6.20 ± 1.66	6.15 ± 1.55

Data are expressed as mean ± S.D. (n = 65). Means with * in each column are significantly different ($p < 0.05$).

4. Conclusions

An increased loading amount and decreased electrode plate gap caused a higher RF output power and faster heating rate. Under the treatment conditions of 1 kg almonds with a 10 cm distance between the RF electrode plates, it only took 3.5 min to reach the target roasting temperature of 120 °C, while traditional oven roasting required 120 min. In addition, the moisture content, water activity, acid value, and DPPH radical scavenging ability of the RF-roasted almonds were better than those of the commercial products. The results of the GC–IMS aroma analysis showed that the almonds produced more aromatic components after roasting, and the aroma of the RF-roasted almonds was different from that of the commercial almonds according to PCA analysis. The overall score of the RF-roasted almonds was higher for the sensory evaluation. Therefore, the RF roasting of almonds is a time-saving, effective, and quality-enhancing technique.

Author Contributions: S.-D.C.—supervision, writing (review and editing), and project administration; T.-Y.L.—investigation, formal analysis, data analysis, and writing (original drift manuscript). All authors have read and agreed to the published version of the manuscript.

Funding: This research received no external funding.

Institutional Review Board Statement: Not applicable.

Informed Consent Statement: Not applicable.

Data Availability Statement: The data presented in this study are available in this article.

Acknowledgments: We thank Yue-Wen Chen for providing the GC–IMS equipment for the almond aroma analysis.

Conflicts of Interest: The authors declare no conflict of interest.

References

1. Xu, Y.; Liao, M.; Wang, D.; Jiao, S. Physicochemical quality and volatile flavor compounds of hot air-assisted radio frequency roasted almonds. *J. Food Process. Preserv.* **2020**, *44*, e14376. [CrossRef]
2. Yang, J.; Bingol, G.; Pan, Z.; Brandl, M.T.; McHugh, T.H.; Wang, H. Infrared heating for dry-roasting and pasteurization of almonds. *J. Food Eng.* **2010**, *101*, 273–280. [CrossRef]
3. Agila, A.; Barringer, S. Effect of roasting conditions on color and volatile profile including HMF level in sweet almonds (*Prunus dulcis*). *J. Food Sci.* **2012**, *77*, C461–C468. [CrossRef] [PubMed]
4. Kosoko, S.B.; Oluwole, O.B.; Daramola, A.O.; Adepoju, M.A.; Oyelakin, A.J.; Tugbobo-Amisu, A.O.; Alagbe, G.O.; Elemo, G.N. Comparative quality evaluation of roasted cashew nut kernel: Effect of roasting methods. *Adv. J. Food Sci. Technol.* **2014**, *6*, 1362–1371. [CrossRef]
5. Liao, M.; Zhao, Y.; Gong, C.; Zhang, H.; Jiao, S. Effects of hot air-assisted radio frequency roasting on quality and antioxidant activity of cashew nut kernels. *LWT* **2018**, *93*, 274–280. [CrossRef]
6. Jiao, S.; Zhu, D.; Deng, Y.; Zhao, Y. Effects of hot air-assisted radio frequency heating on quality and shelf-life of roasted peanuts. *Food Bioproc. Technol.* **2016**, *9*, 308–319. [CrossRef]
7. Liao, M.; Zhao, Y.; Xu, Y.; Gong, C.; Jiao, S. Effects of hot air-assisted radio frequency roasting on nutritional quality and aroma composition of cashew nut kernels. *LWT* **2019**, *116*, 108551. [CrossRef]
8. Marra, F.; Zhang, L.; Lyng, J.G. Radio frequency treatment of foods: Review of recent advances. *J. Food Eng.* **2009**, *91*, 497–508. [CrossRef]
9. Xu, B.J.; Chang, S.K.C. A comparative study on phenolic profiles and antioxidant activities of legumes as affected by extraction solvents. *J. Food Sci.* **2007**, *72*, 159–166. [CrossRef] [PubMed]
10. *CNS 3647:2021*; Methods of Test for Edible Oils and Fats—Determination of Acid Value No. 3647 (N6082). Chinese National Standard: Taipei, Taiwan, 1996.

11. Thomas, C.F.; Zeh, E.; Dörfel, S.; Zhang, Y.; Hinrichs, J. Studying dynamic aroma release by headspace-solid phase microextraction-gas chromatography-ion mobility spectrometry (HS-SPME-GC-IMS): Method optimization, validation, and application. *Anal. Bioanal. Chem.* **2021**, *413*, 2577–2586. [CrossRef] [PubMed]
12. Chen, Y.H.; Yen, Y.F.; Chen, S.D. Effects of radio frequency heating on the stability and antioxidant properties of rice bran. *Foods* **2021**, *10*, 810. [CrossRef] [PubMed]
13. Wang, W.; Tang, J.; Zhao, Y. Investigation of hot-air assisted continuous radio frequency drying for improving drying efficiency and reducing shell cracks of inshell hazelnuts: The relationship between cracking level and nut quality. *Food Biopro. Proces.* **2021**, *125*, 46–56. [CrossRef]
14. Tu, X.H.; Wu, B.F.; Xie, Y.; Xu, S.L.; Wu, Z.Y.; Lv, X.; Wei, W.; Du, L.; Chen, H. A comprehensive study of raw and roasted macadamia nuts: Lipid profile, physicochemical, nutritional, and sensory properties. *Food Sci. Nutr.* **2021**, *9*, 1688–1697. [CrossRef]

Article

Effects of Radio Frequency Tempering on the Texture of Frozen Tilapia Fillets

Jiwei Jiang [1,2,3], Fen Zhou [1,2,3], Caining Xian [1,2,3], Yuyao Shi [1,2,3] and Xichang Wang [1,2,3,*]

1. College of Food Science and Technology, Shanghai Ocean University, Shanghai 201306, China; m180300676@st.shou.edu.cn (J.J.); d170202044@st.shou.edu.cn (F.Z.); m190300801@st.shou.edu.cn (C.X.); m190310917@st.shou.edu.cn (Y.S.)
2. Shanghai Aquatic Product Processing and Storage Engineering Technology Research Center, Shanghai 201306, China
3. Laboratory of Quality and Safety Risk Assessment of Aquatic Products Storage and Preservation, Ministry of Agriculture, Shanghai 201306, China
* Correspondence: xcwang@shou.edu.cn

Abstract: Radio frequency (RF) tempering has been proposed as a new alternative method for tempering frozen products because of its advantages of rapid and volumetric heating. In this study, the texture of RF-tempered frozen tilapia fillets was determined under different RF conditions, the effects of related factors on the texture were analyzed, and the mechanisms by which RF tempering affected the texture of the tempered fillets were evaluated. The results show that the springiness (from 0.84 mm to 0.79 mm), cohesiveness (from 0.64 mm to 0.57 mm), and resilience (from 0.33 mm to 0.25 mm) decreased as the electrode gap was increased and the power remained at 600 W, while the shear force increased as the power was increased for the 12 cm electrode gap (from 15.18 N to 16.98 N), and the myofibril fragmentation index (MFI) values were markedly higher at 600 W than at 300 W or 900 W ($p < 0.05$). In addition, the tempering uniformity had a positive effect on hardness and chewiness. The statistical analysis showed that the texture after RF tempering under different RF conditions correlated relatively strongly with the free water content, cooking loss, and migration of bound water to immobilized water. The decrease in free water and bound water migration to immobilized water resulted in a significant increase in cohesiveness and resilience.

Keywords: tempered fillets; electrode gap; springiness; cohesiveness; resilience; hardness

Citation: Jiang, J.; Zhou, F.; Xian, C.; Shi, Y.; Wang, X. Effects of Radio Frequency Tempering on the Texture of Frozen Tilapia Fillets. *Foods* **2021**, *10*, 2663. https://doi.org/10.3390/foods10112663

Academic Editors: Shaojin Wang and Rui Li

Received: 6 October 2021
Accepted: 27 October 2021
Published: 2 November 2021

Publisher's Note: MDPI stays neutral with regard to jurisdictional claims in published maps and institutional affiliations.

Copyright: © 2021 by the authors. Licensee MDPI, Basel, Switzerland. This article is an open access article distributed under the terms and conditions of the Creative Commons Attribution (CC BY) license (https://creativecommons.org/licenses/by/4.0/).

1. Introduction

"Tilapia" is the name given to several genera and species of fish in the family Cichlidae, which have been proven to be among the most important food fish in the world and are popular because of their high protein content [1]. Tilapia are generally processed into frozen fillets for transportation and sales. Fish are among the most highly perishable food products [2], and further quality deterioration in freshness occurs during fishing, storage, transport, and processing due to the presence of rich endogenous enzymes and microorganisms, and therefore, freezing is a common method of fish preservation [3]. For further processing, frozen tilapia fillets need to be defrosted, which is also a key operation for meeting the needs of consumers. However, traditional tempering methods affect the quality, water and juice loss, texture, and organizational structure of fillets by causing varying degrees of damage due to changes in physical properties, chemical reactions, and microorganisms [4], which lead to increased resource consumption due to longer tempering times. Air tempering has disadvantages, such as a slow tempering rate and greater susceptibility to microbial contamination and lipid oxidation. Tempering with running tap water easily causes secondary pollution, considerable juice loss, and tissue collapse. Low-temperature–high-humidity tempering needs a larger processing site, and the investment cost is large [5,6].

New tempering technologies, such as high-voltage electrostatic field (HVEF) tempering, radio frequency (RF) tempering, ohmic tempering, microwave tempering, pressure-assisted tempering, and acoustic tempering, are under consideration by those people working to solve the problems with the tempering process mentioned above [7–10]. The emerging tempering technologies have advantages such as shorter tempering times, low energy consumption, high energy utilization rates, effective inhibition of the growth of microorganisms and lipid oxidation during the tempering process, reduced production of spoilage substances, improved freshness, and extended storage time. Electrical treatments play important roles as alternative methods for processing foods with minimal damage. According to Yang et al. [11], RF tempering technology has led to shorter tempering times and lower energy consumption. RF technology is particularly suitable for heating samples compared with other novel tempering technologies because of its low frequency and deep penetration depth [12]. The volumetric nature of RF tempering shortens the processing time and increases the heating rate, which can improve the sample tissue characteristics in ways that are different from conventional tempering methods. The change in temperature in different parts (tempering uniformity) of a sample and the thermal and non-thermal effects during the RF tempering process can affect the microstructure and macrostructure of the product and cause a variety of phenomena (including water migration and texture changes), depending on the tempering conditions [13].

Water exists inside and in the interstices of myofibrils in muscle tissue and forms compartments, which can complicate the internal change process during tempering, affect the steady state of the complex meat system, and, in turn, affect the texture after tempering [14]. According to some reports, the loss of water retention capacity can lead to the destruction of muscle fiber structure, and the shrinkage of muscles during processing also has a relationship with the distribution of water [15,16]. Therefore, tempering uniformity, water loss, and moisture migration may also affect the final texture. Bedane et al. [13] reported that increasing the electrode gap can improve the hardness of chicken after RF tempering. Farag et al. [17] found that RF tempering reduced the loss of micronutrients and water in beef during the tempering process compared with conventional methods. Zhang et al. [18] demonstrated that a higher L* value was achieved in tilapia fillets tempered by RF. Although the researchers have confirmed that RF tempering can improve some qualities compared with conventional tempering [13,17–19], there was less research on texture, and the mechanism has also not been studied and remains to be further clarified. Therefore, the purposes of this study were to analyze the texture of frozen tilapia fillets under different RF tempering conditions (power and electrode gaps) and to explore the mechanism of the changes in texture for different RF treatments based on tempering uniformity, water loss, and moisture migration, which will provide a useful reference for exploring quality improvements in RF-tempered products.

2. Materials and Methods

2.1. Sample Preparation, Radio Frequency Tempering, and Temperature Measurement

Vacuum-packed frozen tilapia fillets (the weight of each fillet was 155 ± 10 g and the length was 18.8 ± 1.0 cm) were purchased from TongWei Aquatic Products Company (Hainan, China) and stored at −19 ± 1.0 °C. A piece of fillet was tempered in each experiment, which was repeated in at least 3 batches; at least 200 pieces of fillet in total were used in this study. A 50-ohm RF heater (27.12 MHz) (Labotron12, Sairem, Lyon, France) (Figure 1) was used for the RF treatments. Three electrode gaps of parallel-plate RF equipment were selected (10, 12, and 14 cm), and the three power inputs were 300, 600, and 900 W.

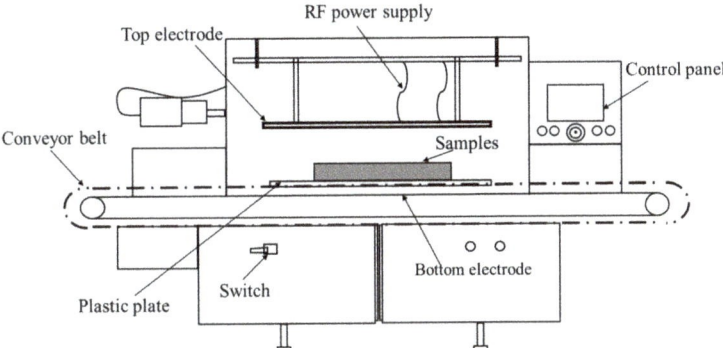

Figure 1. Schematic diagram of radio frequency (RF)-tempering experimental device.

In the stationary experiments, the fish block was placed at the horizontal center of the electrodes. For recording changes in the temperatures of the samples during RF tempering, a 6-channel signal conditioner was used and fiber optic probes (HQ-FTSD120, HeQi Technologies Inc., Xi'an, China) were inserted into the center of its thickest part for each sample. The tempering process was terminated when the temperature reached −4 °C. In the present investigation, the endpoint temperature for tempering was set to approximately −4 °C, because this temperature is typically used in the meat and fish industry for the handling and manipulation of tempered meat blocks [20]. The surface temperature was measured after tempering using an infrared thermal image system (FLIR A655SC, Portland, OR, USA), and surface temperature distribution analysis was carried out using an image analysis system (FLIR Tools). The sample was positioned to ensure that the camera recorded an image of the whole upper surface.

2.2. Texture Determination

2.2.1. Texture Profile Analysis (TPA)

The texture profile analysis test was performed using a model TA-XT2i texture analyzer (Stable Micro Systems, Surrey, UK) with a 5 kg compression load and 20% deformation. The pre-test speed was 1 mm/s, the test speed was 1 mm/s, and the post-test speed was 10 mm/s. The textural parameters of hardness (the hardness generally represents the force exerted by the molars to bite the sample for the first time), springiness (the springiness refers to the ability of the sample to return to its original state after extrusion), chewiness (the chewiness indicates the resistance of the sample to chewing, which is the product of hardness, springiness, and cohesiveness), and cohesiveness (the cohesiveness reflects the internal binding ability of the sample to resist damage and maintain its integrity) were derived from a force–time curve [21].

2.2.2. Warner–Bratzler Shear Test

The fish samples were trimmed into 3 cm × 2 cm × 1.5 cm (length × width × height) pieces. Measurements were taken immediately, and the shear blade was set vertical to the direction of the meat fibers using a Warner–Bratzler shear force blade attached to a model TA-XT2i texture analyzer (Stable Micro Systems, Surrey, UK) [22]. The pre-test speed was 10 mm/s, the test speed was 2 mm/s, and the post-test speed was 10 mm/s. The distance and trigger force were 30 mm and 50 g, respectively.

2.2.3. Measurement of the Myofibril Fragmentation Index (MFI)

The MFI was determined according to the method of Zou et al. [22]: 3.00 g of the minced fish sample was accurately weighed and homogenized for 1 min in 30 mL of phosphate buffer (100 mmol/L KCl, 7 mmol/L KH_2PO_4, 18 mmol/L K_2HPO_4, 1 mmol/L EDTA, and 1 mmol/L $MgCl_2$, at pH 7.0, stored at 4 °C). Then, the sample was centrifuged at 10,614× g for 30 min at 4 °C, and the precipitate was resuspended in 30 mL of phosphate buffer and centrifuged again. The precipitate was dissolved in 15 mL of extraction solution and then filtered with a single layer of gauze to remove connective tissue and cell debris. The filtrate was the myofibril extract. The protein concentration of the myofibril extract was diluted to 0.5 mg/mL and measured spectrophotometrically at 540 nm, and the results were calculated according to Equation (1):

$$MFI = A540 \text{ nm} \times 200 \qquad (1)$$

2.3. Determination of Drip Loss and Cooking Loss

2.3.1. Drip Loss

Tilapia fillets (frozen fillets and fillets that had been tempered and treated to remove surface water) were accurately weighed according to the method of Choi et al. [19], and the drip loss was calculated according to Equation (2):

$$Driploss(\%) = \frac{M_0 - M_t}{M_0} \times 100 \qquad (2)$$

M_0 is the weight of frozen tilapia fillets before tempering (g) and M_t is the weight of fillets after they were tempered and treated to remove the surface water (g).

2.3.2. Cooking Loss

The cooking loss was determined as described by Li et al. [3]. Based on the weight of the meat before and after cooking, the cooking loss of the sample was calculated according to Equation (3):

$$Cookingloss(\%) = \frac{M_e - M_f}{M_e} \times 100 \qquad (3)$$

M_e is the the initial weight of the sample before cooking and M_f is the final weight of the sample after cooking

2.4. Moisture Distribution and Migration

The distribution and migration of moisture in the tilapia fillet samples were analyzed by the relaxation time (T_2) of the GCMPs gel determined using a Niumag Benchtop Pulsed NMR Analyser PQ001 (Niumag Electric Corporation, Shanghai, China). The main parameters were: P_1 = 18 μs, P_2 = 36 μs, NS = 6 and NECH = 3000.

2.5. Statistical Analysis

All the experiments were performed in triplicate. The data were reported as mean ± standard deviation (SD) for each triplicate treatment. Analysis of significant differences ($p < 0.05$) was performed using one-way analysis of variance (ANOVA) calculated by SPSS 20.0 software (SPSS Inc., Chicago, IL, USA). Origin 2018, Origin 2021b and Microsoft PowerPoint-2016 were used to plot and combine figures.

3. Results and Discussion

3.1. Tempering Curve

Figure 2 shows the temperature variation in the tempered tilapia fillets subjected to different methods. The tempering times of the 10 cm and 12 cm samples remained lower than those of the 14 cm group. The tempering times of the 300 W samples significantly exceeded those of the 600 W and 900 W groups. The tempering times of 511 s and 1162 s for 900 W-10 cm and 300 W-14 cm were ensured, and were the shortest and longest, respectively. Increasing the power or reducing the electrode gap increased the tempering rate.

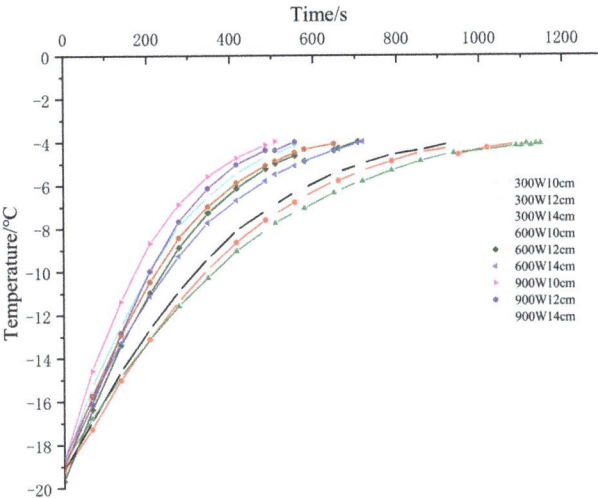

Figure 2. Internal center temperature curve of RF tempering treatment group with different electrode gaps and powers.

3.2. Texture

3.2.1. Texture Profile Analysis (TPA)

The textures of the tilapia fillet samples after RF treatment are presented in Table 1. The springiness, cohesiveness, and resilience decreased as the power increased for the 10 cm electrode gap. There was a decreasing trend in the textural characteristics of springiness, cohesiveness, and resilience when the power was retained at 600 W and the electrode gap was increased. These results agree with those reported by Bedane [13], who found that the springiness and chewiness of melted chicken breasts did not increase significantly with increases in the electrode gap during RF tempering. The hardness values first increased and then decreased after treatment with increasing power when the electrode gap remained unchanged.

Table 1. Effects of power and electrode gap on the texture profile analysis (TPA) results and shear force of frozen tilapia fillets.

Electrode Gap (cm)	Power (W)	Hardness (g)	Springiness (mm)	Cohesiveness (mm)	Chewiness (mJ)	Resilience (mm)	Shear Force (N)
10	300	162.01 ± 8.15 Cb	0.85 ± 0.01 Aa	0.67 ± 0.01 Aa	91.35 ± 7.08 Bb	0.34 ± 0.01 Aa	11.77 ± 0.42 Bb
	600	216.22 ± 16.83 Ab	0.84 ± 0.01 Aa	0.64 ± 0.03 Aa	116.22 ± 11.14 Ab	0.33 ± 0.03 Aa	13.36 ± 0.53 Ab
	900	188.01 ± 6.15 Bc	0.83 ± 0.00 Aa	0.64 ± 0.01 Aa	100.92 ± 1.43 ABb	0.32 ± 0.01 Aa	13.27 ± 0.15 Ab
12	300	137.66 ± 8.14 Cc	0.82 ± 0.06 Ac	0.63 ± 0.00 Ab	71.41 ± 8.83 Cc	0.30 ± 0.01 Aa	15.18 ± 1.30 Aa
	600	293.81 ± 30.72 Aa	0.80 ± 0.00 Ab	0.61 ± 0.01 Aab	142.36 ± 15.85 Aa	0.31 ± 0.03 Aa	16.45 ± 1.68 Aa
	900	221.38 ± 13.45 Bb	0.81 ± 0.03 Aa	0.64 ± 0.03 Aa	114.68 ± 3.26 Ba	0.33 ± 0.04 Aa	16.98 ± 0.60 Aa
14	300	218.08 ± 17.47 Ba	0.81 ± 0.03 Ab	0.64 ± 0.03 Aab	113.20 ± 9.44 Aa	0.33 ± 0.03 Aa	14.06 ± 1.12 Ba
	600	263.95 ± 13.58 Aa	0.79 ± 0.01 Ab	0.57 ± 0.01 Ab	119.35 ± 3.62 Ab	0.25 ± 0.02 Ab	16.51 ± 0.43 Aa
	900	247.02 ± 12.92 ABa	0.81 ± 0.05 Aa	0.61 ± 0.05 Aa	122.06 ± 8.23 Aa	0.30 ± 0.05 Aa	15.76 ± 1.41 ABa

A–C Mean values in the same column with the same electrode gap followed by different letters are significantly different according to Duncan's multiple range test ($p < 0.05$). a–c Mean values in the same column with the same power followed by different letters are significantly different according to Duncan's multiple range test ($p < 0.05$).

This result was related to the shorter exposure time (to the electric field and air) for the 600 W than 300 W samples. Small ice crystals were not completely melted, and there was less cell loss that, together with the relatively intact muscle tissue and slower protein denaturation, made the tissue not loose. However, at 900 W, the energy accumulation increased, and the temperature distribution was uneven due to the strong electric field deflection, which caused the tissue to collapse with increased juice loss from the fillet edge [23]. Therefore, all of these factors led to the aforementioned trends. The hardness gradually increased with the increasing electrode gap at 900 W, possibly because the electric field was gradually deflected with the increasing electrode gap, which reduced the electric field concentration. The temperature increased slowly per unit time, and the ice crystals melted more slowly, so the tissue was less damaged, and the tissue firmness of the fillets was better maintained [13]. For the 10 cm and 12 cm electrode gaps, the chewiness and hardness of the fillets tempered at different powers showed the same trend, and the chewiness of the fillets tempered in the 600 W group was better than that of fillets subjected to other powers.

3.2.2. Warner–Bratzler Shear Test

There is an inverse relationship between the shear force exerted on muscle and tenderness. The shear force upon the tilapia fillets tempered with a 10 cm electrode gap was significantly lower than those of the 12 cm and 14 cm groups at the same power ($p < 0.05$), as shown in Table 1. These results are related to the increases in thermal conductivity and the loss of some soluble substances with the decrease in electrode gap. The findings show that the shear force increased as the power increased under the 12 cm electrode gap. During heating, different muscle proteins denature and cause structural changes, such as the destruction of cell membranes, the transversal and longitudinal shrinkage of meat fibers, aggregation, the gel formation of sarcoplasmic proteins, and the shrinkage and solubilization of connective tissue [16]. However, the tempering time decreases as the power increases, and there is less damage to mitochondria, sarcoplasmic reticulum, and lysosome membrane and less release of salt-soluble proteins and tenderizing enzymes. Therefore, a lower tempering time reduces the enrichment of salt-soluble proteins on the surface of the meat, and plays a role in reducing meat tenderness [22].

3.2.3. Myofibril Fragmentation Index (MFI)

The MFI is a useful indicator of the extent of proteolysis, indicating both the degree of rupturing of the I-bands and the breakage of inter-myofibril linkages [24]. The MFI reflects the integrity of muscle fibers and their skeletal proteins. A higher MFI value corresponds to more severe myofibril breakage and greater meat tenderness. The effects of the electrode gap and power on the degradation of myofibrils are shown in Figure 3. Compared with the 300 W and 900 W samples, the MFI values of the 600 W samples were markedly higher ($p < 0.05$). The MFI in the 10 cm group was significantly higher than that in the 12 cm and 14 cm groups at 300 W ($p < 0.05$), indicating better tenderness at 10 cm, which is consistent with the tenderness reflected by the shear force measurements. In the present study, it is likely that the observed increase or decrease in MFI for the RF-treated samples at different powers for the same electrode gap was related to the chemical and physical changes occurring in the meat due to RF. The MFI in the 300 W-14 cm group was the lowest of all the RF groups, possibly because the slowest electric field deflection and the smallest energy accumulation occurred when the electrode gap was the largest and the power was the highest, resulting in the slow degradation of myofibrils [25].

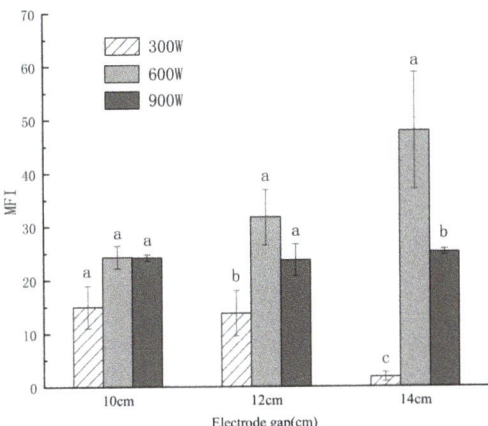

Figure 3. Effects of different power and electrode gaps on the value of myofibril fragmentation index (MFI) of frozen tilapia fillets. "a–c" letters indicate significant differences ($p < 0.05$). Error bars show standard deviation.

3.3. Temperature Distribution

As shown in Figure 4, the lowest temperature of the sample in the nine RF processing groups was frequently between −3 °C and −4 °C. When the tempering process was terminated, all samples showed a gradual increase in temperature from the center to the edge, with the temperature near the edge of the sample being higher. The part with higher temperature gradually increased as the power increased under the 10 cm electrode gap; the distribution was more uneven, and local overheating was more pronounced in the 900 W samples. This finding is contrary to the trends of springiness, cohesiveness, and resilience under the same conditions. As the temperature increased, the frozen fillet absorbed energy and started to exhibit a phase change. The frozen water at the surface of the fillet began to melt, the water molecules underwent a gradual transition to a less structured state, and a certain amount of water loss resulted; at this point, the internal structure was no longer compact [11].

Alfaifi et al. [26] also reported that, with increasing power, the temperature increased more rapidly in the corners and edges than the center. The edge of the fillet melted faster, and the transition speed of the water molecules was faster because the edge was thinner, and a thermal runaway phenomenon occurred due to an increase in the dielectric loss factor as the temperature increased, which caused a large area of melting at the edge [20]. Farag [27] also indicated that an increase in the power level resulted in a higher average temperature in the fillet within a constant tempering terminal range. The changing trend of the average temperature for different electrode gap groups at the same power was contrary to the trend of increasing hardness, which was consistent with the report of Yang [11]. In addition, the tempering uniformity of the 600 W group was relatively better than that of the other two power groups due to the smaller numbers of overheated and excessively cold areas, and the temperature difference was relatively small, which is consistent with the states of hardness and chewiness under the same conditions. Therefore, this study shows that there is a certain relationship between tempering uniformity and the changes in meat texture during tempering.

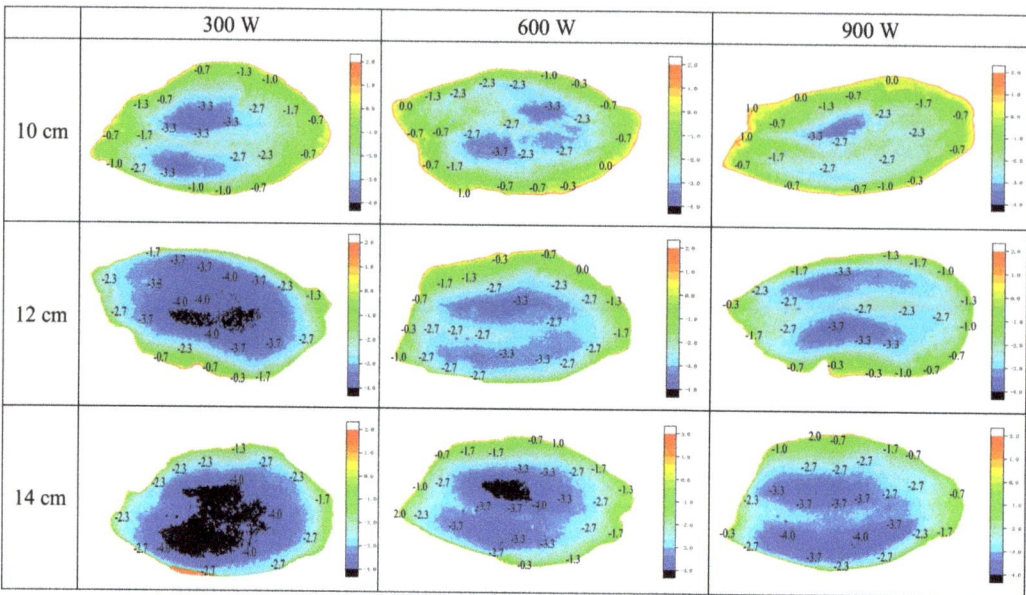

Figure 4. Contour map of the temperature distribution of different RF processing groups.

3.4. Drip Loss and Cooking Loss

The structure of muscle tissue and the degree of shrinkage of myofibrils may be affected by changes in the water loss during tempering, which in turn affects the stability of meat texture [19]. The drip loss increased with increasing electrode gap when the power was 900 W (Table 2). This outcome may be because the tempering time gradually increased as the electrode gap increased, resulting in the release of a large number of oxidases and pro-oxidants, which promoted protein carbonylation reactions and led to the destruction of cell structures, gradually increasing the drip loss. In addition, the drip loss increased with increasing power and a 12 cm electrode gap, which was consistent with the changing trend in shear force under the same conditions. Studies have reported that high power promoted water flow from intra-myofibrillar spaces into the extra-myofibrillar spaces, increasing the water content of extra-myofibrillar spaces, where it was more easily lost and thus increased the drip loss [28]. Therefore, the loss of water from the intra-myofibrillar spaces under this condition destroyed the tissue structure and increased the shear force on the fillets.

Table 2. Effects of power and electrode gap on drip loss and cooking loss of frozen tilapia fillets.

Electrode Gap (cm)	Power (W)	Drip Loss (%)	Cooking Loss (%)
10	300	0.43 ± 0.06 [BY]	15.71 ± 0.31 [AY]
	600	0.84 ± 0.06 [AX]	17.84 ± 0.99 [AX]
	900	0.31 ± 0.09 [BY]	15.90 ± 0.05 [AY]
12	300	0.16 ± 0.01 [BY]	17.34 ± 1.03 [AX]
	600	0.17 ± 0.00 [CY]	15.94 ± 0.37 [AXY]
	900	0.57 ± 0.10 [ABX]	14.20 ± 0.07 [CY]
14	300	0.85 ± 0.19 [AX]	16.55 ± 0.44 [AX]
	600	0.38 ± 0.01 [BY]	11.08 ± 0.49 [BY]
	900	0.76 ± 0.08 [AXY]	15.44 ± 0.03 [BX]

A–C Mean values in the same column with the same power followed by different letters are significantly different according to Duncan's multiple range test ($p < 0.05$). X, Y Mean values in the same column with the same electrode gap followed by different letters are significantly different according to Duncan's multiple range test ($p < 0.05$).

Cooking loss includes the loss of moisture and water-soluble ingredients from the flesh. A gradual compression of the meat structure occurs during cooking, as the melting of fat and the denaturation of proteins lead to the release of chemically bound water, which reflects the specific water-holding capacity of the muscles [29]. The results of this study show that cooking loss decreased with the increasing electrode gap for the 600 W samples. The reason may be that the protein solubility was slowly diminished with the increasing electrode gap at a constant voltage, which led to the denatured proteins slowly losing their ability to retain water, with a minor portion released after tempering and another portion released after cooking [30]. These results are consistent with the changing trends of springiness, cohesiveness, and resilience observed under the same conditions.

3.5. Moisture Distribution and Migration

The flow of water destroys cells, causes tissue degradation, and affects the texture of muscles [31]. The water fluidity and the bonding forces between water and meat tissue can be described by the transverse relaxation time (T_2), as measured by LF-NMR. As shown in Figure 5a,b, the three peaks (T_{21}, T_{22}, T_{23}) were assigned to three key populations of water in muscle: T_{21} (0–10 ms), T_{22} (10–100 ms), and T_{23} (100–1000 ms) represent bound water (bound to macromolecules), immobilized water (in spaces with a high density of myofibrils), and free water (in the myofibril lattice), respectively [30].

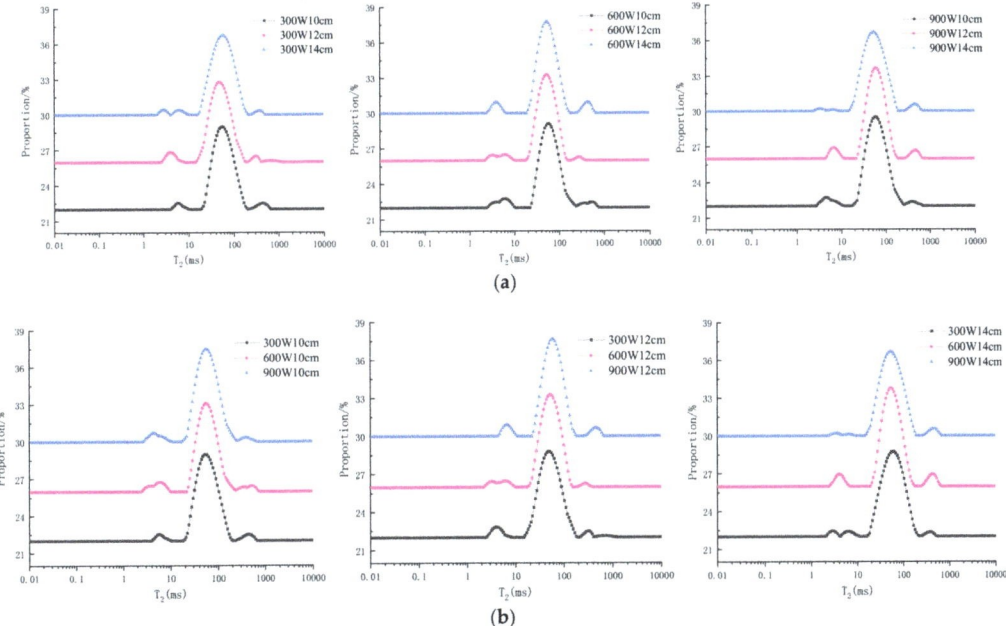

Figure 5. Moisture distribution and migration of frozen tilapia fillets with different RF tempering groups. (**a**) Moisture distribution and migration of frozen tilapia fillets with RF tempering at different electrode gaps (when power is the same). (**b**) Moisture distribution and migration of frozen tilapia fillets with RF tempering at different powers (when electrode gap is the same).

In Figure 5a, the bound water relaxation time under the 12 and 14 cm electrode gaps tends to move more to the right than that under the 10 cm gap when the power is 300 W. This suggests that the fraction of water was weakly bound by RF tempering under the 12 and 14 cm electrode gaps, which is consistent with the changing trend in cohesiveness under the same conditions. This result may be related to the protease and microorganisms being more active when the energy conduction was faster due to higher power, causing the decomposition of proteins, which led to an increase in water fluidity. Shao et al. [32] also pointed out that the T_2 relaxation time of water increased, which indicates that the binding between water and macromolecules was loose and that the water was more fluid.

As shown in Figure 5b, the free water relaxation time at 900 W tended to move more to the right than the 300 W and 600 W groups when the electrode gap was 12 cm. The primary cause for this movement was that the ice crystals gradually melted as the power increased, the free water migrated to the external environment, and the physical adsorption by the meat gradually decreased. The transverse relaxation time (T_2) typically reflects the bonding force between water and meat tissue, and the area of the peak represents the moisture content [33]. As shown in Table 3, A_{21} gradually decreased as the electrode gap increased for the 600 W group, which was consistent with the changes in springiness and cohesiveness. Other studies have reported similar results; when the binding ability of water was lower, the muscles were looser [34]. The peak area of immobilized water for the 12 cm electrode gap group was higher than that of the other electrode gap groups at 600 W, and the peak area of immobilized water for the 600 W group was higher than that of other power groups under the 12 cm electrode gap, suggesting that the free water refluxed and increased the content of immobilized water under these conditions, and that the sample maintained a satisfactory microstructure, stable conformation, and excellent physicochemical properties [31].

Table 3. Peak areas of moisture content of frozen tilapia fillets tempered by different RF groups.

Electrode Gap (cm)	Power (W)	A_{21}	A_{22}	A_{23}
10	300	2.73 ± 0.26 [Bb]	94.05 ± 0.90 [Aa]	3.22 ± 0.65 [Aa]
	600	6.98 ± 1.13 [Aa]	89.17 ± 2.93 [Aa]	3.85 ± 1.80 [Aa]
	900	5.25 ± 1.54 [ABa]	92.63 ± 1.42 [Aa]	2.12 ± 0.12 [Aa]
12	300	5.27 ± 1.03 [Aa]	91.67 ± 0.73 [Ab]	3.06 ± 1.75 [Aa]
	600	5.52 ± 0.70 [Aa]	93.21 ± 0.25 [Aa]	1.27 ± 0.45 [Aa]
	900	5.15 ± 1.04 [Aa]	91.18 ± 1.74 [Aa]	3.67 ± 0.70 [Aa]
14	300	4.11 ± 0.46 [Aab]	94.36 ± 0.36 [Aa]	1.53 ± 0.10 [Aa]
	600	5.13 ± 0.13 [Aa]	89.80 ± 1.02 [Ba]	5.07 ± 1.15 [Aa]
	900	1.67 ± 0.69 [Ba]	95.22 ± 1.41 [Aa]	3.11 ± 2.11 [Aa]

A–C Mean values in the same column with the same electrode gap followed by different letters are significantly different according to Duncan's multiple range test ($p < 0.05$). a, b Mean values in the same column with the same power followed by different letters are significantly different according to Duncan's multiple range test ($p < 0.05$).

3.6. Relationship between Texture and Moisture Index and the Principal Component Analysis (PCA)

The correlations between moisture characteristics and TPA characteristics, shear force, and MFI for different RF treatments were determined (Figure 6). The shear force and MFI showed a significant positive correlation with A_{23}. The springiness, cohesiveness, and resilience showed a negative correlation with A_{23} and a positive correlation with T_{21}. The hardness, shear force, and MFI showed a negative correlations with cooking loss. These results indicate that the contents of immobilized water and free water and the cooking loss may have a certain effect on the texture, and that a close correlation exists between textural properties, water migration and water loss during RF tempering, which requires further analysis and confirmation. Some people have reported that the lower hardness values after tempering could be an indicator of the juiciness of the meat and could have an effect on overall consumer acceptability of the meat [13]. However, in this study, the drip loss was weakly correlated with the texture index, and this result may be related to the short tempering time.

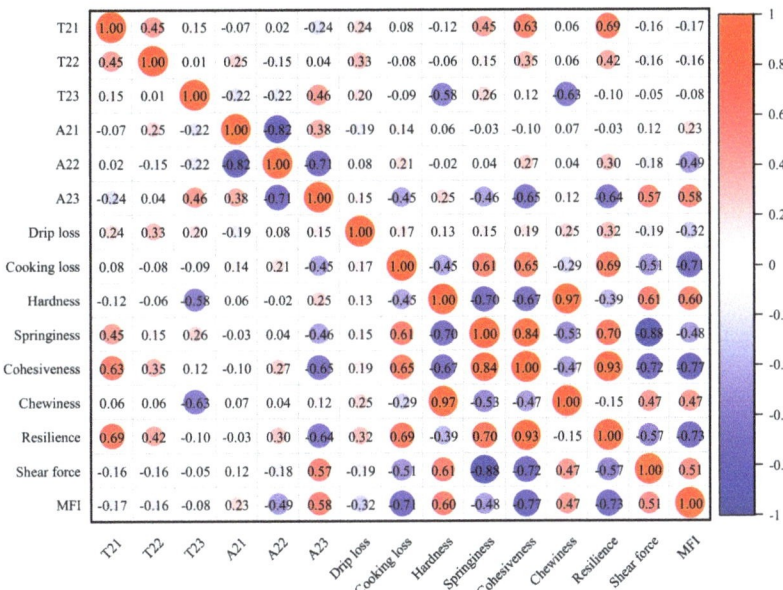

Figure 6. Pearson plot between TPA index and moisture index, shear force, and MFI.

The PCA biplot shown in Figure 7 simultaneously represents the observations and variables in the new space, where the first two principal components (PCs) explain 57.60% of the total variation (PC1 and PC2 were 40.70 and 16.90%, respectively). The first major component (PC1), which represents 40.70% of the variability, most often describes resilience, springiness, cooking loss, and cohesiveness variables, with a negative association with chewiness, hardness, shear force, MFI, and A_{23} variables, as shown in Figure 6. In the first component, only samples in the 300 W-10 cm RF group were associated with higher resilience, springiness and cohesiveness, indicating that the better springiness and cohesiveness gain associated with the 300 W-10 cm RF treatment resulted in a compact myofibrillar structure, which is consistent with the results shown in Table 1. PC2 was defined by chewiness and hardness, which were negatively correlated with A_{23} and T_{23}, indicating that the content of free water caused a change in the texture of the tempered fillets. For component 2 (positive side of graphic), only the 14 cm group with 900 W was associated with greater hardness, chewiness, A_{22} and cohesiveness, which indicates that the hardness and chewiness increased as the electrode gap increased to 14 cm for the 900 W group, and the immobilized water can be prevented from becoming free water, as shown in Figure 6.

This study investigated the texture of fillets after RF tempering and the associated mechanisms that caused their differences. Previous studies had observed that RF tempering can affect the texture and other qualities of products; however, most of these were focused on exploring the effects on color, water loss, and fat oxidation, and less attention was paid to the texture [11,13,17–19]. Bedane et al. [13] reported that a higher hardness of chicken after RF tempering was achieved by increasing the electrode gap. In this study, we also proved that the hardness of fillets gradually increased with the increasing electrode gap at 900 W. Zhang et al. [18] found that there was no significant difference ($p > 0.05$) in hardness, chewiness, springiness, or resilience in RF-tempered fillets with alterations in the power and electrode gap. However, our results indicate a decreasing trend in springiness, cohesiveness, and resilience as the electrode gap increased, which is consistent with Bedane et al.'s result [13]. Compared to previous research, the variations in temperature distribution, water holding capacity, and moisture distribution after tempering were not only observed in our study, and we also assessed the mechanisms for difference in texture after tempering. Finally, the pivotal factors that led to the difference in texture after tempering were confirmed by the correlation analysis and principal component analysis.

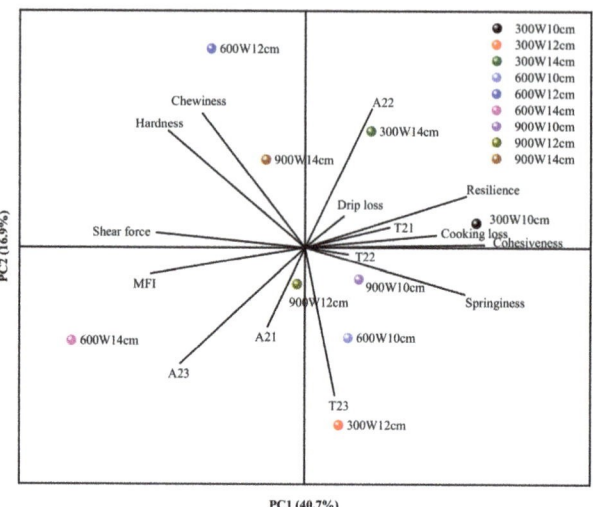

Figure 7. Analysis of principal components (PCA) of the texture index (hardness, cohesiveness, chewiness, resilience, springiness), moisture index (NMR index (T_{21}, T_{22}, T_{23}, A_{21}, A_{22}, A_{23}), drip loss and cooking loss), shear force and MFI (myofibril fragmentation index) of all samples.

4. Conclusions

This study focused on exploring the texture change mechanism of RF-tempered frozen food products under different RF conditions by analyzing the factors that affect the texture. The results show that the power or the electrode gap can affect the texture of tempered fillets. At 600 W, the springiness decreased from 0.84 mm to 0.79 mm, the cohesiveness decreased from 0.64 mm to 0.57 mm, and the resilience decreased from 0.33 mm to 0.25 mm as the electrode gap was increased. The shear force was increased from 15.18 N to 16.98 N by increasing the power, while the electrode gap remained at 12 cm. Besides this, good tempering uniformity can improve the hardness and chewiness of tempered fillets. The free water content, cooking loss and migration of bound water to immobilized water may be the main factors in the texture change observed after RF tempering. However, deeper and more exact mechanisms of texture change during RF tempering still need to be determined.

Author Contributions: J.J.: conceptualization, methodology, investigation, data curation, writing—original draft, writing—review and editing. F.Z.: investigation. C.X.: investigation. Y.S.: investigation. X.W.: conceptualization, methodology, resources, supervision. All authors have read and agreed to the published version of the manuscript.

Funding: This research was funded by National Key R&D Program of China, grant number (2018YFD0901006).

Institutional Review Board Statement: Not applicable.

Informed Consent Statement: Not applicable.

Data Availability Statement: The datasets generated for this study are available on request to the corresponding author.

Acknowledgments: This study was supported by the National Key R&D Program of China (2018YFD0901006).

Conflicts of Interest: The authors declare no conflict of interest.

References

1. Britz, P. Tilapia: Biology, culture and nutrition. *Afr. J. Aquat. Sci.* **2010**, *33*, 103. [CrossRef]
2. Roiha, I.S.; Tveit, G.M.; Backi, C.J.; Jónsson, Á.; Karlsdóttir, M.; Lunestad, B.T. Effects of controlled thawing media temperatures on quality and safety of pre-rigor frozen Atlantic cod (*Gadus morhua*). *LWT* **2018**, *90*, 138–144. [CrossRef]
3. Li, D.; Jia, S.; Zhang, L.; Li, Q.; Pan, J.; Zhu, B.; Prinyawiwatkul, W.; Luo, Y. Post-thawing quality changes of common carp (*Cyprinus carpio*) cubes treated by high voltage electrostatic field (HVEF) during chilled storage. *Innov. Food Sci. Emerg. Technol.* **2017**, *42*, 25–32. [CrossRef]
4. Ramaswamy, H.S.; Marcotte, M.; Sastry, S.; Abdelrahim, K.J.C.P. *Ohmic Heating in Food Processing*; CRC Press: Boca Raton, FL, USA, 2014.
5. Icier, F.; Izzetoglu, G.T.; Bozkurt, H.; Ober, A. Effects of ohmic thawing on histological and textural properties of beef cuts. *J. Food Eng.* **2010**, *99*, 360–365. [CrossRef]
6. Mousakhani-Ganjeh, A.; Hamdami, N.; Soltanizadeh, N. Thawing of frozen tuna fish (*Thunnus albacares*) using still air method combined with a high voltage electrostatic field. *J. Food Eng.* **2016**, *169*, 149–154. [CrossRef]
7. Hsieh, C.W.; Lai, C.H.; Ho, W.J.; Huang, S.C.; Ko, W.C. Effect of thawing and cold storage on frozen chicken thigh meat quality by high-voltage electrostatic field. *J. Food Sci.* **2010**, *75*, M193–M197. [CrossRef]
8. Li, B.; Sun, D.W. Novel methods for rapid freezing and thawing of foods—A review. *J. Food Eng.* **2002**, *54*, 175–182. [CrossRef]
9. Zhu, S.M.; Su, G.M.; He, J.S.; Ramaswamy, H.S.; Le Bail, A.; Yu, Y. Water phase transition under pressure and its application in high pressure thawing of agar gel and fish. *J. Food Eng.* **2014**, *125*, 1–6. [CrossRef]
10. Shazman, A.; Mizrahi, S.; Cogan, U.; Shimoni, E. Examining for possible non-thermal effects during heating in a microwave oven. *Food Chem.* **2006**, *103*, 444–453. [CrossRef]
11. Yang, H.; Chen, Q.; Cao, H.; Fan, D.; Huang, J.; Zhao, J.; Yan, B.; Zhou, W.; Zhang, W.; Zhang, H. Radio frequency thawing of frozen minced fish based on the dielectric response mechanism. *Innov. Food Sci. Emerg. Technol.* **2018**, *52*, 80–88. [CrossRef]
12. Nagaraj, G.; Singh, R.; Hung, Y.C.; Mohan, A. Effect of radio-frequency on heating characteristics of beef homogenate blends. *LWT-Food Sci. Technol.* **2015**, *60*, 372–376. [CrossRef]
13. Bedane, T.F.; Altin, O.; Erol, B.; Marra, F.; Erdogdu, F. Thawing of frozen food products in a staggered through-field electrode radio frequency system: A case study for frozen chicken breast meat with effects on drip loss and texture. *Innov. Food Sci. Emerg. Technol.* **2018**, *50*, 139–147. [CrossRef]
14. Leygonie, C.; Britz, T.J.; Hoffman, L.C. Impact of freezing and thawing on the quality of meat: Review. *Meat Sci.* **2012**, *91*, 93–98. [CrossRef] [PubMed]
15. Tornberg, E. Engineering processes in meat products and how they influence their biophysical properties. *Meat Sci.* **2013**, *95*, 871–878. [CrossRef] [PubMed]
16. Rincon, A.M.; Singh, R.K.; Stelzleni, A.M. Effects of endpoint temperature and thickness on quality of whole muscle non-intact steaks cooked in a Radio Frequency oven. *LWT-Food Sci. Technol.* **2015**, *64*, 1323–1328. [CrossRef]
17. Farag, K.W.; Duggan, E.; Morgan, D.J.; Cronin, D.A.; Lyng, J.G. A comparison of conventional and radio frequency defrosting of lean beef meats: Effects on water binding characteristics. *Meat Sci.* **2009**, *83*, 278–284. [CrossRef] [PubMed]
18. Zhang, Y.; Li, S.; Jin, S.; Li, F.; Tang, J.; Jiao, Y. Radio frequency tempering multiple layers of frozen tilapia fillets: The temperature distribution, energy consumption, and quality. *Innov. Food Sci. Emerg. Technol.* **2021**, *68*, 102603. [CrossRef]
19. Choi, E.J.; Park, H.W.; Chung, Y.B.; Park, S.H.; Kim, J.S.; Chun, H.H. Effect of tempering methods on quality changes of pork loin frozen by cryogenic immersion. *Meat Sci.* **2017**, *124*, 69–76. [CrossRef] [PubMed]
20. Li, Y.; Li, F.; Tang, J.; Zhang, R.; Wang, Y.; Koral, T.; Jiao, Y. Radio frequency tempering uniformity investigation of frozen beef with various shapes and sizes. *Innov. Food Sci. Emerg. Technol.* **2018**, *48*, 42–55. [CrossRef]
21. Barekat, S.; Soltanizadeh, N. Improvement of meat tenderness by simultaneous application of high-intensity ultrasonic radiation and papain treatment. *Innov. Food Sci. Emerg. Technol.* **2017**, *39*, 223–229. [CrossRef]
22. Zou, Y.; Shi, H.; Xu, P.; Jiang, D.; Zhang, X.; Xu, W.; Wang, D. Combined effect of ultrasound and sodium bicarbonate marination on chicken breast tenderness and its molecular mechanism. *Ultrason. Sonochemistry* **2019**, *59*, 104735. [CrossRef]
23. Romano, V.; Marra, F. A numerical analysis of radio frequency heating of regular shaped foodstuff. *J. Food Eng.* **2007**, *84*, 449–457. [CrossRef]
24. Volpelli, L.A.; Failla, S.; Sepulcri, A.; Piasentier, E. Calpain system in vitro activity and myofibril fragmentation index in fallow deer (Dama dama): Effects of age and supplementary feeding. *Meat Sci.* **2005**, *69*, 579–582. [CrossRef] [PubMed]
25. Wang, A.; Kang, D.; Zhang, W.; Zhang, C.; Zou, Y.; Zhou, G. Changes in calpain activity, protein degradation and microstructure of beef m. semitendinosus by the application of ultrasound. *Food Chem.* **2018**, *245*, 724–730. [CrossRef]
26. Alfaifi, B.; Tang, J.; Jiao, Y.; Wang, S.; Rasco, B.; Jiao, S.; Sablani, S. Radio frequency disinfestation treatments for dried fruit: Model development and validation. *J. Food Eng.* **2014**, *120*, 268–276. [CrossRef]
27. Farag, K.W.; Marra, F.; Lyng, J.G.; Morgan, D.J.; Cronin, D.A. Temperature changes and power consumption during radio frequency tempering of beef lean/fat formulations. *Food Bioprocess Technol.* **2010**, *3*, 732–740. [CrossRef]
28. Jia, G.; Liu, H.; Nirasawa, S.; Liu, H. Effects of high-voltage electrostatic field treatment on the thawing rate and post-thawing quality of frozen rabbit meat. *Innov. Food Sci. Emerg. Technol.* **2017**, *41*, 348–356. [CrossRef]
29. Vieira, C.; Diaz, M.T.; Martínez, B.; García-Cachán, M.D. Effect of frozen storage conditions (temperature and length of storage) on microbiological and sensory quality of rustic crossbred beef at different states of ageing. *Meat Sci.* **2009**, *83*, 398–404. [CrossRef]

30. Li, D.; Jia, S.; Zhang, L.; Wang, Z.; Pan, J.; Zhu, B.; Luo, Y. Effect of using a high voltage electrostatic field on microbial communities, degradation of adenosine triphosphate, and water loss when thawing lightly-salted, frozen common carp (*Cyprinus carpio*). *J. Food Eng.* **2017**, *212*, 226–233. [CrossRef]
31. Cao, M.; Cao, A.; Wang, J.; Cai, L.; Regenstein, J.; Ruan, Y.; Li, X. Effect of magnetic nanoparticles plus microwave or far-infrared thawing on protein conformation changes and moisture migration of red seabream (Pagrus Major) fillets. *Food Chem.* **2018**, *266*, 498–507. [CrossRef] [PubMed]
32. Shao, J.-H.; Deng, Y.-M.; Song, L.; Batur, A.; Jia, N.; Liu, D.-Y. Investigation the effects of protein hydration states on the mobility water and fat in meat batters by LF-NMR technique. *LWT-Food Sci. Technol.* **2016**, *66*, 1–6. [CrossRef]
33. Guo, L.Y.; Shao, J.H.; Liu, D.Y.; Xu, X.L.; Zhou, G.H. The distribution of water in pork meat during wet-curing as studied by low-field NMR. *Food Sci. Technol. Res.* **2014**, *20*, 393–399. [CrossRef]
34. Bertram, H.C.; Donstrup, S.; Karlsson, A.H.J.M.S. Continuous distribution analysis of T(2) relaxation in meat-an approach in the determination of water-holding capacity. *Meat Sci.* **2002**, *60*, 279–285. [CrossRef]

Article

Effects of Radio Frequency Tempering on the Temperature Distribution and Physiochemical Properties of Salmon (*Salmo salar*)

Rong Han [1,2], Jialing He [1,2], Yixuan Chen [1,2], Feng Li [1,2], Hu Shi [1,2] and Yang Jiao [1,2,*]

[1] College of Food Science and Technology, Shanghai Ocean University, Shanghai 201306, China; sxczhr@163.com (R.H.); m13974501128_1@163.com (J.H.); yxchen@shou.edu.cn (Y.C.); fli@shou.edu.cn (F.L.); hshi@shou.edu.cn (H.S.)
[2] Engineering Research Center of Food Thermal-Processing Technology, Shanghai 201306, China
* Correspondence: yjiao@shou.edu.cn; Tel.: +86-21-6190-8758

Citation: Han, R.; He, J.; Chen, Y.; Li, F.; Shi, H.; Jiao, Y. Effects of Radio Frequency Tempering on the Temperature Distribution and Physiochemical Properties of Salmon (*Salmo salar*). *Foods* **2022**, *11*, 893. https://doi.org/10.3390/foods11060893

Academic Editor: Oscar Martinez-Alvarez

Received: 25 January 2022
Accepted: 1 March 2022
Published: 21 March 2022

Publisher's Note: MDPI stays neutral with regard to jurisdictional claims in published maps and institutional affiliations.

Copyright: © 2022 by the authors. Licensee MDPI, Basel, Switzerland. This article is an open access article distributed under the terms and conditions of the Creative Commons Attribution (CC BY) license (https://creativecommons.org/licenses/by/4.0/).

Abstract: Salmon (*Salmo salar*) is a precious fish with high nutritional value, which is perishable when subjected to improper tempering processes before consumption. In traditional air and water tempering, the medium temperature of 10 °C is commonly used to guarantee a reasonable tempering time and product quality. Radio frequency tempering (RT) is a dielectric heating method, which has the advantage of uniform heating to ensure meat quality. The effects of radio frequency tempering (RT, 40.68 MHz, 400 W), water tempering (WT + 10 °C, 10 ± 0.5 °C), and air tempering (AT + 10 °C, 10 ± 1 °C) on the physiochemical properties of salmon fillets were investigated in this study. The quality of salmon fillets was evaluated in terms of drip loss, cooking loss, color, water migration and texture properties. Results showed that all tempering methods affected salmon fillet quality. The tempering times of WT + 10 °C and AT + 10 °C were 3.0 and 12.8 times longer than that of RT, respectively. AT + 10 °C produced the most uniform temperature distribution, followed by WT + 10 °C and RT. The amount of immobile water shifting to free water after WT + 10 °C was higher than that of RT and AT + 10 °C, which was in consistent with the drip and cooking loss. The spaces between the intercellular fibers increased significantly after WT + 10 °C compared to those of RT and AT + 10 °C. The results demonstrated that RT was an alternative novel salmon tempering method, which was fast and relatively uniform with a high quality retention rate. It could be applied to frozen salmon fillets after receiving from overseas catches, which need temperature elevation for further cutting or consumption.

Keywords: tempering; radio frequency; quality; salmon (*Salmo salar*); fish

1. Introduction

Salmon (*Salmo salar*) is a good source of n-3 polyunsaturated fatty acids, essential amino acids and bioactive peptides [1,2], which makes it a great supplement to people's daily diets. The consumption of salmon has increased gradually thanks to its characteristics of high nutrition and delicious taste. However, fish and fish products are susceptible to spoilage and deterioration; thus, their shelf-lives are usually short. Freezing is a convenient and effective method for fish preservation, while tempering is usually a necessary process for the further processing, marketing, and cooking of frozen fish while restoring its quality. Improper tempering methods, which usually result in a longer tempering time and uncontrollable temperature, may cause physical and chemical damages and changes in the fish, such as higher drip loss [3,4], color change [5,6], protein denaturation and lipid oxidation [6–10] and texture change [11]. Conventional tempering processes currently applied in the food industry, including air and water tempering, were characterized as time-consuming and causes of food quality losses, while new tempering technologies may present the advantages of a fast tempering rate, energy saving and good quality [12].

Thus, more novel tempering technologies have been emerging and studied to replace the traditional technologies.

Radio frequencies (RF) are electromagnetic waves with a frequency range of 3 kHz to 300 MHz, and the commonly used frequencies in the food industry are 13.56, 27.12, and 40.68 MHz [13–16]. RF heating is a dielectric heating method with the characteristics of a high heating rate, volumetric heating, uniform heating, and a controllable process to ensure food quality and hygiene [17,18]. The RF generator produces an alternating electromagnetic field, and when the food is exposed to this alternating electromagnetic field, the dipole molecules, atoms, and ions in the food material rotate and rearrange, causing friction that generates heat inside the food [19]. Due to the longer wavelength compared to microwave, RF technology is more suitable for heating large volume samples and minimizing non-uniform heating because of its relatively higher penetration depth [20].

RF tempering of food products has been researched for its application to various products in recent years. Farag et al. [21] thawed lean beef using RF and discovered that RF could reduce the drip and micronutrient losses during the tempering process compared with conventional tempering methods. It was also found that the tempering rate of RF was 85 folds reduced compared to air tempering. Llave et al. [22] discovered the RF tempering rate of tuna fish was three times higher than that with air tempering and that the best uniformity occurred with a top electrode projection similar in size to the sample. Kim et al. [23] investigated RF tempering of pork with a curved-shape top electrode and found that temperature uniformity was improved by using a top electrode with a projection area smaller than that of the sample. Koray Palazoğlu et al. [24] compared RF (27.12 MHz) and microwave (915 MHz) tempering of frozen blocks of shrimp (1.75 kg) and observed that localized overheating occurred in samples subjected to microwave tempering. In contrast, RF tempering resulted in a relatively uniform overall temperature distribution. Palazoğlu and Miran [25] studied the effect of conveyor movement and sample vertical position on RF tempering of frozen lean beef (2.0 kg). It was demonstrated that tempering beef at the mid position in stationary mode and at high position on an upwardly inclined conveyor in continuous mode could achieve the best temperature uniformity. Li et al. [26] investigated the effects of shapes (cuboid, trapezoidal prism, step) and sizes (small: $160 \times 102 \times 60$ mm^3, medium: $220 \times 140 \times 60$ mm^3, large: $285 \times 190 \times 60$ mm^3) of frozen beef on RF tempering uniformity. It was reported that the worst heating uniformity occurred in the samples with sharp edges and steps ($STUI$ 0.282) ($STUI$: simulated temperature uniformity index), followed by the trapezoidal prism ($STUI$ 0.209) and cuboid shape samples ($STUI$ 0.194). Additionally, heating uniformity increased as base area of the sample increased. Furthermore, Zhang et al. [27] studied the effects of different tempering methods and freeze-thaw cycles on melanosis and quality parameters of Pacific white shrimp. It was found that RF tempering effectively inhibited melanosis and reduced protein oxidation in Pacific white shrimp with its fast and uniform heating characteristics. Previous research explored not only the attributes of RF tempering but also methods to improve tempering uniformity on selected meat and fish products. However, not much research has been conducted on exploring the effectiveness of RF tempering of fish fillets and analyzing the fish quality. The lack of information hinders the application of RF tempering in the fish industry. A comprehensive comparison of the tempering processes among radio frequency tempering, water tempering and air tempering with an analysis of their tempering times, temperature distributions, and resulting fish quality attributes would provide new information to both researchers and industry people for accelerating the commercialization of RF tempering technology.

Dielectric properties are the essential factors affecting the electromagnetic wave absorption and energy conversion of materials [11]. Dielectric constant (ε') and dielectric loss factor (ε'') are often used to characterize material dielectric properties [28]. The former represents the storage capacity of electromagnetic energy, while the latter reflects the ability of materials to absorb electromagnetic energy or convert electromagnetic energy into heat [29]. Furthermore, the dielectric properties of food are affected by many factors such as temperature and frequency, and also vary with food composition and quality. Many studies have been conducted using dielectric properties as a non-destructive method for predicting the quality and maturity of fruits and [30–32], eggs [33], milk [34] and meat maturity [35] and quality [36].

In this study, we selected salmon (*Salmo salar*) as a target fish for exploring the effectiveness of radio frequency tempering. The objective of this study was to compare RF tempering with water tempering and air tempering on the tempering rate, temperature distribution and physiochemical properties of salmon fillets. The investigated quality parameters included drip loss, cooking loss, color, water migration and texture properties.

2. Materials and Methods

2.1. Sample Preparation

Salmon fillets (*Salmo salar*), which were deep frozen in Norway, were purchased from a local supermarket in China (Lingang, Shanghai, China). The fillets were kept in a cooler filled with ice and transported to the laboratory within 0.5 h. The initial transport temperature of the salmon was $-18\ °C$ and there was no significant temperature fluctuation within the process. Salmon fillets were firstly cut and trimmed into cuboid shape (10 × 8 × 2.5 cm³), weighed to 240 ± 5 g, and then kept frozen at $-20\ °C$ for 24 h. After freezing, a Φ2 mm hole was drilled at the geometrical center of each fillet for temperature sensor insertion. The prepared samples were frozen at $-20\ °C$ for 24 h again before tempering experiments. Salmon fillets without freezing-thawing treatments were used as control samples for comparison. The initial moisture content of the sample was measured as 61.3% (AOAC 950.46).

2.2. Dielectric Properties

The dielectric properties of salmon were measured with an open-ended coaxial probe (Agilent N1501A, Agilent Technologies Inc., San Jose, CA, USA) connected with a network analyzer (Agilent E5071C, Agilent Technologies Inc., San Jose, CA, USA). The system was equipped with a custom-made cylindrical sample container (d = 2.5 cm, h = 10.0 cm) and an oil bath with temperature control device [37]. Before starting, the vector network analyzer was turned on firstly to preheat for 1 h, and the scanning type of the software was set to linear scanning with a scanning frequency range of 1 and 300 MHz. Probe calibration was conducted with air, short circuit and deionized water at 25.0 ± 0.5 °C. After that, the thawed salmon samples were minced by hand and stuffed into the sample container for measurement. Subsequent measurements were taken with a temperature range from -20 to $20\ °C$ with a temperature interval of $5\ °C$ by controlling the oil bath temperature. Each group of experiments was repeated three times. Detailed measuring procedures can be found in Chen et al. [38].

2.3. Penetration Depth

The penetration depth of an electromagnetic wave into a material refers to the vertical distance of the electromagnetic wave passing through the food when the intensity of the electromagnetic wave is reduced to $1/e$ (e = 2.71828) of the strength on the food surface. The equation used to determine penetration depth is defined as:

$$d_p = \frac{c}{2\sqrt{2}\pi f \left[\varepsilon' \left(\sqrt{1 + \left(\frac{\varepsilon''}{\varepsilon'}\right)^2} - 1\right)\right]^{\frac{1}{2}}} \tag{1}$$

where d_p is the penetration depth (m), c is the speed of light in free space (3×10^8 m/s), f stands for the working frequency (MHz), ε' represents the dielectric constant (F/m) and ε'' represents the dielectric loss factor (F/m).

2.4. Tempering Experiments

The prepared frozen salmon fillets were tempered from −20 to −4 °C with three methods: radio frequency tempering (RT), water immersion tempering (WT + 10 °C), and air tempering (AT + 10 °C); all tempering experiments were replicated three times in parallel.

2.4.1. Radio Frequency Tempering (RT)

A 40.68 MHz, 400 W radio frequency oven (D20 Plus, Dotwil Intelligent Technology Co., Ltd., Shanghai, China) (Figure 1) with a fixed electrode gap of 122 mm was used for tempering experiment. A pre-calibrated fiber optic sensor (HQ-FTS-D1F00, Heqi guangdian Technology Co., Ltd., Xi'an, China) was inserted into the pre-drilled hole of one piece of salmon fillet for automatically monitoring and recording temperature histories during the tempering process. The sample was placed in the oven cavity for the tempering experiment. The experiment was stopped right away once the sample's center temperature reached −4 °C. The treated sample was taken out of the oven immediately to obtain the top and bottom surface temperature distribution under a thermal infrared imager (FLIR A655sc, Wilsonville, OR, USA). The experiment was performed in triplicate with three different pieces of salmon fillets from different batches.

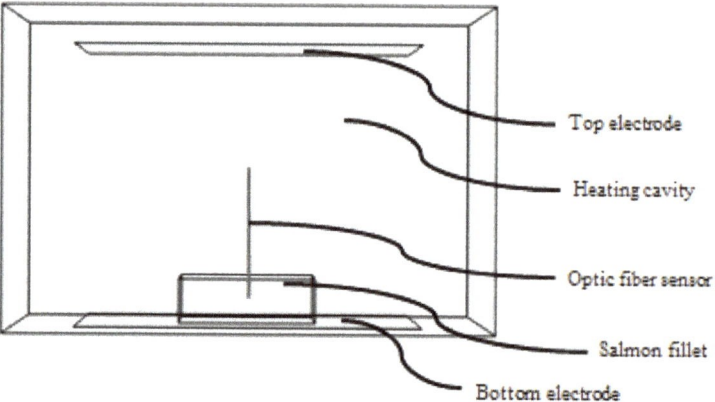

Figure 1. Layout of RF system and placement of salmon fillet sample and fiber optic sensor.

2.4.2. Water Tempering (WT) and Air Tempering (AT)

Prepared frozen salmon fillets were placed in a thermostatic water bath (HH-S, Titan Technology Co., Ltd., Shanghai, China), and the water temperature was set to 10 ± 0.5 °C for the WT + 10 °C experiment. For the AT + 10 °C experiment, one piece of frozen salmon fillet was placed in atmospheric air at 10 ± 1 °C. Before the experiments started, thermocouple temperature sensors (OMEGA Engineering, Norwalk, CT, USA) were inserted into the centers of samples to monitor and record the temperature history until reaching −4 °C. After reaching the target temperature, the samples were taken out immediately to obtain the surface temperature picture using a thermal infrared imager (FLIR A655sc, Wilsonville, OR, USA). Each experiment with WT and AT was performed in triplicate with three different pieces of salmon fillets from different batches.

2.5. Drip Loss and Cooking Loss

The frozen samples were weighed before tempering experiments. After tempering, the salmon fillet samples were transferred to a 4 °C refrigerator for temperature equilibration for 4 h until fully thawed. After equilibration, samples were wiped dry with absorbent paper and weighed again. The drip loss was expressed as a percentage of water loss in the initial sample weight [39].

$$\text{Drip loss }(\%) = \frac{weight\ of\ frozen\ sample\ (g) - weight\ of\ thawed\ sample\ (g)}{weight\ of\ frozen\ sample\ (g)} \times 100 \qquad (2)$$

Thawed salmon samples were cut into six cuboid pieces ($3 \times 2 \times 1$ cm^3), weighed, and placed into polyethylene bags individually. The samples with bags were labeled and then immersed in a thermostatic water bath (HH-S, Titan Technology Co., Ltd., Shanghai, China) at 85 °C for 30 min. The salmon samples were weighed again after cooking, and the cooking loss was expressed as follow:

$$\text{Cooking loss }(\%) = \frac{weight\ of\ thawed\ sample\ (g) - weight\ of\ cooked\ sample\ (g)}{weight\ of\ uncooked\ sample\ (g)} \times 100 \qquad (3)$$

2.6. Color

The surface color of control and thawed samples were both measured using a handheld colorimeter (Minolta CR-400, Tokyo, Japan). Before measurements, the colorimeter was calibrated with a white standard calibration plate. The L^* (lightness), a^* (redness-greenness) and b^* (yellowness-blueness) values were obtained by contacting the probe to the surfaces of samples vertically. Each experiment was performed in triplicate at randomly selected locations on the sample surfaces.

2.7. Lipid Oxidation

The lipid oxidation degree of salmon fillets was determined by measuring the thiobarbituric acid reactive substances (TBARS) based on the methods of Zhu et al. [40] and Xuan et al. [41] with slight modifications. Around 10.0 g minced sample was placed in a conical flask and mixed with 50 mL 7.5% trichloroacetic acid with 0.1% EDTA. Then the mixture was shocked in a thermostatic water bath oscillator (SHZ-B, Titan Technology Co., Ltd., Shanghai, China) at a speed of 130 r/min for 30 min and then filtered. Five milliliter (5 mL) filtrate was transported into a glass test tube and mixed thoroughly with 5 mL 0.02 mol/L 2-thiobarbituric acid. The above tubes were placed in a water bath at 90 °C for 40 min, and then cooled down in 25 °C water for 30 min. After cooling down, 5 mL chloroform was added into the mixture solution, which was then shaken thoroughly by hand. Then, the supernatant was used to determine the absorbance value at 532 and 600 nm using a UV-Visible Spectrophotometer (Evolution 220, Thermo Fisher Scientific Inc., Waltham, MA, USA). Analyses were performed in $n = 6$ for each fish sample. The TBARS values were expressed in mg of malonaldehyde/kg of sample and calculated using the following equation:

$$\text{TBARS (mg MDA/kg)} = \frac{A_{532\ nm} - A_{600\ nm}}{155} \times \frac{1}{10} \times 72.6 \times 1000 \qquad (4)$$

2.8. Protein Denaturation

Differential scanning calorimetry (DSC) is a common method for measuring the protein denaturation temperature and enthalpy to evaluate the protein denaturation degree of meat and fish products [42]. The experiment was conducted according to Rahbari et al. [43] with slight modifications. Around 10.0~15.0 mg samples were placed into a hermetically sealed aluminum pan, and thermal transition was assessed with a differential scanning calorimeter (Q2000, TA Instrument, New Castle, DE, USA). An empty pan was used as a reference. Samples were equilibrated at 20 °C for 2 min before heating at a rate of 5 °C/min till 100 °C. After each run, the maximum denaturation temperature (T_{max}) and denaturation enthalpy (ΔH) were determined by analyzing the heat flow curves [44]. Experiments were performed for each treatment replicate.

2.9. Water Migration

Low-field nuclear magnetic resonance (LF-NMR) was used to analyze the water states and migration in salmon fillet samples. Each control and thawed sample was cut into 3 cubes (2 × 2 × 2 cm^3) and wiped gently with absorbent paper. Each individual sample was placed in an NMR tube (Φ = 70 mm) at room temperature. T_2 transverse relaxation measurements were performed using an LF-NMR analyzer (MesoMR23-060H.I, Niumag Electronic Technology Co., Ltd., Shanghai, China) with a magnetic field strength of 0.4 T corresponding to a proton resonance frequency of 21 MHz. T_2 was measured at 32 °C using the Carr-Purcell-Meiboom-Gill (CPMG) pulse sequence [45,46]. The T_2 measurements were conducted with a time delay between the 90° and 180° pulse (τ) of 150 µs. For each sample, 16 scans were acquired at 2 s intervals with 8000 echoes. Detailed experimental procedure can be found in Aursand et al. [47]. Each measurement was performed in triplicate.

2.10. Microstructure

Tissues of salmon fillet sample were cut into 3 cubes (5 × 5 × 5 mm^3) and fixed at 4 °C for 24 h in 10% phosphate-buffered (pH 6.9~7.1) formaldehyde and dehydrated with a gradient series of ethanol solutions, and then embedded in paraffin wax. Samples were then cut into 5 µm thick slices and dried for 24 h, then dewaxed and stained with hematoxylin and eosin [48]. The specimens were observed and photographed by a light microscope (BZ-9000, Keyence, Osaka, Japan). Three measurements were carried out for each treatment.

2.11. Statistical Analysis

Three independent trials were performed to test the effects of tempering methods on physicochemical properties of salmon, including drip loss, cooking loss, TBARS, denaturation temperatures and enthalpy, color, and water migration. One-way analysis of variance (ANOVA) was used to analyze the previously mentioned physiochemical parameters following a Duncan's multiple range test, expressed as mean ± standard deviation (SD) (SPSS 25.0, Chicago, IL, USA). Significant difference ($p < 0.05$) was used to compare treatment means. All figures were plotted with OriginPro 9.0 (OriginLab Co., Northampton, MA, USA).

3. Results and Discussions

3.1. Dielectric Properties

Table 1 shows that the dielectric constant and dielectric loss factor of salmon increase with temperatures within the range of −20 to 20 °C. At all selected frequencies, the dielectric constant and dielectric loss factor increased slowly from −20 to −10 °C, and then increased rapidly from −10 to −5 °C due to the melting of ice crystals (Figure 2). In general, both dielectric constant and loss factor increased as temperature increased from −20 to 20 °C which was attributed to the increasing water molecule dipole rotation and ionic conductivity [49]. This would possibly result in significant temperature non-uniformity at the sample edges, since the edges usually tend to absorb more energy and the localized heating results in rapid melting of ice crystals, which further aggravates the non-uniform heating. Table 1 also lists the penetration depth of salmon at selected frequencies and temperatures between −20 and 20 °C. The penetration depth decreased with increasing temperature at three different frequencies, and the lower the frequency was, the greater the decrease was. The tendency was similar to those reported in literature, namely that the penetration depth decreases as the temperature increases for most foods [36]. In addition, the penetration depth of the sample at 13.56 MHz was the largest, followed by 27.12 and 40.68 MHz, which is consistent with the results of some researchers who considered that the penetration depth usually decreases with increases in frequency [50,51]. According to the results, it can be predicted that the thawing uniformity of the sample will be better at lower frequency.

Table 1. Dielectric properties and penetration depth (dp) of salmon at 13.56, 27.12, and 40.68 MHz and temperatures between −20 and 20 °C.

T (°C)	Parameters	13.56 MHz	27.12 MHz	40.68 MHz
−20	ε'	11.25 ± 0.31 [Ad]	7.54 ± 0.02 [Bb]	2.64 ± 0.03 [Cc]
	ε''	6.10 ± 0.17 [Af]	2.82 ± 0.09 [Bh]	1.65 ± 0.04 [Ch]
	dp/cm	200.41 [Aa]	174.53 [Ba]	120.56 [Ca]
−15	ε'	15.39 ± 0.06 [Ad]	10.16 ± 0.02 [Bb]	4.79 ± 0.11 [Cbc]
	ε''	14.17 ± 0.26 [Af]	6.81 ± 0.12 [Bh]	3.24 ± 0.56 [Ch]
	dp/cm	105.97 [Ab]	86.56 [Bb]	83.45 [Bb]
−10	ε'	24.13 ± 0.08 [Ad]	13.76 ± 3.23 [Bb]	11.41 ± 3.45 [Bb]
	ε''	45.05 ± 4.37 [Af]	15.74 ± 6.78 [Bg]	9.93 ± 5.09 [Bg]
	dp/cm	47.96 [Ac]	46.58 [ABc]	43.09 [Bc]
−5	ε'	97.36 ± 5.99 [Ac]	71.42 ± 4.88 [Ba]	63.91 ± 4.04 [Ba]
	ε''	432.35 ± 23.38 [Ae]	203.89 ± 8.03 [Bf]	140.43 ± 4.94 [Cf]
	dp/cm	13.40 [Ad]	10.36 [ABd]	8.73 [Bd]
0	ε'	101.15 ± 8.35 [Abc]	72.07 ± 5.54 [Ba]	64.55 ± 4.52 [Ba]
	ε''	498.56 ± 27.76 [Ade]	233.50 ± 8.69 [Be]	159.66 ± 5.30 [Ce]
	dp/cm	12.34 [Ad]	9.49 [ABd]	8.00 [Bd]
5	ε'	104.74 ± 9.50 [Abc]	73.43 ± 6.08 [Ba]	65.69 ± 4.72 [Ba]
	ε''	572.69 ± 45.01 [Acd]	266.26 ± 5.47 [Bd]	182.33 ± 2.95 [Cd]
	dp/cm	11.40 [Ad]	8.75 [Ad]	7.34 [Ad]
10	ε'	111.48 ± 9.21 [Aabc]	75.22 ± 6.37 [Ba]	66.91 ± 4.57 [Ba]
	ε''	648.70 ± 53.84 [Abc]	300.79 ± 5.09 [Bc]	205.87 ± 2.40 [Cc]
	dp/cm	10.65 [Ad]	8.13 [Ad]	6.79 [Ad]
15	ε'	115.91 ± 11.57 [Aab]	77.10 ± 7.08 [Ba]	68.21 ± 4.92 [Ba]
	ε''	722.13 ± 61.82 [Aab]	335.76 ± 3.99 [Bb]	229.94 ± 1.74 [Cb]
	dp/cm	10.04 [Ad]	7.62 [Ad]	6.34 [Ad]
20	ε'	122.97 ± 13.03 [Aa]	79.27 ± 8.01 [Ba]	69.65 ± 5.44 [Ba]
	ε''	804.56 ± 74.99 [Aa]	373.25 ± 2.63 [Ba]	256.32 ± 0.80 [Ca]
	dp/cm	9.48 [Ad]	7.16 [Ad]	5.93 [Ad]

Different lowercase letters in the same column indicate significant differences ($p < 0.05$). Different uppercase letters in the same row indicate significant differences ($p < 0.05$). ε': dielectric constant, ε'': dielectric loss factor, dp: penetration depth.

3.2. Tempering Rate

The tempering rates of salmon fillets with different tempering methods are shown in Figure 3. The tempering rate of RT was higher than that of AT + 10 °C and WT + 10 °C, and the tempering time of WT (15 min) and AT (64 min) were 3 and 12.8 times longer than RT (5 min), respectively. Choi et al. [52] also reported that the tempering rate of RT (at 400 W) of pork loin (100 × 100 × 70 mm³) was 100 times faster than that of AT. The reason is that WT and AT elevate the sample temperature mainly by heat convection and conduction, and the heating rate relies on the limited convective heat transfer coefficient and thermal conductivity, while radio frequency generates heat within the food samples rapidly and volumetrically [53,54]. Meanwhile, the tempering rate from −20 to −5 °C was much faster than from −5 to −4 °C. This is because the thermal conductivity of ice is four times that of water [55], and the immobilized water started to shift to the mobilized state when the temperature reached −5 °C, causing the slow rise in temperature.

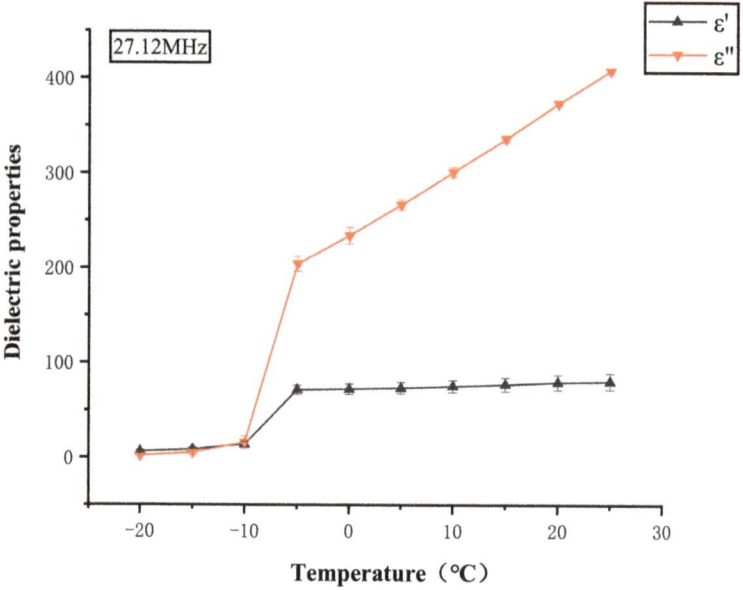

Figure 2. Dielectric constant (ε') and dielectric loss factor (ε'') of salmon as a function of temperature at 27.12 MHz.

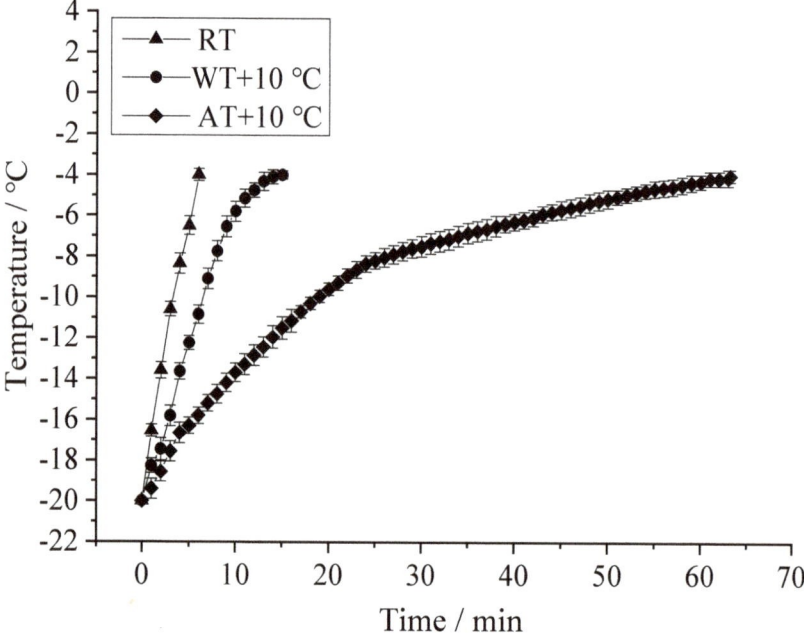

Figure 3. Temperature–time histories of salmon samples during different tempering treatments. (RT: Radio frequency tempering; WT + 10 °C: Water tempering at 10 ± 0.5 °C; AT + 10 °C: Air tempering at 10 ± 1 °C).

3.3. Surface Temperature Distribution

The temperature distribution on the top and bottom surfaces of salmon samples after different tempering methods are shown in Figure 4. For RT, the temperature distribution on the top surfaces of salmon fillets were more uniform than for the other two tempering methods, and the temperature over most of the area was controlled below 0 °C. The result was similar to the results reported by Zhang et al. [27]. However, overheating at the sample edges and corners was observed since edges and corners are a convergence of many surfaces, and the electromagnetic field intensity at these locations is higher than in the rest of the area, thus resulting in more severe heating [26,56]. However, the bottom surfaces showed WT and AT produced more uniform heating, and RT showed a much higher temperature than on the top surfaces. This is possibly because water migration occurred during the tempering process and free water transferred to the bottom surface, resulting in higher a heating rate on the bottom surface.

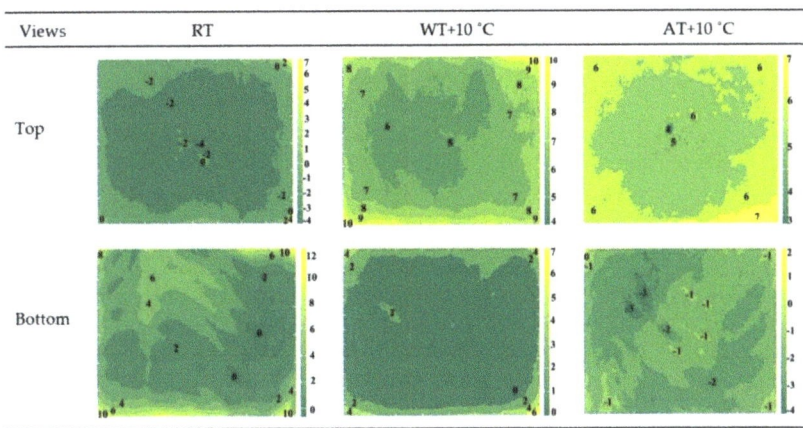

Figure 4. Surface temperature distribution of salmon fillets after different tempering treatments when the sample center reached −4 °C. (RT: Radio frequency tempering; WT + 10 °C: Water tempering at 10 ± 0.5 °C; AT + 10 °C: Air tempering at 10 ± 1 °C).

From Table 2, the maximum temperature on the surfaces of RT, WT + 10 °C and AT + 10 °C samples were 7.3 °C, 11.8 °C, 7.2 °C and 13.4 °C, 8.4 °C, 2.2 °C for the top and bottom surfaces, respectively. The smallest differences in surface temperature were found in AT + 10 °C, and also the standard deviation of AT + 10 °C was the smallest, indicating that AT + 10 °C was the most uniform tempering method. Similar results were reported by Zhu et al. [40]. This is because the convective heat transfer coefficient between water and the salmon fillet is larger than that between air and the [57], resulting in a faster temperature increase but worse tempering uniformity than with air tempering. Combining the heating uniformity results with tempering rates, it was noted that although AT had the best tempering uniformity, its tempering rate was the lowest. RT had the highest tempering rate, but the tempering uniformity could be better controlled before extending it to industrial use by reducing the edge effect and absorbing the drip loss during tempering. In most industrial radio frequency equipment, the heating rate can be further regulated by varying the electrode gap, which also influences the heating uniformity.

Table 2. Top and bottom surface temperature of salmon after different tempering treatments. (RT: Radio frequency tempering; WT + 10 °C: Water tempering at 10 ± 0.5 °C; AT + 10 °C: Air tempering at 10 ± 1 °C).

Treatments	Views	Maximum Temperature/(°C)	Minimum Temperature/(°C)	Standard Deviation/(°C)
RT	top	7.3	−5.2	±1.1
	bottom	13.4	−0.6	±2.2
WT + 10 °C	top	11.8	3.7	±0.9
	bottom	8.4	−0.5	±1.1
AT + 10 °C	top	7.2	3.3	±0.5
	bottom	2.2	−3.9	±0.6

3.4. Drip and Cooking Loss

The drip and cooking loss of salmon fillet under different tempering conditions are presented in Table 3. The drip and cooking loss usually contain water and water-soluble substances. A lower drip and cooking loss represent a higher water holding capacity of raw and cooked meat, respectively. It could be observed that the tempering methods had a significant effect on both drip and cooking loss ($p < 0.05$). For tempering loss, AT + 10 °C produced the least amount (0.43%) of drip loss while WT + 10 °C produced the most (1.03%). Meanwhile, RT, WT and AT resulted in a cooking loss of 13.58%, 15.75% and 12.26%, respectively, after a freezing and tempering cycle, while the control sample exhibited a cooking loss of 11.34%. The higher drip and cooking loss of WT + 10 °C samples were also possibly a result of increased myosin denaturation [58]. Xia et al. [58] demonstrated that freezing and thawing cycles decreased the stability of myosin and actin, changed the microstructure of myofibrillar protein, and finally affected the physical attributes of meat, such as juiciness and texture. Damaged structure and denatured proteins reduced the ability of muscle to retain water, resulting in a lower water holding capacity [8]. This result agreed with Farag et al. [21], who observed that air tempering resulted in higher drip loss (18.0%) than RT (9.0%) in whole lean beef samples. The results were in accordance with the temperature distribution in the fish fillets, where WT samples had the worst tempering uniformity and resulted in the highest amount of drip and cooking loss. This indicates that temperature is a key factor affecting drip and cooking loss.

Table 3. Effects of different tempering methods on drip loss and cooking loss of salmon. (RT: Radio frequency tempering; WT + 10 °C: Water tempering at 10 ± 0.5 °C; AT + 10 °C: Air tempering at 10 ± 1 °C).

Treatments	Drip Loss/(%)	Cooking Loss/(%)
Control	-	11.34 ± 0.02 [b]
RT	0.66 ± 0.052 [b]	13.58 ± 0.63 [ab]
WT + 10 °C	1.029 ± 0.050 [a]	15.75 ± 0.06 [a]
AT + 10 °C	0.43 ± 0.021 [b]	12.26 ± 0.48 [b]

The superscript of the numbers (a,b) indicate significant differences ($p < 0.05$) between samples treated with different methods.

3.5. Color

Studies have shown that consumers usually relate flesh color with the freshness and quality of fish [59]. Table 4 shows the color (L^*, a^*, b^* and ΔE values) of salmon fillets before and after different tempering methods. The total color difference parameter (ΔE) showed that treated salmon fillets had significant differences with the control sample. The AT + 10 °C sample differed the least from the control sample compared with the WT + 10 °C and RT samples. The L^* value of thawed samples increased significantly ($p < 0.05$) compared to the control sample, and the highest increase in the L^* value was discovered in the WT + 10 °C sample, while the lowest was observed in the AT + 10 °C sample. It was discovered that all tempering methods had significant ($p < 0.05$) effects on the lightness (L^*) of salmon fillets, but the WT + 10 °C and AT + 10 °C methods showed the most and least negative effect on the lightness (L^*) of salmon sample, respectively. Because of the long tempering time and the small temperature differences between salmon fillet and air during the final tempering stage, the surface of fillet flesh was protected from negative change. Additionally, the higher tempering rate and relatively uniform tempering of RT resulted in a tempered sample color close to that of the original salmon fillet.

Table 4. Effects of different tempering methods on the color of salmon flesh. (RT: Radio frequency tempering; WT + 10 °C: Water tempering at 10 ± 0.5 °C; AT + 10 °C: Air tempering at 10 ± 1 °C; L^*: lightness; a^*: redness-greenness; b^*: yellowness-blueness; ΔE: the total color difference).

Treatments	L^*	a^*	b^*	ΔE
Control	42.27 ± 0.48 [b]	14.29 ± 0.33 [a]	13.88 ± 0.32 [a]	-
RT	44.46 ± 0.94 [a]	13.57 ± 0.83 [a]	12.14 ± 0.85 [b]	2.89
WT + 10 °C	44.74 ± 0.32 [a]	13.35 ± 0.73 [a]	12.80 ± 0.52 [ab]	2.85
AT + 10 °C	43.34 ± 1.00 [a]	14.41 ± 0.23 [a]	12.10 ± 0.60 [b]	2.08

The superscript of the numbers (a,b) indicate significant differences ($p < 0.05$) between samples treated with different methods.

As for the a^* value, no significant ($p > 0.05$) difference in redness (a^*) was found between the control sample and the samples treated with three different tempering methods. The reason is possibly that the dominant pigment in salmon fillet is astaxanthin, which is fat-soluble and not easily lost with drip loss. However, the redness of the sample treated with AT + 10 °C increased, which may be because the low temperature resulted in less oxidation reaction. For the b^* value, no significant difference ($p > 0.05$) was found between the WT + 10 °C sample and control sample, and AT + 10 °C and RT had a lower b^* value. The change in yellowness of salmon fillet was possibly due to protein and lipid oxidation [22].

3.6. TBARS

The lipid oxidation degree was quantified by the TBARS values of the fish samples. The higher the TBARS value, the higher the degree of lipid oxidation. The TBARS values of the control and thawed salmon are shown in Figure 5. It could be seen that the TBARS values increased significantly after all tempering processes ($p < 0.05$) except for AT + 10 °C. Ke et al. [60] suggested that levels below 8 mg MDA/kg fish flesh could be indicative of good quality. The TBARS value of control salmon was 1.027 mg MDA/kg, while that of WT + 10 °C was the highest (3.102 mg MDA/kg), followed by RT (2.135 mg MDA/kg) and AT + 10 °C (1.054 mg MDA/kg). This result was similar to that of Xia et al. [39], who reported that AT and WT resulted in TBARS value increases of 56.2% and 71.1%, respectively, compared to fresh longissimus muscle flesh. RT fish samples showed a higher tempering rate (Figure 3) and less lipid oxidation than WT samples (Figure 5). Comparatively, WT samples were exposed directly to water, and the differences in pressure between water and the samples caused an oxidation reaction and damage to the samples' microstructure.

3.7. Protein Denaturation

Three heat flow peaks were observed in salmon fillets representing myosin, sarcoplasmic proteins and actin. The peak temperature and endothermic values, T_{max} and ΔH, were the denaturation temperatures and enthalpy of myofibrillar protein, respectively. The values of T_{max} and ΔH reflect the degree of protein denaturation compared to that of the control sample. The smaller the T_{max} and ΔH values, the higher the degree of protein denaturation [61].

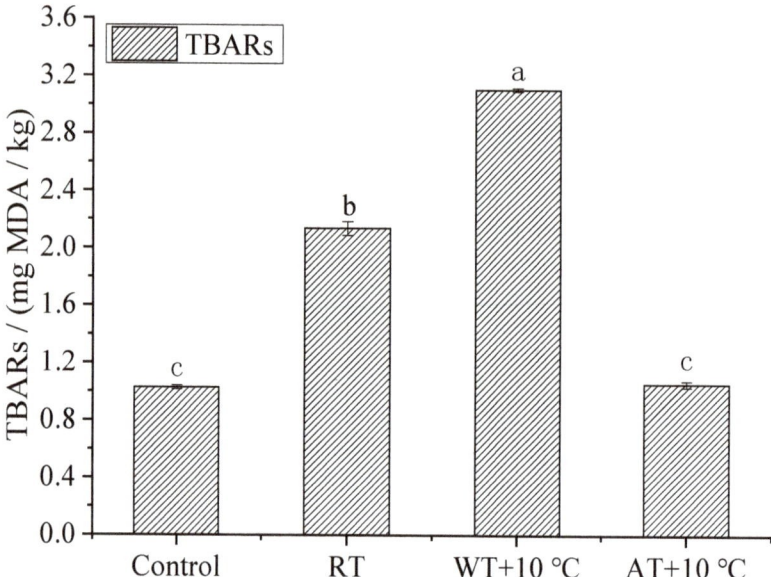

Figure 5. Effects of different tempering methods on TBARS of salmon. (RT: Radio frequency tempering; WT + 10 °C: Water tempering at 10 ± 0.5 °C; AT + 10 °C: Air tempering at 10 ± 1 °C). Different letters on the bar (a,b,c) indicate significant differences ($p < 0.05$) among samples treated with different methods.

As shown in Table 5, no significant differences ($p > 0.05$) were observed among control sample and thawed samples in their T_{2max}, T_{3max} ΔH_2, and ΔH_3 values. However, for T_{1max} and ΔH_1, a significant decrease ($p < 0.05$) in peak temperature was found in the WT + 10 °C and RT samples, indicating these two tempering methods had a stronger negative effect on myosin than AT + 10 °C. This result illustrated that among the three tempering methods, AT + 10 °C was the most appropriate tempering method for protein preservation, followed by RT and WT + 10 °C. Further, it was found that myosin was more easily damaged and denatured in freeze-temper cycles. The denaturation of myofibrillar protein was also associated with the lipid oxidation and microstructure damage [48].

Table 5. Effects of different tempering methods on T_{max} and ΔH (the denaturation temperature and enthalpy of myofibrillar protein, respectively) of salmon. (RT: Radio frequency tempering; WT + 10 °C: Water tempering at 10 ± 0.5 °C; AT + 10 °C: Air tempering at 10 ± 1 °C).

Treatment	T_{1max}/°C	ΔH_1/(J/g)	T_{2max}/°C	ΔH_2/(J/g)	T_{3max}/°C	ΔH_3/(J/g)
Control	58.18 ± 0.82 [a]	0.0779 ± 0.002 [a]	67.88 ± 0.38 [a]	0.0962 ± 0.001 [a]	77.27 ± 0.06 [a]	0.2315 ± 0.003 [a]
RT	57.24 ± 0.36 [b]	0.0662 ± 0.004 [b]	67.51 ± 0.23 [a]	0.0928 ± 0.001 [ab]	76.56 ± 0.12 [a]	0.2309 ± 0.002 [a]
WT + 10 °C	56.93 ± 0.14 [b]	0.0623 ± 0.005 [b]	67.04 ± 0.19 [a]	0.0890 ± 0.003 [b]	76.55 ± 0.23 [a]	0.2200 ± 0.006 [a]
AT + 10 °C	57.88 ± 0.17 [a]	0.0718 ± 0.003 [a]	67.75 ± 0.21 [a]	0.0941 ± 0.002 [ab]	77.10 ± 0.24 [a]	0.2307 ± 0.004 [a]

The superscripts on the numbers (a,b) indicate significant differences ($p < 0.05$) between samples treated with different methods.

3.8. Moisture Migration

Low-field nuclear magnetic resonance (LF-NMR) was used to study the water migration mobility in food materials [62], which was expressed as transverse relaxation time, T_2. A longer T_2 indicates a stronger mobility of water. As shown in Figure 6, there were three peaks in salmon fillets corresponding to T_2 of the three states of water in the samples: T_{21} (0–10 ms), representing bound water with the strongest binding force among three states; T_{22} (10–100 ms), representing immobilized water or intracellular water that was entrapped in the myofibrillar network and was usually the predominant water component in flesh; and T_{23} (100–1000 ms), representing free water that was restricted by capillary forces and could be lost by heating and mechanical damage [63].

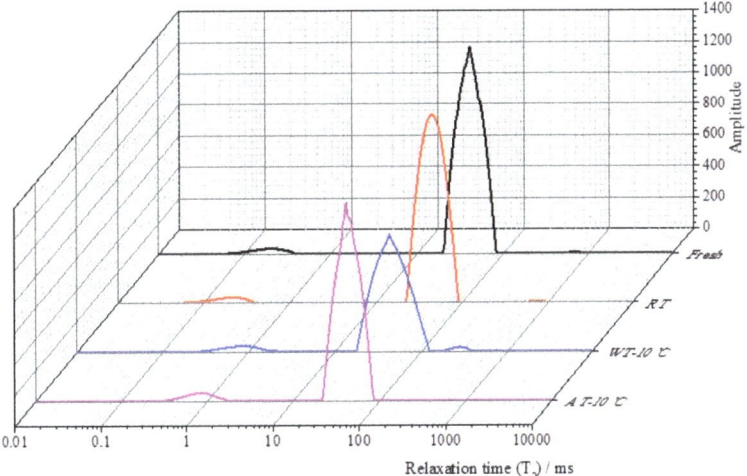

Figure 6. Transverse relaxation time (T_2) curves of salmon samples after tempering, evaluated by LF-MNR. (RT: Radio frequency tempering; WT + 10 °C: Water tempering at 10 ± 0.5 °C; AT + 10 °C: Air tempering at 10 ± 1 °C).

From Figure 6, it can be seen that the changes in bound water (T_{21}) in all samples were subtle. This is possibly because the bound water was tightly bound to proteins and other macromolecular substances and was barely affected by the tempering processes. It was observed that WT + 10 °C led to longer T_{21} and T_{22} and larger amplitudes compared to the control sample and the other two treated samples, which revealed that the bound and immobile water shifted to free water. The P_{21}, P_{22} and P_{23} of fresh sample were 2.9%, 97.0% and 0.1%, respectively (Figure 7). There were significant differences ($p < 0.05$) in the peak area (P_{21}, P_{22} and P_{23}) of samples after different tempering treatments. After tempering, a significant decrease in P_{22} (the peak area of T_{22}, immobile water) was observed in the sample treated with WT + 10 °C, followed by that of RT and AT + 10 °C. The increase in P_{23} of the WT + 10 °C sample indicated a possible migration from tightly bound (T_{22}) to loosely bound (T_{23}) water as a result of microstructure damage and protein denaturation. McDonnell et al. [63] also found that WT had a significant destructive effect on pork, resulting in severe myofibrillar protein denaturation.

3.9. Microstructure

The microstructures of the control and thawed tissue tempered with different methods are shown in Figure 8. After tempering, myofibrils with clear boundaries and a larger intercellular space were observed in tempered samples, especially the WT + 10 °C sample. From the severely deformed myofibrils observed in the WT + 10 °C sample, it could be speculated that the sample tissue suffered significant mechanical injury from water tempering and released the moisture and cellular substances to the intercellular spaces or as tempering loss. The microstructures of RT and AT + 10 °C samples were found to be less damaged since the spaces between the muscle fiber bundles were smaller than those of WT + 10 °C sample. This was consistent with the results of tempering loss, cooking loss and moisture migration of salmon flesh. RF tempering shortened the tempering time and reduced the damage to muscle fiber and fragmentation of cell structures [10]. Zhang et al. [64] also reported that improper tempering methods resulted in protein denaturation and structural damage to muscle tissue and cells.

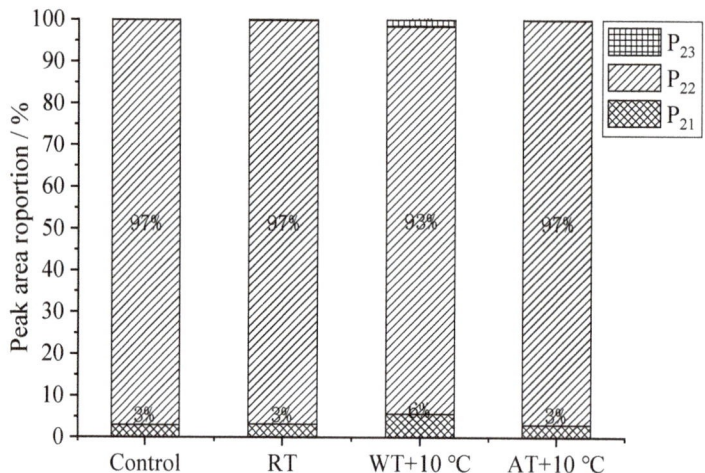

Figure 7. The relaxation time (T_2) corresponding to relative peak areas (P_{21}, P_{22} and P_{23}) of salmon fillets after different tempering treatments. (RT: Radio frequency tempering; WT + 10 °C: Water tempering at 10 ± 0.5 °C; AT + 10 °C: Air tempering at 10 ± 1 °C).

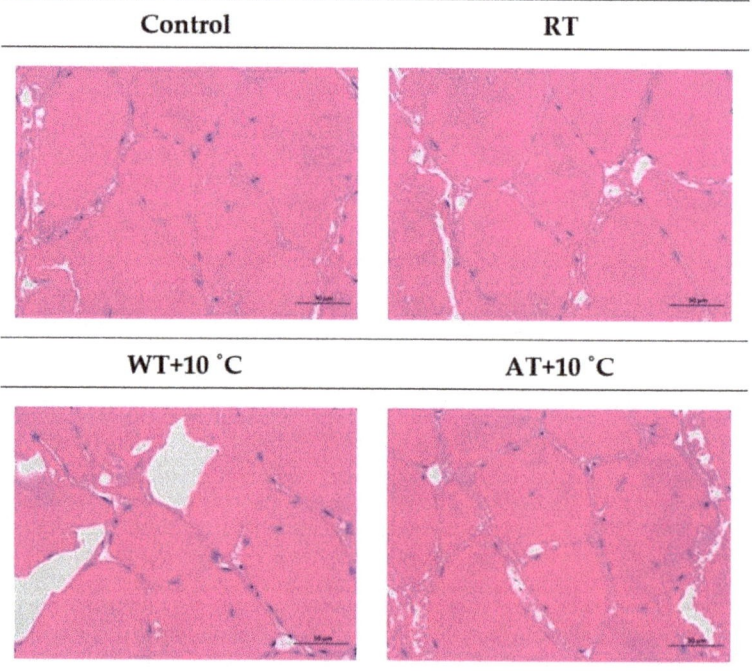

Figure 8. Microstructure of salmon fillet tissues after different tempering methods under light microscope (magnification ×400). (RT: Radio frequency tempering; WT + 10 °C: Water tempering at 10 ± 0.5 °C; AT + 10 °C: Air tempering at 10 ± 1 °C).

4. Conclusions

The tempering rate, temperature distribution and physiochemical properties of frozen salmon fillets subjected to radio frequency (RT), water (WT + 10 °C) and air (AT + 10 °C) tempering were evaluated. RT presented the shortest tempering time, less tempering and cooking loss, and a lower degree of lipid oxidation and protein denaturation compared to WT + 10 °C. These results were consistent with the water mobility results from LF-NMR and muscle histology. Although WT + 10 °C produced a faster tempering rate than AT + 10 °C, it was an undesirable method for salmon fillet tempering because the physiochemical properties were not better than those obtained with AT + 10 °C. In conclusion, the best overall quality was found in the salmon fillet treated with AT + 10 °C, but with the drawback of long tempering time. RT is a promising technology and has been proposed as an alternative novel salmon tempering method for the food industry, offering fast and relatively uniform tempering with a higher quality retention rate. RT throughput can be further enhanced with continuous processing.

This study provided a general evaluation of salmon fillet quality after various tempering methods, and showed that RF is a better method of thawing frozen salmon fillets. However, industrial-scale tempering of frozen salmon fillets in bulk or of whole salmon fish has not been studied yet. Thus, the industrial processing parameters for fishery product tempering with RF should be further explored, including analyzing the influencing factors that contribute to tempering uniformity, quality and energy consumption. Furthermore, whole fish and fish fillets could present significantly different effects in parallel-plate RF tempering and will require further investigation of the influences on process development from their complex shapes and composition. In addition, the endpoint tempering temperature should be optimized on the basis of salmon fish processing procedures.

Author Contributions: Conceptualization, R.H., J.H. and Y.J.; methodology, J.H., Y.J.; analysis, R.H., J.H., Y.C., F.L.; writing—original draft preparation, R.H., J.H., Y.C.; writing—review and editing, F.L., H.S., Y.J.; supervision, H.S., Y.J. All authors have read and agreed to the published version of the manuscript.

Funding: This research was funded by China National Science Foundation (31801613), China Postdoctoral Science Foundation (2018M632299).

Institutional Review Board Statement: Not applicable.

Informed Consent Statement: Not applicable.

Data Availability Statement: Not applicable.

Conflicts of Interest: The authors declare no conflict of interest.

References

1. Calder, P.C. Fatty acids and inflammation: The cutting edge between food and pharma. *Eur. J. Pharmacol.* **2011**, *668*, S50–S58. [CrossRef] [PubMed]
2. Mozaffarian, D.; Wu, J.H. Omega-3 fatty acids and cardiovascular disease: Effects on risk factors, molecular pathways, and clinical events. *J. Am. Coll. Cardiol.* **2011**, *58*, 2047–2067. [CrossRef] [PubMed]
3. Boonsumrej, S.; Chaiwanichsiri, S.; Tantratian, S.; Suzuki, T.; Takai, R. Effects of freezing and tempering on the quality changes of tiger shrimp (*Penaeus monodon*) frozen by air-blast and cryogenic freezing. *J. Food Eng.* **2007**, *80*, 292–299. [CrossRef]
4. Li, D.; Jia, S.; Zhang, L.; Wang, Z.; Pan, J.; Zhu, B.; Luo, Y. Effect of using a high voltage electrostatic field on microbial communities, degradation of adenosine triphosphate, and water loss when tempering lightly-salted, frozen common carp (*Cyprinus carpio*). *J. Food Eng.* **2017**, *212*, 226–233. [CrossRef]
5. Cui, Y.; Xuan, X.; Ling, J.; Liao, X.; Zhang, H.; Shang, H.; Lin, X. Effects of high hydrostatic pressure-assisted tempering on the physicohemical characteristics of silver pomfret (*Pampus argenteus*). *Food Sci. Nutr.* **2019**, *7*, 1573–1583. [CrossRef]
6. Ersoy, B.; Aksan, E.; Ozeren, A. The effect of tempering methods on the quality of eels (*Anguilla anguilla*). *Food Chem.* **2008**, *111*, 377–380. [CrossRef]

7. Cai, L.; Zhang, W.; Cao, A.; Cao, M. Effects of different tempering methods on the quality of largemouth bass (*Micropterus salmonides*). *LWT Food Sci. Technol.* **2020**, *120*, 108908. [CrossRef]
8. Mousakhani-Ganjeh, A.; Hamdami, N.; Soltanizadeh, N. Impact of high voltage electric field tempering on the quality of frozen tuna fish (*Thunnus albacares*). *J. Food Eng.* **2015**, *156*, 39–44. [CrossRef]
9. Mousakhani-Ganjeh, A.; Hamdami, N.; Soltanizadeh, N. Effect of high voltage electrostatic field tempering on the lipid oxidation of frozen tuna fish (*Thunnus albacares*). *Innov. Food Sci. Emerg. Technol.* **2016**, *36*, 42–47. [CrossRef]
10. Li, X.; Sun, P.; Ma, Y.; Cai, L.; Li, J. Effect of ultrasonic thawing on the water holding capacity, physico-chemical properties, and structure of frozen tuna fish (*Thunnus tonggol*) myofibrillar proteins. *J. Sci. Food Agric.* **2019**, *99*, 5083–5091. [CrossRef]
11. Zhu, S.; Ramaswamy, H.; Simpson, B. Effect of high-pressure versus conventional tempering on color, drip loss and texture of Atlantic salmon frozen by different methods. *LWT Food Sci. Technol.* **2004**, *37*, 291–299. [CrossRef]
12. Anderson, B.A.; Singh, R.P. Modeling the tempering of frozen foods using air impingement technology. *Int. J. Refrig.* **2006**, *29*, 294–304. [CrossRef]
13. Hafezparast-Moadab, N.; Hamdami, N.; Dalvi-Isfahan, M.; Farahnaky, A. Effects of radiofrequency-assisted freezing on microstructure and quality of rainbow trout (*Oncorhynchus mykiss*) fillet. *Innovative Food Sci. Emerg. Technol.* **2018**, *47*, 81–87. [CrossRef]
14. Guo, C.; Wang, X.; Wang, Y. Dielectric properties of soy protein isolate dispersion and its temperature profile during radio frequency heating. *J. Food Process. Preserv.* **2018**, *42*, e13659. [CrossRef]
15. Li, Y.; Zhou, L.; Chen, J.; Subbiah, J.; Chen, X.; Fu, H.; Wang, Y. Dielectric properties of chili powder in the development of radio frequency and microwave pasteurisation. *Int. J. Food Prop.* **2018**, *20* (Suppl. S3), S3373–S3384. [CrossRef]
16. Mao, Y.; Wang, P.; Wu, Y.; Hou, L.; Wang, S. Effects of various radio frequencies on combined drying and disinfestation treatments for in-shell walnuts. *LWT Food Sci. Technol.* **2021**, *144*, 111246. [CrossRef]
17. Bedane, T.F.; Chen, L.; Marra, F.; Wang, S. Experimental study of radio frequency (RF) tempering of foods with movement on conveyor belt. *J. Food Eng.* **2017**, *201*, 17–25. [CrossRef]
18. Nagaraj, G.; Singh, R.; Hung, Y.-C.; Mohan, A. Effect of radio-frequency on heating characteristics of beef homogenate blends. *LWT Food Sci. Technol.* **2015**, *60*, 372–376. [CrossRef]
19. Piyasena, P.; Dussault, C.; Koutchma, T.; Ramaswamy, H.S.; Awuah, G.B. Radio frequency heating of foods: Principles, applications and related properties—A review. *Crit. Rev. Food Sci. Nutr.* **2003**, *43*, 587–606. [CrossRef]
20. Yang, H.; Chen, Q.; Cao, H.; Fan, D.; Huang, J.; Zhao, J.; Zhang, H. Radiofrequency tempering of frozen minced fish based on the dielectric response mechanism. *Innov. Food Sci. Emerg. Technol.* **2019**, *52*, 80–88. [CrossRef]
21. Farag, K.W.; Duggan, E.; Morgan, D.J.; Cronin, D.A.; Lyng, J.G. A comparison of conventional and radio frequency defrosting of lean beef meats: Effects on water binding characteristics. *Meat Sci.* **2009**, *83*, 278–284. [CrossRef]
22. Llave, Y.; Terada, Y.; Fukuoka, M.; Sakai, N. Dielectric properties of frozen tuna and analysis of defrosting using a radio-frequency system at low frequencies. *J. Food Eng.* **2014**, *139*, 1–9. [CrossRef]
23. Kim, J.; Park, J.W.; Park, S.; Choi, D.S.; Choi, S.R.; Kim, Y.H.; Cho, B.-K. Study of Radio Frequency Tempering for Cylindrical Pork Sirloin. *J. Biosyst. Eng.* **2016**, *41*, 108–115. [CrossRef]
24. Palazoglu, T.K.; Miran, W. Experimental comparison of microwave and radio frequency tempering of frozen block of shrimp. *Innov. Food Sci. Emerg. Technol.* **2017**, *41*, 292–300. [CrossRef]
25. Palazoğlu, T.K.; Miran, W. Experimental investigation of the effect of conveyor movement and sample's vertical position on radio frequency tempering of frozen beef. *J. Food Eng.* **2018**, *219*, 71–80. [CrossRef]
26. Li, Y.; Li, F.; Tang, J.; Zhang, R.; Wang, Y.; Koral, T.; Jiao, Y. Radio frequency tempering uniformity investigation of frozen beef with various shapes and sizes. *Innov. Food Sci. Emerg. Technol.* **2018**, *48*, 42–55. [CrossRef]
27. Zhang, Y.; Li, F.; Yao, Y.; He, J.; Tang, J.; Jiao, Y. Effects of freeze-thaw cycles of Pacific white shrimp (*Litopenaeus vannamei*) subjected to radio frequency tempering on melanosis and quality. *Innov. Food Sci. Emerg. Technol.* **2021**, *74*, 102860. [CrossRef]
28. Cai, L.; Zhang, W.; Cao, A.; Li, J. Effects of ultrasonics combined with far infrared or microwave tempering on protein denaturation and moisture migration of *Sciaenops ocellatus* (red drum). *J. Ultrason. Sonochem.* **2019**, *55*, 96–104. [CrossRef]
29. Jiao, Y.; Tang, J.; Wang, Y.; Koral, T.L. Radio-Frequency Applications for Food Processing and Safety. *J. Annu. Rev. Food Sci. Technol.* **2018**, *9*, 105–127. [CrossRef]
30. An, H.; Ma, H.; Ren, X. Relationship between electrical parameters and quality indexes during apple fruit storage. *J. Food Sci.* **2013**, *34*, 298–302.
31. Zhou, S.; Zhang, H.; Li, H.; Ma, Q. Review of fruit nondestructive testing technology research based on the dielectric characteristics. *Food Res. Dev.* **2015**, *36*, 131–135.
32. Li, X.; Kang, N.; Li, D.; Liu, X.; Jia, Y.; Li, F.; Wang, Y.; Pan, Z. Correlation between dielectric characteristics and storage time of goji berry fruit. *Agric. Mech. Res.* **2015**, *6*, 136–139.
33. Jin, Y.; Wang, H.; Yang, N. Effect of low frequency band on dielectric properties of eggs during storage. *Chin. J. Food Sci.* **2015**, *15*, 220–225.
34. Guo, W.; Zhu, X.; Liu, H.; Yue, R.; Wang, S. Effects of milk concentration and freshness on microwave dielectric properties. *J. Food Eng.* **2010**, *99*, 344–350. [CrossRef]

35. Castro-Giráldez, M.; Toldrá, F.; Fito, P. Low frequency dielectric measurements to assess post-mortem ageing of pork meat. *J. LWT Food Sci. Technol.* **2011**, *44*, 1465–1472. [CrossRef]
36. Wang, Y.; Tang, J.; Rasco, B.; Kong, F.; Wang, S. Dielectric properties of salmon fillets as a function of temperature and composition. *J. Food Eng.* **2008**, *87*, 236–246. [CrossRef]
37. Lau, S.K.; Subbiah, J. An automatic system for measuring dielectric properties of foods: Albumen, yolk, and shell of fresh eggs. *J. Food Eng.* **2018**, *223*, 79–90. [CrossRef]
38. Chen, Y.; He, J.; Li, F.; Tang, J.; Jiao, Y. Model food development for tuna (*Thunnus Obesus*) in radio frequency and microwave tempering using grass carp mince. *J. Food Eng.* **2021**, *292*, 110267. [CrossRef]
39. Xia, X.; Kong, B.; Liu, J.; Diao, X.; Liu, Q. Influence of different tempering methods on physicochemical changes and protein oxidation of porcine longissimus muscle. *LWT Food Sci. Technol.* **2012**, *46*, 280–286. [CrossRef]
40. Zhu, Y.; Li, F.; Tang, J.; Wang, T.; Jiao, Y. Effects of radio frequency, air and water tempering, and different end-point tempering temperatures on pork quality. *J. Food Process Eng.* **2019**, *42*, e13026. [CrossRef]
41. Xuan, X.; Cui, Y.; Lin, X.; Yu, J.; Liao, X.; Ling, J.; Shang, H. Impact of high hydrostatic pressure on the shelling efficacy, physicochemical properties, and microstructure of fresh razor clam (*Sinonovacula constricta*). *J. Food Sci.* **2018**, *83*, 284–293. [CrossRef]
42. Siró, I.; Vén, C.; Balla, C.; Jónás, G.; Zeke, I.; Friedrich, L. Application of an ultrasonic assisted curing technique for improving the diffusion of sodium chloride in porcine meat. *J. Food Eng.* **2009**, *91*, 353–362. [CrossRef]
43. Rahbari, M.; Hamdami, N.; Mirzaei, H.; Jafari, S.M.; Kashaninejad, M.; Khomeiri, M. Effects of high voltage electric field tempering on the characteristics of chicken breast protein. *J. Food Eng.* **2018**, *216*, 98–106. [CrossRef]
44. McDonnell, C.K.; Allen, P.; Morin, C.; Lyng, J.G. The effect of ultrasonic salting on protein and water-protein interactions in meat. *Food Chem.* **2014**, *147*, 245–251. [CrossRef]
45. Carr, H.Y.; Purcell, E.M. Effects of Diffusion on Free Precession in Nuclear Magnetic Resonance Experiments. *Phys. Rev.* **1954**, *94*, 630–638. [CrossRef]
46. Meiboom, S.; Gill, D. Modified Spin-Echo Method for Measuring Nuclear Relaxation Times. *Rev. Sci. Instrum.* **1958**, *29*, 688–691. [CrossRef]
47. Ida, G.; Aursand, L.G.-J.; Ulf, E.D.; Axelson, E.; Turid, R. Water Distribution in Brine Salted Cod (*Gadus morhua*) and Salmon (*Salmo salar*): A Low-Field 1H NMR Study. *J. Agric. Food Chem.* **2008**, *56*, 6252–6260.
48. Jiang, Q.; Jia, R.; Nakazawa, N.; Hu, Y.; Osako, K.; Okazaki, E. Changes in protein properties and tissue histology of tuna meat as affected by salting and subsequent freezing. *Food Chem.* **2019**, *271*, 550–560. [CrossRef]
49. Kannan, S.; Dev, S.R.S.; Gariepy, Y.; Raghavan, G.S.V. Effect of radiofrequency heating on the dielectric and physical properties of eggs. *Prog. Electr. Res.* **2013**, *51*, 201–220. [CrossRef]
50. Wang, Y.; Tang, J.; Rasco, B.; Wang, S.; Alshami, A.A.; Kong, F. Using whey protein gel as a model food to study dielectric heating properties of salmon (*Oncorhynchus gorbuscha*) fillets. *LWT Food Sci. Technol.* **2009**, *42*, 1174–1178. [CrossRef]
51. Guan, D.; Cheng, M.; Wang, Y.; Tang, J. Dielectric properties of mashed potatoes relevant to Microwave and Radio frequency Pasteurization and Sterilization Processes. *J. Food Sci.* **2004**, *69*, FEP30–FEP37.
52. Choi, E.J.; Park, H.W.; Chung, Y.B.; Park, S.H.; Kim, J.S.; Chun, H.H. Effect of tempering methods on quality changes of pork loin frozen by cryogenic immersion. *Meat Sci.* **2017**, *124*, 69–76. [CrossRef]
53. Brown, T.; James, S.J. The effect of air temperature, velocity and visual lean (VL) composition on the tempering times of frozen boneless beef blocks. *Meat Sci.* **2006**, *73*, 545–552. [CrossRef]
54. Uyar, R.; Bedane, T.F.; Erdogdu, F.; Koray Palazoglu, T.; Farag, K.W.; Marra, F. Radio-frequency tempering of food products—A computational study. *J. Food Eng.* **2015**, *146*, 163–171. [CrossRef]
55. Li, X.; Sun, P.; Jia, J.; Cai, L.; Li, J.; Lv, Y. Effect of low frequency ultrasound tempering method on the quality characteristics of Peru squid (*Dosidicus gigas*). *Food Sci. Technol. Int.* **2018**, *25*, 171–181. [CrossRef]
56. Jiao, Y.; Tang, J.; Wang, S.; Koral, T. Influence of dielectric properties on the heating rate in free-running oscillator radio frequency systems. *J. Food Eng.* **2014**, *120*, 197–203. [CrossRef]
57. Singh, R.P.; Heldman, D.R. Chapter 4—Heat Transfer in Food Processing. In *Introduction to Food Engineering*, 5th ed.; Singh, R.P., Heldmann, D.R., Eds.; Academic Press: Cambridge, MA, USA, 2014; pp. 265–419.
58. Xia, X.; Kong, B.; Xiong, Y.; Ren, Y. Decreased gelling and emulsifying properties of myofibrillar protein from repeatedly frozen-thawed porcine longissimus muscle are due to protein denaturation and susceptibility to aggregation. *Meat Sci.* **2010**, *85*, 481–486. [CrossRef]
59. He, X.; Liu, R.; Nirasawa, S.; Zheng, D.; Liu, H. Effect of high voltage electrostatic field treatment on tempering characteristics and post-tempering quality of frozen pork tenderloin meat. *J. Food Eng.* **2013**, *115*, 245–250. [CrossRef]
60. Ke, P.J.; Cervantes, E.; Robles-Martinez, C. Determination of thiobarbituric acid reactive substances (TBARS) in fish tissue by an improved distillationspectrophotometric method. *J. Sci. Food Agric.* **1984**, *35*, 1248–1254. [CrossRef]
61. Cao, M.; Cao, A.; Wang, J.; Cai, L.; Regenstein, J.; Ruan, Y.; Li, X. Effect of magnetic nanoparticles plus microwave or far-infrared tempering on protein conformation changes and moisture migration of red seabream (*Pagrus Major*) fillets. *Food Chem.* **2018**, *266*, 498–507. [CrossRef]

62. Zheng, H.; Xiong, G.; Han, M.; Deng, S.; Xu, X.; Zhou, G. High pressure/thermal combinations on texture and water holding capacity of chicken batters. *Innov. Food Sci. Emerg. Technol.* **2015**, *30*, 8–14. [CrossRef]
63. McDonnell, C.K.; Allen, P.; Duggan, E.; Arimi, J.M.; Casey, E.; Duane, G.; Lyng, J.G. The effect of salt and fibre direction on water dynamics, distribution and mobility in pork muscle: A low field NMR study. *Meat Sci.* **2013**, *95*, 51–58. [CrossRef] [PubMed]
64. Zhang, M.; Li, F.; Diao, X.; Kong, B.; Xia, X. Moisture migration, microstructure damage and protein structure changes in porcine longissimus muscle as influenced by multiple freeze-thaw cycles. *Meat Sci.* **2017**, *133*, 10–18. [CrossRef] [PubMed]

MDPI
St. Alban-Anlage 66
4052 Basel
Switzerland
Tel. +41 61 683 77 34
Fax +41 61 302 89 18
www.mdpi.com

Foods Editorial Office
E-mail: foods@mdpi.com
www.mdpi.com/journal/foods